Sustainable Modelling, Processes and Applications for Societal Development

Sustainable Modelling, Processes and Applications for Societal Development

Editors

Yoshimitsu Uemura
Jun-Wei Lim
Worapon Kiatkittipong

MDPI

Basel • Beijing • Wuhan • Barcelona • Belgrade • Novi Sad • Cluj • Manchester

Editors

Yoshimitsu Uemura
Universiti Teknologi Petronas
Seri Iskandar
Malaysia

Jun-Wei Lim
Universiti Teknologi
PETRONAS
Seri Iskandar
Malaysia

Worapon Kiatkittipong
Silpakorn University
Nakhon Pathom
Thailand

Editorial Office
MDPI
St. Alban-Anlage 66
4052 Basel, Switzerland

This is a reprint of articles from the Special Issue published online in the open access journal *Processes* (ISSN 2227-9717) (available at: https://www.mdpi.com/journal/processes/special_issues/sustainable_modelling).

For citation purposes, cite each article independently as indicated on the article page online and as indicated below:

Lastname, A.A.; Lastname, B.B. Article Title. *Journal Name* **Year**, *Volume Number*, Page Range.

ISBN 978-3-7258-0419-1 (Hbk)
ISBN 978-3-7258-0420-7 (PDF)
doi.org/10.3390/books978-3-7258-0420-7

Contents

About the Editors

Yoshimitsu Uemura

In 1990, Dr. Yoshimitsu Uemura obtained his Ph.D. in Chemical Engineering from the Tokyo Institute of Technology, Japan. He was a Professor of Chemical Engineering (2009–2019) and the Head of the Center for Biofuel and Biochemical Research (CBBR) (2010–2018) at the Universiti Teknologi PETRONAS (UTP) in Malaysia. For 23 years, he also worked for the Kagoshima University in Japan as a Faculty Member. During this period, he spent two years (July 1993–July 1995) at Erindale College at the University of Toronto, Canada, as a Postdoctoral Researcher. Recently, he has also worked for the KPR Institute of Engineering and Technology, India, between Sep. and Nov. 2023 as an International Visiting Professor. His expertise revolves around catalytic reaction engineering, catalyst preparation engineering, polymer engineering, and biomass conversion. Across his professional career, he has published around 300 peer-reviewed journal papers.

Jun-Wei Lim

Dr. Jun-Wei Lim received his Ph.D. qualification in Environmental Chemistry from the Universiti Sains Malaysia in 2013. He is affiliated with Universiti Teknologi PETRONAS, serving as the Senior Lecturer and Cluster Head of the Applied Chemistry program. His major research interests are insect-based biological compounds, the bioremediation of solid wastes and wastewaters, microalgal biofuels, and green hydrogen. Accordingly, he has published over 300 research papers, including book chapters. He is a Member of The Royal Society of Chemistry (MRSC), an Associate Member of the Institution of Chemical Engineers (AMIChemE) at an international level, and a Professional Chemist registered with the Malaysian Institute of Chemistry and Professional Technologists who are registered with the Malaysia Board of Technologists at a national level. He is also a Chartered Scientist registered with the Science Council, UK, Chartered Environmentalist registered with Society for the Environment, UK, and Chartered Chemist registered with The Royal Society of Chemistry, UK. He is a member of the Distinguished Adjunct Faculty at Saveetha Institute of Medical and Technical Sciences and Visiting Professor at Prince of Songkla University.

Worapon Kiatkittipong

Dr. Worapon Kiatkittipong is an Associate Professor in the Department of Chemical Engineering at Silpakorn University, Thailand. He received a B.Eng. degree from Kasetsart University and M.Eng and D.Eng. degrees from Chulalongkorn University in 2005, supported by The Royal Golden Jubilee Ph.D. The program is funded by The Thailand Research Fund and supervised by Prof. Suttichai Assabumrungrat. His research focuses on the implications of biofuel and biochemical production integrated with biorefinery and the concept of process intensification. He has published more than 150 ISI/SCOPUS-indexed journals, receiving 3000+ citations and an H index of 32. He has been a Guest Editor for many journals and proceedings, such as *Biomass and Bioenergy*, *Energies*, and *Processes*. He previously served as the Vice Dean for Research and Academic Services at the Faculty of Engineering and Industrial Technology, Silpakorn University (2019–2022). He is currently the Associate Editor of Science Engineering and Health Studies.

processes

Editorial

Special Issue on "Sustainable Modellings, Processes and Applications for Societal Development"

Jun Wei Lim [1,*] and Worapon Kiatkittipong [2,*]

[1] HICoE—Centre for Biofuel and Biochemical Research, Institute of Self-Sustainable Building, Department of Fundamental and Applied Sciences, Universiti Teknologi PETRONAS, Seri Iskandar 32610, Perak Darul Ridzuan, Malaysia
[2] Department of Chemical Engineering, Faculty of Engineering and Industrial Technology, Silpakorn University, Nakhon Pathom 73000, Thailand
* Correspondence: junwei.lim@utp.edu.my (J.W.L.); kiatkittipong_w@su.ac.th (W.K.)

Global society has experienced a tremendous development since the institution of civilization. Indeed, the societal development is further intensified together with the new-age transformation associated with the Fourth Industrial Revolution [1]. The upward ascending society development is encompassing greater levels of energy, efficiency, productivity, comprehension, creativity and innovation in order to spur the intrinsic accomplishments [2]. In particular, the societal development is a continuous unabated process, interlocking from one generation to other generations. Although the development is necessary to achieve a decent enjoyment, various sustainable approaches have been explored and exploited recently in targeting the rapid transformation relevant to societal development. The approaches include modellings to deal with simulation, change of matters, time and space, analytical and statistical analyses and real case studies for future prediction [3,4]. The enhancement of the overall processes is undoubtedly essential in accounting for the escalation of various demands, stemming from the population growth to approximately 10 billion by 2050 [5]. The processes include physical, biological and chemical modes for sustainable technicality, treatment, bioremediation, control, production and development [6]. Ahead of all that, the advancement of applications corresponding to the feedstock and products, new materials, operations, systems, theories to know how and managing tangible and intangible resources are the indispensable prerequisites in attaining a sustainable societal development [7,8]. Therefore, the prime intention of this Special Issue is to document a novel "Sustainable Modellings, Processes and Applications for Societal Development". The gaps among societies would be eventually narrowed in creating a global harmony whilst enriching the natural environment.

Citation: Lim, J.W.; Kiatkittipong, W. Special Issue on "Sustainable Modellings, Processes and Applications for Societal Development". *Processes* **2022**, *10*, 1153. https://doi.org/10.3390/pr10061153

Received: 30 May 2022
Accepted: 6 June 2022
Published: 8 June 2022

Author Contributions: Conceptualization, J.W.L. and W.K.; resources, J.W.L. and W.K.; writing—original draft preparation, J.W.L. and W.K.; writing—review and editing, J.W.L. and W.K. All authors have read and agreed to the published version of the manuscript.

Funding: This research received no external funding.

Conflicts of Interest: The authors declare no conflict of interest.

References

1. Khalili, N.; Othman, M.; Abu Bakar, M.; Abdullah, L. Modelling of a Single Passage Air PV/T Solar Collector: Experimental and Simulation Design. *Processes* **2020**, *8*, 763. [CrossRef]
2. Leong, W.; Kiatkittipong, K.; Kiatkittipong, W.; Cheng, Y.; Lam, M.; Shamsuddin, R.; Mohamad, M.; Lim, J. Comparative Performances of Microalgal-Bacterial Co-Cultivation to Bioremediate Synthetic and Municipal Wastewaters Whilst Producing Biodiesel Sustainably. *Processes* **2020**, *8*, 1427. [CrossRef]
3. Wong, C.; Kiatkittipong, K.; Kiatkittipong, W.; Ntwampe, S.; Lam, M.; Goh, P.; Cheng, C.; Bashir, M.; Lim, J. Black Soldier Fly Larval Valorization Benefitting from Ex-Situ Fungal Fermentation in Reducing Coconut Endosperm Waste. *Processes* **2021**, *9*, 275. [CrossRef]
4. Raksasat, R.; Kiatkittipong, K.; Kiatkittipong, W.; Wong, C.; Lam, M.; Ho, Y.; Oh, W.; Suryawan, I.; Lim, J. Blended Sewage Sludge–Palm Kernel Expeller to Enhance the Palatability of Black Soldier Fly Larvae for Biodiesel Production. *Processes* **2021**, *9*, 297. [CrossRef]
5. Wong, C.; Kiatkittipong, K.; Kiatkittipong, W.; Lim, J.; Lam, M.; Wu, T.; Show, P.; Daud, H.; Goh, P.; Sakuragi, M. Elfis Rhizopus oligosporus-Assisted Valorization of Coconut Endosperm Waste by Black Soldier Fly Larvae for Simultaneous Protein and Lipid to Biodiesel Production. *Processes* **2021**, *9*, 299. [CrossRef]
6. Othman Zailani, N.; Yunus, N.; Ab Rahim, A.; Bustam, M. Thermophysical Properties of Newly Synthesized Ammonium-Based Protic Ionic Liquids: Effect of Temperature, Anion and Alkyl Chain Length. *Processes* **2020**, *8*, 742. [CrossRef]
7. Isiyaka, H.; Jumbri, K.; Sambudi, N.; Lim, J.; Saad, B.; Ramli, A.; Zango, Z. Experimental and Modeling of Dicamba Adsorption in Aqueous Medium Using MIL-101(Cr) Metal-Organic Framework. *Processes* **2021**, *9*, 419. [CrossRef]
8. Kumakiri, I.; Yokota, M.; Tanaka, R.; Shimada, Y.; Kiatkittipong, W.; Lim, J.; Murata, M.; Yamada, M. Process Intensification in Bio-Ethanol Production–Recent Developments in Membrane Separation. *Processes* **2021**, *9*, 1028. [CrossRef]

processes

Article

Removal of Nitrate Nitrogen by *Rhodotorula graminis* Immobilized in Alginate Gel for Groundwater Treatment

Masashi Nishida [1,*], Syuhei Matsuo [1], Karin Yamanari [1], Masayoshi Iwahara [2] and Katsuki Kusakabe [1]

[1] Department of Nanoscience, Sojo University, 4-22-1 Ikeda, Nishi-ku, Kumamoto 860-0082, Japan; g1512053@m.sojo-u.ac.jp (S.M.); g1812048@m.sojo-u.ac.jp (K.Y.); kusakabe@nano.sojo-u.ac.jp (K.K.)
[2] Department of Applied Microbial Technology, Sojo University, 4-22-1 Ikeda, Nishi-ku, Kumamoto 860-0082, Japan; iwahara@bio.sojo-u.ac.jp
* Correspondence: nishida@nano.sojo-u.ac.jp

Abstract: Groundwater is the source of all tap water in Kumamoto City, Japan. However, the concentration of nitrate nitrogen (NO_3^--N) tends to increase every year due to the influences of over-fertilization, field disposal of livestock manure, and inflow of domestic wastewater. A heterotrophic nitrification–aerobic denitrification (HN-AD) system is an attractive approach for nitrate-nitrogen removal. In this study, *Rhodotorula graminis* NBRC0190, a naturally occurring red yeast that shows high nitrogen removal performance in glucose, was immobilized on calcium alginate hydrogel beads. NO_3^--N removal efficiency exceeded 98% in the region of NO_3^--N concentration below 10 mg/L in the model groundwater. Even after the same treatment was repeated five times, the denitrification performance of the *R. gra* immobilized alginate hydrogel beads was maintained. Finally, when this treatment method was applied to actual groundwater in Kumamoto City, it was possible to make the water of even higher quality.

Keywords: *Rhodotorula graminis*; nitrate nitrogen; groundwater; heterotrophic nitrification; aerobic denitrification; nitrogen removal

Citation: Nishida, M.; Matsuo, S.; Yamanari, K.; Iwahara, M.; Kusakabe, K. Removal of Nitrate Nitrogen by *Rhodotorula graminis* Immobilized in Alginate Gel for Groundwater Treatment. *Processes* **2021**, *9*, 1657. https://doi.org/10.3390/pr9091657

Academic Editors: Worapon Kiatkittipong, Yoshimitsu Uemura and Jun-Wei Lim

Received: 4 August 2021
Accepted: 11 September 2021
Published: 14 September 2021

1. Introduction

Kumamoto City, with a population of 0.7 million, uses groundwater as its water source for all its tap water. The water quality of groundwater sources in the city is comparable to commercial mineral water, except for nitrate nitrogen [1]. The concentration of nitrate nitrogen in the groundwater tends to increase every year due to the influences of overfertilization, field disposal of livestock manure, and inflow of domestic wastewater. The removal of nitrate nitrogen from groundwater is an important issue to be solved from the perspective of maintaining sustainability. Excess organic nitrogen entering the soil decomposes into ammonium, nitrite, and nitrate ions in the soil. These ions return to nitrogen molecules and organic nitrogen through denitrification and assimilation, respectively. In this way, the nitrogen cycle proceeds, but excess nitrate nitrogen pollutes groundwater. Various treatments have been applied to remove nitrate nitrogen from water sources; i.e., biological removal [2–5], ion exchange [6,7], reverse osmosis [8], and chemical reduction [9–11].

Biological nitrification and denitrification are widely applied because they do not pollute the environment and have low maintenance costs. Traditionally, nitrification reactions under aerobic conditions and denitrification reactions under anaerobic conditions have been treated in independent processes. Robertson and Kuenen [12] reported that *Thiosphaera pantotropha*, a heterotrophic bacterium, acts on the denitrification reaction even under aerobic conditions. The heterotrophic–aerobic denitrification system using such bacteria has attracted much attention because of their advantages such as high tolerance in acidic condition, high growth rate, and simultaneous nitrification and denitrification [13]. Bacteria with the ability to remove nitrogen by HN-AD include *Agrobacterium* sp. LAD9 [14], *Acinetobacter junii* YB [15], *Pseudomonas stutzeri* T1 [16], *Pseudomonas tolaasii* Y-11 [17], and

Sporidiobolus pararoseus Y1 [4]. These studies have focused on wastewater treatment, and there have been no reports of denitrification reactions in the dilute concentration range of nitrate nitrogen below 10 mg/L.

Rhodotorula species have been found in soil, air, and plant-associated organisms in a variety of environments ranging from the deep sea to deserts [18,19]. *Rhodotorula graminis* has properties suitable for the production of L-Phenylalanine ammonia-lyase [20] and lipids [21]. However, water treatment using *Rhodotorula graminis* has not yet been studied.

In the previous research [22], we found that the environmental-friendly and naturally occurring yeast *Rhodotorula graminis* NBRC0190 (hereinafter referred to as *R. gra*) has a high nitrate assimilation performance. Evaluated by *R. gra* flask culture experiments, it was found that the system could remove 97% and 99% of nitrate nitrogen from dairy well water (nitrate-nitrogen concentration = 23 mg/L) and treated water of a water purification center (nitrate-nitrogen concentration = 45 mg/L), respectively. For practical application, the immobilization of *R. gra* on porous substrates is necessary. The advantages of the yeast immobilization system include increased activity due to high accumulation, reduced risk of coexisting bacterial contamination, and recycling of yeast. Calcium alginate hydrogel beads are commonly used carriers for the immobilization of biocatalysts [23] due to their advantages of low cost, high porosity, and simplicity of preparation.

In this study, *R. gra* was immobilized in calcium alginate hydrogel beads. A leakage test of *R. gra* from *R. gra*-immobilized alginate beads was carried out. The immobilized *R. gra* beads were added to model water containing 10 mg/L nitrate nitrogen and incubated with shaking to evaluate the nitrate-nitrogen removal, ammonia-nitrogen removal, and repeated removal performance. Finally, nitrate-nitrogen removal tests were conducted using groundwater from two locations in Kumamoto City.

2. Materials and Methods

2.1. Materials

R. gra was used after thawing a lyophilized product purchased from the Biological Resource Center of the National Institute of Technology and Evaluation (NITE), Japan. Glucose was used as the carbon source due to the low concentration of organic carbon in groundwater. First, 0.6 g of sodium nitrate and 2 g of glucose were dissolved in 200 mL of Czapek liquid medium and sterilized in an autoclave at 110 °C for 10 min. After sterilization, the inoculating loop of *R. gra* was dipped into this medium. Shaking the culture was performed at 20 °C and 50 rpm for 3–4 days. The free organisms of *R. gra* were obtained by repeatedly centrifuging the pre-cultured suspension at 3000 rpm for 15 min and then added to the Czapek liquid medium. For immobilized yeast, *R. gra* suspension was washed by centrifugation with sterilized water and added to 0.5, 1.0, 1.5, and 2.0% sodium alginate solution, and then, 1% $CaCl_2 \cdot H_2O$ solution was added dropwise. The beads solidified into a spherical shape at all concentrations of sodium alginate except 0.5%. After the sample was allowed to stand for 30 min, the immobilized yeasts were collected by filtering and added to Czapek liquid medium.

Model water containing 10 mg/L nitrate nitrogen was prepared using ion-exchange water and $NaNO_3$. In the previous study [22], we used a nitrogen concentration of 25–200 mg/L and a glucose carbon concentration of 500–4000 mg/L to remove nitrate nitrogen by *R. gra* yeast. In the process of removing nitrate ions while growing this yeast, we confirmed that a C/N ratio of 20 was sufficient, and therefore, the glucose concentration was set at 200 mg/L in this study. Two types of groundwater were collected from Kumamoto city, Japan. These waters had been sterilized in an autoclave at 110 °C for 10 min for use in the experiments.

2.2. Methods

R. gra suspension (1 mL) or immobilized yeast suspension (10 mL) was added to 100 mL of the water medium containing 10 mg/L nitrate nitrogen, and the nitrate assimilation reaction was carried out. The turbidity of the medium was assessed using a UV-vis spectrometer at 660 nm. It was confirmed that there was a direct proportional relationship between the value of turbidity and the number of *R. gra* yeast counted by the microscopic method. The sample solution was centrifuged at 3000 rpm for 10 min, and the supernatant liquid was filtered through a 0.45 μm membrane filter to measure the nitrate-nitrogen concentration and ammonia-nitrogen concentration. Determination of nitrate nitrogen was performed by ion chromatography (IA-100, DKK-TOA Corporation, Tokyo, Japan). Ammonia nitrogen and residual sugar concentrations were determined using a UV-vis spectrophotometer by the indophenol blue method (630 nm) and Somogyi–Nelson method (660 nm), respectively.

3. Results

3.1. Leak of *R. gra* Yeast from Alginate Hydrogel Beads

Figure 1 shows the change in turbidity caused by *R. gra* leaked from alginate beads during the nitrate assimilation reaction. At a sodium alginate concentration of 2.0%, the turbidity was slightly higher than at 1.0 and 1.5%, suggesting that the leakage of *R. gra* slightly increased due to the volume expansion during the preparation of alginate hydrogel beads. However, the turbidity increased in the early stage of the nitrate assimilation reaction, but it did not change significantly in the subsequent reaction period. This suggests that the initial increase is due to the leakage of *R. gra* attached to the surface of alginate beads. Compared to the turbidity of the free bacterial suspension, the increase in turbidity of the alginate beads was about 5% or less. In the subsequent experiments, alginate beads were prepared with a concentration of 1.5% sodium alginate and 1% calcium chloride.

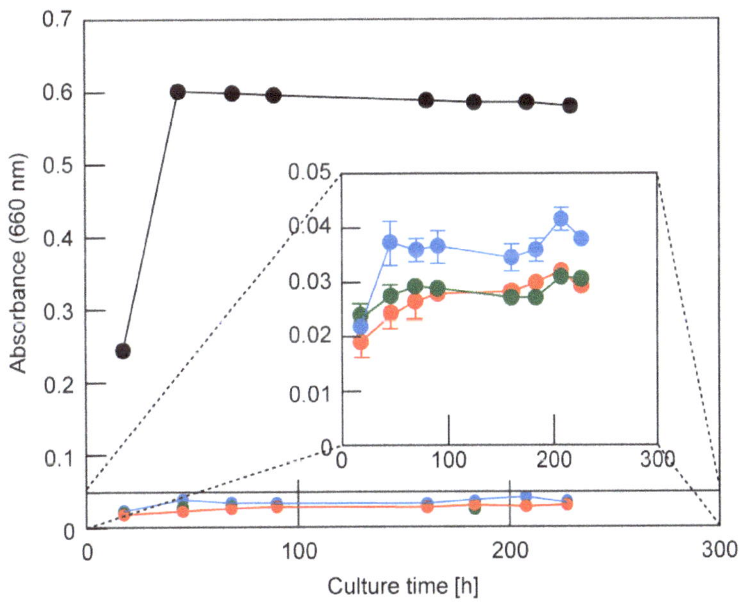

Figure 1. Change of turbidity (absorbance at 660 nm) of the medium caused by *R. gra* leaked from alginate hydrogel beads during the nitrate assimilation reaction: *R. gra* suspension (1 mL) or immobilized yeast suspension (10 mL) was added to 100 mL of the water medium containing 10 mg/L nitrate nitrogen. Sodium alginate concentration in the preparation of alginate beads with *R. gra*, red circle, 1.0%, green circle, 1.5%, blue circle, 2.0%, black circle free *R. gra* suspension.

3.2. Effect of Yeast Density on Nitrate-Nitrogen Removal

Figure 2 shows the effect of yeast density in alginate beads on nitrate-nitrogen removal. The yeast density in the beads was adjusted by changing the volume of 1.5% sodium alginate solution and the volume of *R. gra* suspension during the preparation of alginate beads. The yeast density in alginate beads prepared by adding 1 mL of *R. gra* suspension to 10 mL of sodium alginate solution is equivalent to 1×10^7 cells/mL. To eliminate the effect of yeast leakage, the alginate beads that were re-cultured after one week of nitrate assimilation reaction were used for this experiment. As shown in Figure 2, there was no change in the removal rate of nitrate nitrogen by changing the density in the alginate beads. This suggests that the reaction might be conducted on the surface without diffusing into the beads at low nitrate-nitrogen concentration. In all conditions, the nitrate-nitrogen concentration was below 0.2 mg/L, the detection limit by the ion chromatography method, after 24 h of incubation. There are no data on the growth of *R. gra* in alginate beads, but the beads became darker in pink color due to *R. gra* as the reaction progressed. In this study, the effect of yeast density on the nitrogen assimilation reaction was small. However, the activity of *R. gra* tended to decrease gradually for the high yeast density in alginate beads and might be affected by miscellaneous germs for the low yeast density. Therefore, as a standard preparation condition for alginate beads, the liquid volume of sodium alginate to be mixed with 1 mL of *R. gra* suspension was set at 10 mL.

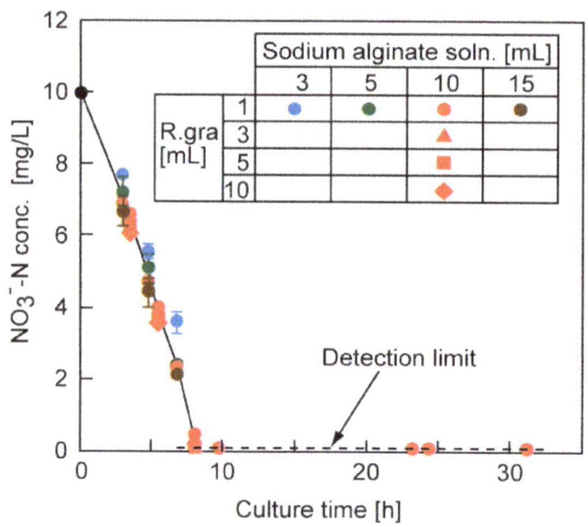

Figure 2. Change in nitrate-nitrogen concentration at various yeast densities in alginate hydrogel bead. The detection limit of nitrate nitrogen by the ion chromatography method is 0.2 mg/L: Red symbols, Volume of *R. gra* suspension was varied from 1, 3, 5, and 10 mL to 1 mL of sodium alginate of 1.5% concentration. Circular symbols, Volume of sodium alginate at a concentration of 1.5% was varied from 3, 5, 10, and 15 mL to 1 mL of *R. gra* suspension.

Residual carbon in drinking water is undesirable because it is a precursor for trihalomethane formation [24]. In these experiments with a C/N ratio of 20, the glucose concentration decreased from 200 mg/L to less than 10 mg/L, which is the qualitative limit of the Somogyi–Nelson method, in 24 h. As a result, it was confirmed that the amount of glucose at C/N ratio 20 was sufficient for *R. gra* immobilized alginate beads as well as for the removal of nitrate nitrogen in *R. gra* suspensions [22].

Zen et al. [4] used a red yeast, *Sporidiobolus pararoseus* Y1, for the denitrification reaction of nitrate nitrogen. The results showed that the nitrogen removal efficiency with glucose was 92.2% at a nitrate-nitrogen concentration of 14 mg/L for 72 h, and the average removal rate was 0.39 mg·L^{-1}·h^{-1}. The results shown in Figure 3 indicate that the immobilized *R. gra* has a high performance in nitrogen removal efficiency and the removal rate of more than 98% and 1.0 mg·L^{-1}·h^{-1}, respectively.

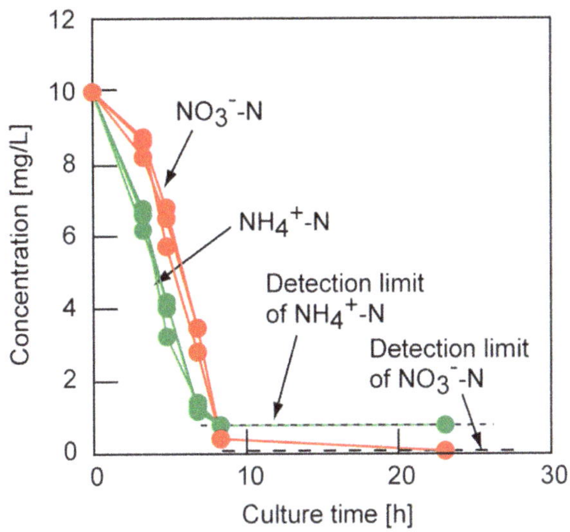

Figure 3. Comparison of nitrogen removal between denitrification of nitrate nitrogen and nitrification of ammonia nitrogen by alginate beads: red circle, nitrate-nitrogen concentration, green circle, ammonia-nitrogen concentration. The detection limit of nitrate nitrogen by the ion chromatography method is 0.2 mg/L. The alginate beads were prepared by adding 1 mL of *R. gra* suspension to 10 mL of 1.5% concentration sodium alginate solution. The detection limit of nitrate nitrogen by the ion chromatography method is 0.2 mg/L, and that of ammonia nitrogen by the indophenol blue method is 0.8 mg/L.

3.3. *Simultaneous Removal of Nitrate Nitrogen and Ammonia Nitrogen*

Nitrate-nitrogen removal and ammonia-nitrogen removal experiments were performed independently using *R. gra*-immobilized alginate beads prepared under standard conditions (10 mL of 1.5% concentration sodium alginate solution and 1 mL of *R. gra* suspension). As shown in Figure 3, ammonia nitrogen was removed at the same rate as nitrate-nitrogen removal, and it reached 0.8 mg/L, the detection limit by the indophenol blue method, in 8 h. In the ammonia-nitrogen removal experiment, no nitrite nitrogen could be detected in the samples. Similar results have been obtained with ammonia-nitrogen removal by the red yeast *Sporidiobolus pararoseus* Y1 [4]. In this study, the removal rate of ammonia nitrogen was almost equal to that of nitrate nitrogen, suggesting that the removal of nitrate nitrogen is the rate-determining step.

Figure 4 shows the performance of *R. gra* yeast in removing ammonia nitrogen and nitrate nitrogen under coexisting conditions. In this case, the total concentration of ammonia nitrogen plus nitrate nitrogen was 20 mg/L, but the elemental ratio of glucose to nitrogen was set at C/N = 20. Comparing Figures 3 and 4, it was found that the removal rate of ammonia nitrogen did not change much. The concentration of nitrate nitrogen decreased simultaneously with ammonia nitrogen in the early stage of the reaction, but it increased at 5 h of incubation due to the formation of nitrate nitrogen by nitrification of ammonia nitrogen. Then, after the ammonia nitrogen reached the detection limit, it gradually decreased and reached the detection limit of nitrate nitrogen in about 20 h. In these experiments, the initial glucose concentration of 400 mg/L was reduced to less than 10 mg/L, the detection limit by the Somogy–Nelson method, after 24 h.

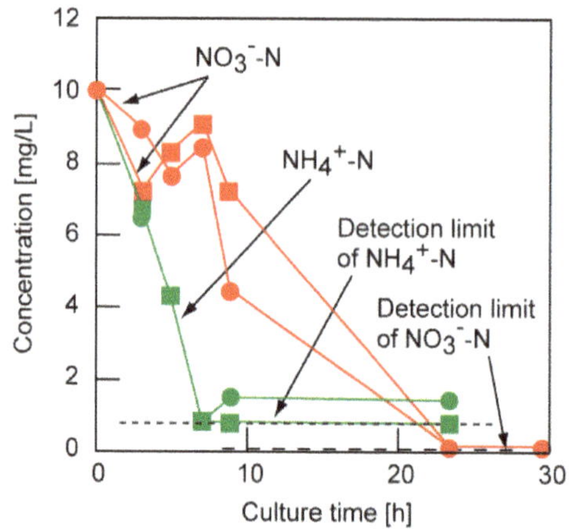

Figure 4. Simultaneous nitrogen removal of nitrate nitrogen and ammonia nitrogen by alginate beads: red symbols, nitrate-nitrogen concentration, green symbols, ammonia-nitrogen concentration, Circle and square symbols indicate first run and second run, respectively. The detection limit of nitrate nitrogen by the ion chromatography method is 0.2 mg/L, and that of ammonia nitrogen by the indophenol blue method is 0.8 mg/L.

3.4. Repeated Use of Alginate Beads Immobilized with R. gra

To examine the durability of *R. gra* immobilized alginate beads, the nitrate assimilation reaction was first carried out. The concentration of nitrate nitrogen reached below the detection limit in 24 h, as shown in Figure 5, and then, the beads were left to stand for 1 week. After one week, the immobilized beads were transferred to a new medium containing nitrate nitrogen and glucose, and the nitrate assimilation reaction was performed again. In total, five nitrate assimilation reactions were performed over a period of 5 weeks. Excess glucose (2000 mg/L) was added for the first incubation, but 200 mg/L glucose concentration was added for the second and subsequent incubations. As shown in Figure 5, there was no significant difference in the removal rate of nitrate nitrogen even after five repetitions. Even when excess glucose was added in the first reaction, the removal rate did not change. These results showed that the activity of the *R. gra*-immobilized alginate gel was maintained even after the glucose in the solution was completely consumed in one week.

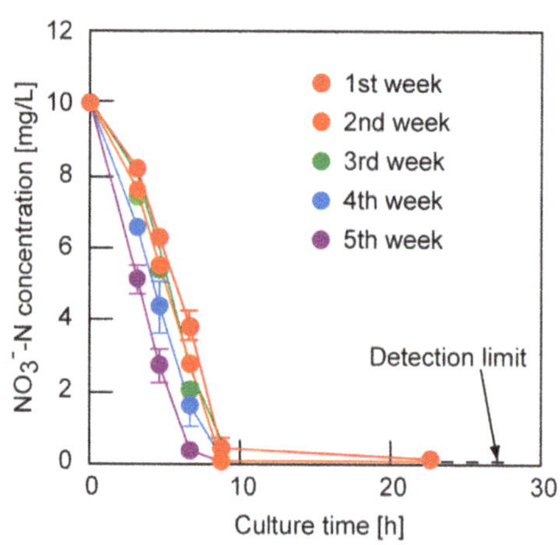

Figure 5. Repeated denitrification properties of alginate hydrogel beads with *R. gra*.

3.5. Denitrification of Ground Water

In the water treatment experiments of groundwater in Kumamoto City, glucose with a mass ratio C/N = 20 was dissolved in 100 mL of the sterilized groundwater, and then, 5 mL of immobilized *R. gra* was added. The mixed solution was incubated at 20 °C and 50 rpm for 5 days with shaking. The ion concentrations were measured after the denitrification reaction.

Table 1 summarizes the ion concentrations in groundwater collected from two wells in Kumamoto City and treated with immobilized *R. gra*. These groundwaters contain relatively high concentrations of nitrate ions, 36.4 mg/L and 32.1 mg/L. These values were converted to nitrate-nitrogen concentrations of 8.22 mg/L and 7.25 mg/L, which are close to the equivalent Japanese environmental quality standard of 10 mg/L. Water treatment with *R. gra* reduced the nitrate ion concentration to below the detection limit for well #1 and to 0.60 mg/L for well #2 at a culture time of 120 h. In actual groundwater, more than 98% of nitrate ions were removed. In the treatment of groundwater, the Ca^{2+} and Cl^- ion concentrations increased due to leaching from the alginate hydrogel beads after 5 days of treatment, but in the actual process, the increase in these concentrations could be controlled by optimizing the denitrification time. In drinking water treatment plants that use groundwater as their source of water, the addition of glucose does not significantly affect the overall cost because the cost of coagulants and other chemicals required for advanced treatment is not required.

Table 1. Changes in water quality due to denitrification of groundwater.

	Well #1 (Tensuiko) [a]		Well #2 (Myoukensan) [b]	
	Before Water Treatment	**After Water Treatment**	**Before Water Treatment**	**After Water Treatment**
$[NO_3^-]$ [mg/L]	36.4	<0.2	32.1	0.60
$[Cl^-]$ [mg/L]	6.5	22.8	9.0	43.2
$[SO_4^{2-}]$ [mg/L]	14.6	13.9	24.0	22.6
$[PO_4^{3-}]$ [mg/L]	<1	<1	<1	<1
$[NH_4^+]$ [mg/L]	<0.1	0.15	<0.1	<0.1
$[Na^+]$ [mg/L]	10.5	10.4	17.4	17.1
$[K^+]$ [mg/L]	2.9	3.6	9.4	10.3
$[Mg^{2+}]$ [mg/L]	6.2	5.0	5.8	4.8
$[Ca^{2+}]$ [mg/L]	14.2	26.7	17.8	37.0
pH	6.7	6.9	6.5	6.8

[a] Kawautimachi, Nishi-ku, Kumamoto, [b] Mitsugumachi, Kitaku, Kumamoto, The alginate beads were prepared by adding 0.5 mL of *R. gra* suspension to 5 mL of 1.5% concentration sodium alginate solution. Experimental condition (Shaking culture was performed at 20 °C and 50 rpm for 5 days., groundwater 100 mL, *R. gra* immobilized alginate hydrogel beads 5 mL, C/N = 20).

4. Conclusions

The leakage of *R. gra* immobilized on alginate beads was evaluated by turbidity; the leakage increased after 50 h of incubation, but thereafter, the change in turbidity was small, indicating that *R. gra* was stably immobilized. The nitrate-nitrogen removal experiments were conducted at an initial nitrate-nitrogen concentration of 10 mg/L, assuming groundwater treatment. The density of *R. gra* in alginate beads was varied by changing the amount of 1.5% sodium alginate solution and the amount of *R. gra* suspension. However, it had little effect on the removal behavior of nitrate nitrogen. The immobilized *R. gra* showed high performance in nitrogen removal efficiency and the removal rate of more than 98% and 1.0 $mg.L^{-1}{\cdot}h^{-1}$, respectively. Ammonia-nitrogen removal by *R. gra* immobilized in the alginate beads alone in a solution with an initial ammonia-nitrogen concentration of 10 mg/L showed almost the same nitrogen removal property as the nitrate nitrogen. However, in a mixed solution containing 10 mg/L of both nitrate nitrogen and ammonia nitrogen, the removal of ammonia nitrogen proceeded regardless of the presence of nitrate nitrogen at the beginning of the reaction, and then, the concentration of nitrate nitrogen slightly increased due to the nitrification reaction, and then the concentration decreased to the detection limit of nitrate nitrogen within 24 h. Even after the same treatment was repeated five times, the denitrification performance of the *R. gra* immobilized alginate hydrogel beads was maintained. Finally, when the treatment method of this study was applied to actual groundwater in Kumamoto City, the nitrate ion concentration was reduced to below the detection limit for well #1 and to 0.60 mg/L for well #2 at a culture time of 120 h. In actual groundwater, more than 98% of nitrate ions were removed. In addition, the Ca^{2+} and Cl^- ion concentrations increased due to leaching from the alginate hydrogel beads after 5 days of treatment.

In order to apply the removal of nitrate nitrogen by *R. gra* immobilized alginate beads in a drinking water treatment plant, the removal rate of nitrate nitrogen in a packed bed bioreactor must be considered. The post-treatment of treated water containing residual *R. gra*, etc. by filtration, adsorption, or disinfection should be investigated in the future.

Author Contributions: Conceptualization, M.N.; Formal analysis, M.I.; Investigation, S.M. and K.Y.; Supervision, M.N.; Validation, K.K.; Writing—original draft, M.N. All authors have read and agreed to the published version of the manuscript.

Funding: The authors thank the financial support from Office Yokoo Company, Japan.

Institutional Review Board Statement: Not applicable.

Informed Consent Statement: Not applicable.

Data Availability Statement: Data is contained within the article.

Acknowledgments: Insightful comments from reviewers are acknowledged.

Conflicts of Interest: The authors declare no conflict of interest.

References

1. Tomie, K.; Iwasa, Y.; Maeda, K.; Otsuzuki, M.; Yunoue, T.; Kakimoto, R.; Kawagoshi, Y. Present status and feature of groundwater contaminations by nitrate-nitrogen in Kumamoto city. *J. Waste Environ. Technol.* **2009**, *7*, 19–28. [CrossRef]
2. Park, J.Y.; Yoo, Y.J. Biological nitrate removal in industrial wastewater treatment: Which electron donor we can choose. *Appl. Microbiol. Biotechnol.* **2009**, *82*, 415–429. [CrossRef]
3. Gómez-Guzmán, A.; Jiménez-Magaña, S.; Guerra-Renteria, A.S.; Gómez-Hermosillo, C.; Parra-Rodriguez, F.J.; Velázquez, S.; Aguilar-Uscanga, B.R.; Solis-Pacheco, J.; González-Reynoso, O. Evaluation of nutrients removal (NO_3-N, NH_3-N and PO_4-P) with *Chlorella vulgaris, Pseudomonas putida, Bacillus cereus* and a consortium of these microorganisms in the treatment of wastewater effluents. *Water Sci. Technol.* **2017**, *76*, 49–56. [CrossRef]
4. Zeng, J.; Liao, S.; Qiu, M.; Chen, M.; Ye, J.; Zeng, J.; Wang, A. Effects of carbon sources on the removal of ammonium, nitrite and nitrate nitrogen by the red yeast *Sporidiobolus pararoseus* Y1. *Bioresour. Technol.* **2020**, *312*, 123593. [CrossRef]
5. Rezvani, F.; Sarrafzadeh, M.-H.; Ebrahimi, S.; Oh, H.-M. Nitrate removal from drinking water with a focus on biological methods: A review. *Environ. Sci. Pollut. Res.* **2019**, *26*, 1124–1141. [CrossRef]
6. Primo, O.; Rivero, M.J.; Urtiaga, A.M.; Ortiz, I. Nitrate removal from electro-oxidized landfill leachate by ion exchange. *J. Hazard. Mater.* **2009**, *164*, 389–393. [CrossRef]
7. Samatya, S.; Kabay, N.; Yüksel, Ü.; Arda, M.; Yüksel, M. Removal of nitrate from aqueous solution by nitrate selective ion exchange resins. *React. Funct. Polym.* **2006**, *66*, 1206–1214. [CrossRef]
8. Schoeaman, J.J.; Steyn, A. Nitrate removal with reverse osmosis in a rural area in South Africa. *Desalination* **2003**, *155*, 15–26. [CrossRef]
9. Marchesini, F.A.; Picard, N.; Miró, E.E. Study of the interactions of Pd, In with SiO_2 and Al_2O_3 mixed supports as catalysts for the hydrogenation of nitrates in water. *Catal. Commun.* **2012**, *21*, 9–13. [CrossRef]
10. Liu, Y.; Gong, X.; Yang, W.; Wang, B.; Yang, Z.; Liu, Y. Selective reduction of nitrate into nitrogen using Cu/Fe bimetal combined with sodium sulfite. *Ind. Eng. Chem. Res.* **2019**, *58*, 5175–5185. [CrossRef]
11. Villora-Picó, J.J.; Belda-Alcázar, V.; Garcia-Fernández, M.J.; Serrano, E.; Sepúlveda-Escribano, A.; Pastor-Blas, M.M. Conducting polymer−TiO_2 hybrid materials: Application in the removal of nitrates from water. *Langmuir* **2019**, *35*, 6089–6105. [CrossRef]
12. Robertson, L.A.; Kuenen, J.G. Aerobic denitrifications: A controversy revived. *Arch. Microbiol.* **1984**, *139*, 351–354. [CrossRef]
13. Yang, J.-R.; Wang, Y.; Chen, H.; Lyu, Y.-K. Ammonium removal characteristics of an acid-resistant bacterium *Acinetobacter* sp. JR1 from pharmaceutical wastewater capable of heterotrophic nitrification-aerobic denitrification. *Bioresour. Technol.* **2019**, *274*, 56–64. [CrossRef]
14. Chen, Q.; Ni, J. Ammonium removal by *Agrobacterium* sp. LAD9 capable of heterotrophic nitrification–aerobic denitrification. *J. Biosci. Bioeng.* **2012**, *113*, 619–623. [CrossRef] [PubMed]

15. Ren, Y.-X.; Yang, L.; Liang, X. The characteristics of a novel heterotrophic nitrifying and aerobic denitrifying bacterium, *Acinetobacter junii* YB. *Bioresour. Technol.* **2014**, *171*, 1–9. [CrossRef]
16. Guo, L.; Chen, Q.; Fang, F.; Hu, Z.; Wu, J.; Miao, A.; Xiao, L.; Chen, X.; Yang, L. Application potential of a newly isolated indigenous aerobic denitrifier for nitrate and ammonium removal of eutrophic lake water. *Bioresour. Technol.* **2013**, *142*, 45–51. [CrossRef]
17. He, T.; Li, Z.; Sun, Q.; Xu, Y.; Ye, Q. Heterotrophic nitrification and aerobic denitrification by *Pseudomonas tolaasii* Y-11 without nitrite accumulation during nitrogen conversion. *Bioresour. Technol.* **2016**, *200*, 493–499. [CrossRef] [PubMed]
18. Margesin, R.; Fonteyne, P.-A.; Schinner, F.; Sampaio, J.P. *Rhodotorula psychrophila* sp. nov., *Rhodotorula psychrophenolica* sp. nov. and *Rhodotorula glacialis* sp. nov., novel psychrophilic basidiomycetous yeast species isolated from alpine environments. *Int. J. Syst. Evol. Microbiol.* **2007**, *57*, 2179–2184. [CrossRef] [PubMed]
19. Zalar, P.; Gunde-Cimerman, N. Cold-adapted yeasts in arctic habitats. In *Cold-Adapted Yeasts: Biodiversity, Adaptation Strategies and Biotechnological Significance*; Buzzini, P., Margesin, R., Eds.; Springer: Berlin/Heidelberg, Germany, 2014; pp. 49–74.
20. Orndorff, S.A.; Costantino, N.; Stewart, D.; Durham, D.R. Strain improvement of *Rhodotorula graminis* for production of a novel L-phenylalanine ammonia-lyase. *Appl. Environ. Microbial.* **1988**, *54*, 996–1002. [CrossRef] [PubMed]
21. Galafassi, S.; Cucchetti, D.; Pizza, F.; Franzosi, G.; Bianchi, D.; Compagno, C. Lipid production for second generation biodiesel by the oleaginous yeast *Rhodotorula graminis*. *Bioresour. Technol.* **2014**, *111*, 398–403. [CrossRef] [PubMed]
22. Nishida, M.; Hatta, T.; Iwahara, M. Removal properties of nitrate nitrogen in underground water by the nitrate assimilation yeast *Rhodotorula* sp. *J. Water Waste* **2020**, *62*, 39–44.
23. Zhou, Z.; Li, G.; Li, Y. Immobilization of *Saccharomyces cerevisiae* alcohol dehydrogenase on hybrid alginate-chitosan beads. *Int. J. Biol. Macromol.* **2010**, *47*, 21–26. [CrossRef] [PubMed]
24. Soares, M.I.M. Biological denitrification of groundwater. *Water Air Soil Poll.* **2020**, *123*, 183–193. [CrossRef]

processes

MDPI

Article

Experimental and Modeling of Dicamba Adsorption in Aqueous Medium Using MIL-101(Cr) Metal-Organic Framework

Hamza Ahmad Isiyaka [1,*], Khairulazhar Jumbri [1,2,*], Nonni Soraya Sambudi [3], Jun Wei Lim [1,4], Bahruddin Saad [1], Anita Ramli [1,4] and Zakariyya Uba Zango [1]

1 Department of Fundamental and Applied Sciences, Universiti Teknologi PETRONAS, Seri Iskandar 32610, Perak, Malaysia; junwei.lim@utp.edu.my (J.W.L.); bahruddin.saad@utp.edu.my (B.S.); anita_ramli@utp.edu.my (A.R.); zakariyya_17008416@utp.edu.my (Z.U.Z.)
2 Centre of Research in Ionic Liquids (CORIL), Institute of Contaminant Management, Universiti Teknologi PETRONAS, Seri Iskandar 32610, Perak, Malaysia
3 Department of Chemical Engineering, Universiti Teknologi PETRONAS, Seri Iskandar 32610, Perak, Malaysia; soraya.sambudi@utp.edu.my
4 HICoE-Centre for Biofuel and Biochemical Research, Institute of Self-Sustainable Building, Universiti Teknologi PETRONAS, Seri Iskandar 32610, Perak, Malaysia
* Correspondence: hamza_18001996@utp.edu.my (H.A.I.); khairulazhar.jumbri@utp.edu.my (K.J.)

Citation: Isiyaka, H.A.; Jumbri, K.; Sambudi, N.S.; Lim, J.W.; Saad, B.; Ramli, A.; Zango, Z.U. Experimental and Modeling of Dicamba Adsorption in Aqueous Medium Using MIL-101(Cr) Metal-Organic Framework. *Processes* **2021**, *9*, 419. https://doi.org/10.3390/pr9030419

Academic Editor: María V. López-Ramón

Received: 28 December 2020
Accepted: 18 January 2021
Published: 26 February 2021

Abstract: Drift deposition of emerging and carcinogenic contaminant dicamba (3,6-dichloro-2-methoxy benzoic acid) has become a major health and environmental concern. Effective removal of dicamba in aqueous medium becomes imperative. This study investigates the adsorption of a promising adsorbent, MIL-101(Cr) metal-organic framework (MOF), for the removal of dicamba in aqueous solution. The adsorbent was hydrothermally synthesized and characterized using N_2 adsorption-desorption isotherms, Brunauer, Emmett and Teller (BET), powdered X-ray diffraction (XRD), Fourier Transformed Infrared (FTIR) and field emission scanning electron microscopy (FESEM). Adsorption models such as kinetics, isotherms and thermodynamics were studied to understand details of the adsorption process. The significance and optimization of the data matrix, as well as the multivariate interaction of the adsorption parameters, were determined using response surface methodology (RSM). RSM and artificial neural network (ANN) were used to predict the adsorption capacity. In each of the experimental adsorption conditions used, the ANN gave a better prediction with minimal error than the RSM model. The MIL-101(Cr) adsorbent was recycled six times to determine the possibility of reuse. The results show that MIL-101(Cr) is a very promising adsorbent, in particular due to the high surface area (1439 $m^2\ g^{-1}$), rapid equilibration (~25 min), high adsorption capacity (237.384 mg g^{-1}) and high removal efficiency of 99.432%.

Keywords: adsorption; dicamba; artificial neural network model; response surface methodology; metal-organic framework

1. Introduction

Anthropogenic activities such as crop cultivation, industrial processes and sewage discharge result in the contamination of surface and ground water resources [1]. Herbicides such as dicamba (3,6-dichloro-2-methoxy benzoic acid) are widely used to selectively kill broad leave weeds that affect crop areas, gardens and road sides [2]. When applied in excess, their residue remains in the environment and can be transported from point source to nonpoint sources through leaching, run-off, subsurface drainage and spray drift [3]. Drift deposition of dicamba to non-intended areas has become a major environmental concern, as it directly affects vulnerable crops even at low concentrations [4]. In the United States of America (USA), an estimate of 1.5 million hectares of non-target soybeans were destroyed by dicamba herbicides in 2017 due to uncontrolled drift and extend to 2018. The

US Environmental Protection Agency (USEPA) implemented restrictio on the application of dicamba in 2018. In 2019 the USEPA canceled the registration of dicamba herbicide that restrict farmers to buy and use the products legally (USEPA, 2020).

The US and European Community Environmental Protection Agencies have listed dicamba as a priority pollutant with possible carcinogenic and mutagenic effects [5]. Dicamba is easily bioaccumulated and biomagnified in the tissues of plants and aquatic animals, which poses a serious health risk to humans and the environment at large [6]. Despite the environmental and health consequences of dicamba, many countries still adopt it as an alternative to pest control. Hence, the removal of this toxic contaminant from water becomes imperative.

Over the years, several physical and chemical treatment techniques such as advanced oxidation process, adsorption, bioremediation, membrane filtration have been applied for the removal of toxic contaminants in water [7]. One promising method that has been singled out and applied for the remediation of recalcitrant contaminants in water is adsorption, due to its low cost, simple operations, high selectivity, environmental benignity, convenient recycling and availability of alternative materials [8,9]. Adsorbents such as mesoporous silica [10], polymer [11] and clay material [12] have also been tested for the removal of dicamba in aqueous medium. Yet, the ideal adsorbent for real world application with high surface area, large pore volume, good water and thermal stability, fast equilibration time and easy regeneration remains elusive [13].

Recently, porous materials such as metal-organic frameworks (MOFs) have received considerable attention from researchers for application in water treatment, catalysis, gas sensing, biomedical imaging and drug delivery [14,15]. MOFs are a new class of advanced and porous materials that consist of a cluster of transition metal ion and organic linkers. The high surface area, porous nature, multifunctionality, tunable pore size make MOFs a unique material of interest in wastewater remediation [16,17]. Among the several MOFs reported, the MIL-101(Cr) is an exceptionally promising material that has been applied for the removal of contaminants such as pharmaceuticals, dyes and heavy metals in wastewaters [18]. The MIL-101(Cr) is formed from a combination of chromium (III) oxide octahedral trimers and dicarboxylate linker, resulting in a high class of hybrid supertetrahedron azeotypic mesoporous material [19].

The majority of adsorption studies only vary one parameter at a time; however, it has been recognized that several parameters often act simultaneously on the adsorption process. This conventional 'one-parameter-at-a-time' optimization approach is not only time consuming, costly for industrial applications, but the shared interactions and impacts of other parameters working together are not considered. In this study, we introduce a mathematical and intelligent algorithm that works like the structure of the human neurons using the central composite design response surface methodology and artificial intelligence. This is done to determine the effects and provide due consideration to the shared interactions of the adsorption process between dicamba and MIL-101(Cr) MOF. Additionally, the kinetics, isotherms and thermodynamic processes that affect the adsorption were also studied.

2. Materials and Methods

Chromium nitrate nonahydrate ($Cr(NO_3)_3{\cdot}9H_2O$, 99%), 1,4-benzene dicarboxylic acid (H_2BDC, 99%), hydrochloric acid (HCl), HF (98%), acetone (98%), N-dimethyl formamide (DMF, 99%), ethanol (99.9%), methanol (99%), and sodium hydroxide were purchased from Avantis Laboratory (Perak, Malaysia) and were used without further purification. Dicamba was sourced from Sigma-Aldrich (St. Louis, MO, USA).

2.1. *Synthesis of MIL-101(Cr) MOF*

The adsorbent was synthesized hydrothermally based on a previously reported process [20]. $Cr(NO_3)_3{\cdot}9H_2O$ (8 g) and H_2BDC (3.32 g) were put in a 100 mL volumetric flask containing deionized water. The solution was stirred using a magnetic stirrer and sonicated for 30 min, respectively, for it to be homogenized. HF (10 mmol) was gradually added to the mixture and stirred for 15 min. The solution was then placed in a stainless-steel Teflon-lined autoclave, sealed, and inserted into a preheated electric oven at 483 K for 8 h. Next, the autoclave was allowed to cool to room temperature and the product was filtered and recovered. The as-synthesized product was further purified using deionized water, DMF and ethanol to remove possible impurities in the pores. The purified product was finally dried overnight, cooled to room temperature, and stored in a desiccator prior to use.

2.2. Characterization of MIL-101(Cr) Adsorbent

The BET surface area and pore size of the MOF were analyzed by Micrometric ASAP 2020 using the N_2 adsorption–desorption isotherm. The crystallinity and structural properties of the adsorbent were recorded on a Bruker D8 advanced X-ray diffraction (XRD). Perkin Elmer FTIR Spectrometer was used to ascertain the functional group of the material, which was scanned from 400 to 4000 cm^{-1}. The morphology of the MOF was determined by field emission scanning electron microscopy (FESEM) using the Zeiss supra 55 VP instrument.

2.3. Batch Adsorption Studies

Adsorption studies were done by preparing a stock solution of dicamba (1000 mg L^{-1}). A total of 100 mg of the analyte was dissolved in a volumetric flask of 1000 mL and was stored at a temperature of 0 °C in a refrigerator before use. From the prepared stock, solutions containing different initial concentrations (5–50 mg L^{-1}) were studied by dispersing 20 mg of MIL-101(Cr) adsorbent in 100 mL conical flask. The total volume of 50 mL was maintained in the experiments. Next, the flask containing different concentrations were then inserted into a temperature regulated (incubator ES 20/60, Biosan, Riga, Latvia) and shaken at 150 rpm for 1 h. At an interval of 5 min, the 2 mL sample solution was taken out and filtered using a 0.45 μm nylon syringe membrane. The absorbance of the analyte solution was measured in a UV-vis spectrophotometer (Shimadzu Lamda 25, Waltham MA, USA). The pH in which the adsorption took palace was studied by varying the pH from 2 to 12, and the effect of temperature was studied from 25 to 50 °C. The dosage was also studied by varying the quantity of adsorbent from 5 to 50 mg. All the adsorption data were recorded in triplicates from which the average values were calculated. The quantity of dicamba adsorbed at equilibrium (q_e), percentage removal (% R) and quantity adsorbed at a time interval (q_t) were calculated using the following equations:

$$q_e = \frac{(C_o - C_e)V}{w} \quad (1)$$

$$\%R = \frac{(C_o - C_t)}{C_o} \times 100 \quad (2)$$

$$q_t = \frac{(C_o - C_t)V}{w} \quad (3)$$

where C_o is the initial concentration, C_t and C_e are the time and equilibrium dicamba concentration (mg g^{-1}), V represents the solution volume (L), and w is the adsorbent weight (g).

2.4. Adsorption Kinetics Studies

Adsorption kinetics is an important model that describes the rate of adsorbate uptake, adsorption mechanism and the equilibrium time for the adsorption process. It is used to determine the effectiveness and efficiency of the adsorbent material as well as the mass transfer, which explains the rate-limiting steps in designing the adsorption system [21]. The kinetics results were fitted using the pseudo-first order, pseudo-second order and intraparticle diffusion model, as described in the equation below [21,22].

Pseudo-first-order model

$$q_t = q_e\left(1 - e^{-k_1 t}\right) \quad (4)$$

Pseudo-second-order model

$$q_t = \frac{K_2 q_e^2 t}{1 + K_2 q_e t} \tag{5}$$

Intraparticle diffusion model

$$q_t = K_P t^{0.5} + C \tag{6}$$

where q_t and q_e are the amount of dicamba adsorbed at certain equilibrium and time, t (mg g^{-1}), K_1 (min^{-1}) is the pseudo-first-order rate constant, K_2 (g mg^{-1} min^{-1}) is the equilibrium rate constant of the pseudo-second-order and the intra-particle diffusion rate constant is represented as K_p (mg g^{-1} min^{-1}).

2.5. Adsorption Isotherm Studies

The isotherm model is used to describe the interaction mechanism that exists between the adsorbate molecules with the adsorbent surface. Three isotherm models (Langmuir, Freundlich and Temkin isotherms) were used to evaluate the experimental data. The Langmuir isotherm depicts a monolayer adsorption interaction. The following equation was used to analyze the model [23].

$$\frac{C_e}{q_e} = \frac{1}{K_L q_m} + \frac{C_e}{q_m} \tag{7}$$

$$R_L = \frac{1}{1 + C_o K_L} \tag{8}$$

where C_e is the concentration at equilibrium (mg g^{-1}), q_e is the quantity of dicamba and dicamba adsorbed at equilibrium (mg g^{-1}), q_m and K_L are the constants representing adsorption capacity and adsorption energy, respectively. R_L depicts the favorability of the adsorption process ($R_L > 1$, unfavorable; $0 < R_L < 1$, favorable; $R_L = 1$, linear).

The Freundlich model describes a multilayer interaction on multiple adsorption sites.

$$\log(q_e) = \log K_F + \frac{1}{n} \log C_e \tag{9}$$

where K_F is the Freundlich constant of adsorption capacity, n is the adsorption intensity and C_e is the equilibrium concentration of dicamba (mg g^{-1}).

The Temkin model is represented by the following equation:

$$q_e = B \ln A_T + B \ln C_e \tag{10}$$

where B is the heat of adsorption (Jmol) and A_T is the Temkin equilibrium binding constant corresponding with the maximum binding energy (L g^{-1}).

2.6. Thermodynamics Studies

Thermodynamic parameters such as Gibbs free energy change (ΔG°), enthalpy change (ΔH°) and entropy change (ΔS°) were studied to assess the feasibility of the adsorption process based on temperature changes. This helps to determine whether the adsorption process is spontaneous, exothermic, or endothermic. The equations are given [24]:

$$\Delta G^\circ = -RT \text{ In } K_C \tag{11}$$

$$\Delta G^\circ = \Delta H^\circ - T\Delta S^\circ \tag{12}$$

where ΔG° is the free energy (JK mol^{-1}), T (K) and R (JK mol^{-1}) are the temperature and universal gas constant for the adsorption, respectively, and K_c is the equilibrium constant.

2.7. Optimization by Response Surface Methodology (RSM)

The mathematical optimization of the shared interactions between the independent and dependent process parameters for the adsorption of dicamba onto MIL-101(Cr) was modeled using the central composite design (CCD) [25]. The data matrix design for the experimental and predicted values is expressed using a second-order polynomial equation, as described in Equation (13). The selected independent variables comprise of pH, initial concentration, temperature, contact time and adsorbent dosage, while dicamba adsorption capacity was designed as the dependent variable. The accuracy and significance of the fitted model was ascertained by the analysis of variance (ANOVA) based on the probability value (p-value) and the Fischer's test value (F-value) at 95% confidence level. In addition, the coefficient of determination (R^2), R^2 adjusted (R^2adj) and predicted R^2 were used as diagnostic analyses to test the model performance [26].

$$y = \beta_0 + \sum_{i=1}^{k} \beta_i x_i + \sum_{i=1}^{k} \sum_{j \geq i}^{k} \beta_{ij} x_i x_j + \varepsilon \tag{13}$$

where β_0 represents the constant term, β_i and β_{ij} describe the linear and interactive coefficient, respectively. x_i, x_j define the independent variables, k is the number of factors, y is the predicted response and ε is the noise or error detected in the reply.

2.8. Artificial Neural Network (ANN) Model

The ANN model for this study was designed using the multilayer-perceptron feed-forward-artificial neural network (MLP-FF-ANN) with a back-propagation algorithm and activation function [27] to determine the dicamba adsorption capacity onto the MOF adsorbent material. The ANN model mimics the functionality of the biological system of the brain in disseminating information. The model can be subjected to learning process that can predict the pattern and correlate the experimental dataset during the training [28]. The method can be used to ascertain the effect of critical adsorption variables in the behavior of a given outcome. The designed model consists of multiple neurons that are structured in layers. The amount of selected hidden neurons were arrived at by trial through a process of weighted connections during the training process [29]. A total of 60% of the datasets were used to train the network, 20% for testing the model and 20% were used to validate the model. The training datasets were used to train the model by modifying the weight of the network through learning, the testing subset was applied to estimate the generalization ability of the network, and the network efficiency was determined using the validation dataset. Using this model, the diagnostic criteria including the root mean square error (RMSE) and Akaike information criteria (AIC), standard square error (SSE) were considered as the best fit to judge the performance of the adsorption process by regression analysis. The following equations were used:

$$R^2 = 1 - \frac{\sum (x_i - y_i)^2}{\sum y_i^2 - \frac{\sum y_i^2}{n}} \tag{14}$$

$$R^2 adj = 1 - \left(1 - R^2\right)\left(\frac{n-1}{n-p}\right) \tag{15}$$

$$RMSE = \sqrt{\frac{1}{n} \sum_{i=1}^{n} (x_i - y_i)^2} \tag{16}$$

$$AIC = n \ln\left(\frac{SSE}{n}\right) + 2n_p + \frac{2n_p(n_p + 1)}{n(n_p + 1)} \tag{17}$$

where x_i is the data observation that was expressed experimentally, y_i represents the data predicted, n and p are the number of observations and parameters.

2.9. Regeneration and Reuse of the Adsorbent

The potential of recycling the MOF material after use is an important index to determine the quality of the adsorbent. After the adsorption experiments, the adsorbent was decanted, washed and filtered with water and acetone severally. The material was then dried in a vacuum at 80 °C for 4 h and reused as adsorbent for the removal of dicamba in water. This was repeated for six cycles.

3. Results

3.1. Characterization of the MOF

The BET surface area of the MOF is 1439 $m^2\ g^{-1}$, as highlighted in Table 1 and Figure 1a, which is typical of highly porous materials. The diffraction pattern of the MIL-101(Cr) (Figure 1b) adsorbent indicates peaks that are in agreement with those reported in previous studies [30,31], confirming a well-formed crystallite structure of the MOF. The functional groups of the MOF are presented in the FTIR spectra in Figure 1c. The band at 567 cm^{-1} can be ascribed to the Cr–O bond that represents the formation of a well-structured material, and the peaks of 746 and 1287 cm^{-1} were attributed to the stretching of C–H [32]. The sharp peak of 1384 cm^{-1} denotes a symmetric vibration that shows the presence of the dicarboxylate group in the MOF [33]. The peak at 1581 cm^{-1} is attributed to C=C stretching vibration [34] and the strong-broad band around 3433 cm^{-1} shows the presence of the O–H group in the material [35]. The FESEM image of the MIL-101(Cr) (Figure 1d) is similar to that of a previous study [32].

Table 1. Surface properties of MIL-101(Cr) metal-organic framework (MOF).

Properties	MIL-101(Cr)
BET surface area ($m^2\ g^{-1}$)	1439
Langmuir surface area ($m^2\ g^{-1}$)	2124
Micropore surface area ($m^2\ g^{-1}$)	182
Pore size (nm)	0.773

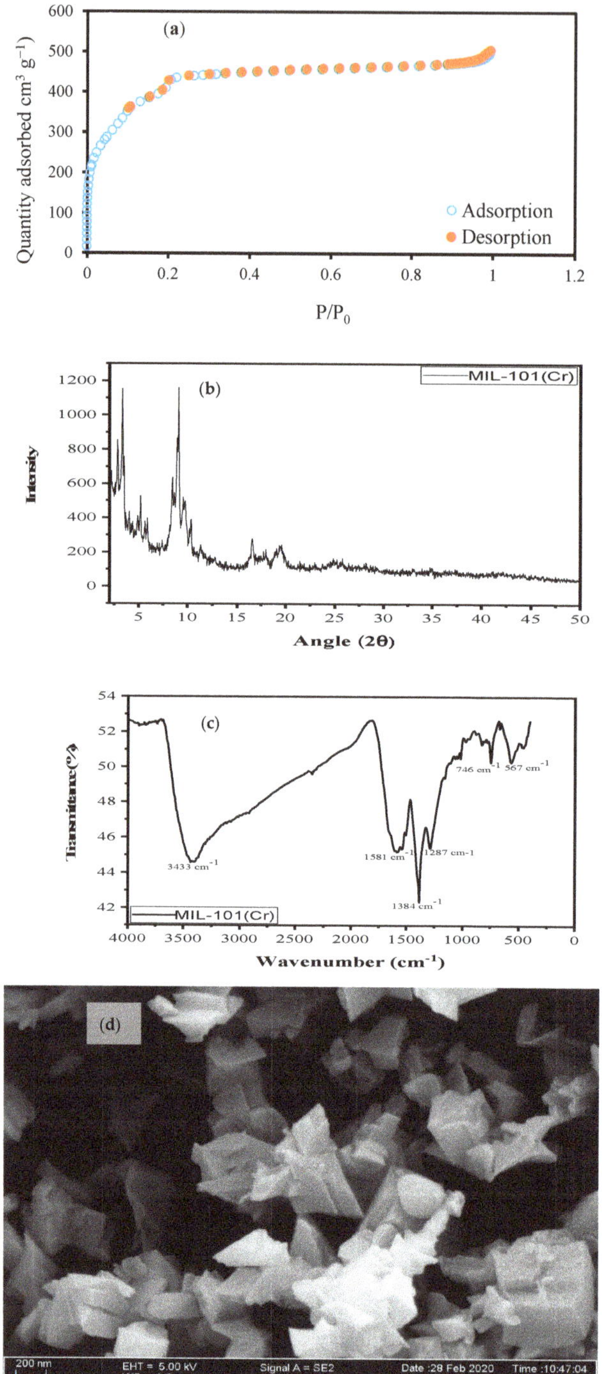

Figure 1. Characterization of the MOF (**a**) N_2 adsorption−desorption isotherm, (**b**) XRD pattern, and (**c**) FTIR spectrum (**d**) FESEM spectrum of MIL-101(Cr).

3.2. Adsorption Kinetics Models

The rate of adsorption uptake and equilibration time were used to determine the adsorption kinetics. Hence, the efficiency of the dicamba removal was ascertained at different initial concentrations (5 to 50 mg L^{-1}), varied time from 5 to 60 min, optimum pH condition (pH 4), dosage (20 mg) and temperature (40 °C). The result is shown in Figure 2. Rapid removal efficiency was recorded within the first 5 to 10 min of contact time, and the adsorption reached equilibrium in ~25 min with high adsorption capacity of 237.384 mg g^{-1} due to favorable interaction, large pores, as well as active and vacant adsorption sites of MIL-101 (Cr). This coincides with the high surface area of the MOF (1439 m^2 g^{-1}). The contact time was extended until 60 min to ensure the maximum interaction of the molecule with the MOF after equilibrium was attained. The optimum condition of the kinetics studies with the highest adsorption capacity was attained with concentration of 50 mg L^{-1}, pH 4, dosage 20 mg, contact time ~25 min and temperature 40 °C.

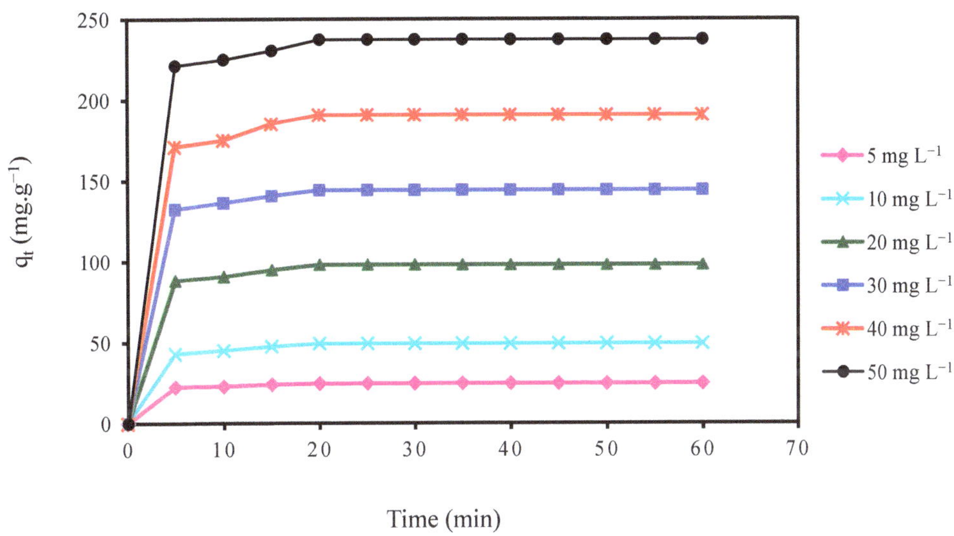

Figure 2. Effect of contact time on dicamba adsorption. Dosage: 20 mg; concentration, 5–50 mg L^{-1}; temp, 40 °C; equilibrium time, 25 min and 150 rpm.

The values obtained for the different kinetic models are displayed in Table 2. The results show that the Pseudo-second order kinetics model best fit the experimental data with the highest coefficient of determination (R^2 = 0.999), R^2adj = 0.997, lowest RMSE = 0.003 and the least AIC value of −133.8. The q_e values calculated for the pseudo-second order is in good agreement with the experimental findings. Hence, the Pseudo-second order model is represented in Figure S1a. The intraparticle diffusion mechanism was also used to describe the kinetics behaviors of the adsorption process based on the interaction and movement of the molecules inside the particles of the MOF adsorbent. The model describes a multiple linear relationship that follows a multistep mechanism. The multistage process is described in Figure S1b that represents an external diffusion of herbicides to the surface of the adsorbent from the bulk phase, and the transport of the molecules from the surface inside the pore of the MOF.

Table 2. Adsorption kinetics parameters for the removal of dicamba.

Pseudo-First Order	(mg L^{-1})	$q_{e, exp}$ (mg g^{-1})	$q_{e, cal}$ (mg g^{-1})	K_1 $(min)^{-1}$	R^2	R^2adj	RMSE	AIC
	5	24.860	10.432	0.152	0.08	0.744	0.753	−1.393
	10	49.504	29.224	0.181	0.945	0.927	0.717	−1.395
	20	98.011	57.512	0.205	0.803	0.737	0.723	−1.769
	30	144.423	79.885	0.223	0.797	0.729	0.795	−0.853
	40	190.903	133.499	0.235	0.832	0.776	0.686	−2.321
	50	237.384	133.686	0.236	0.729	0.639	0.886	0.232
Pseudo-second order	**(mg L^{-1})**	**$q_{e, exp}$ (mg g^{-1})**	**$q_{e, cal}$ (g mg^{-1})**	**K_2 (g mg^{-1} min^{-1})**	**R^2**	**R^2adj**	**RMSE**	**AIC**
	5	24.860	24.875	0.117	0.997	0.995	0.052	−74.97
	10	49.504	49.751	0.041	0.998	0.995	0.027	−92.23
	20	98.011	98.039	0.027	0.999	0.997	0.003	−110.5
	30	144.423	144.927	0.023	0.996	0.995	0.009	−120.6
	40	190.903	192.307	0.012	0.995	0.994	0.007	−127.8
	50	237.384	238.095	0.016	0.995	0.994	0.005	−133.8
Intraparticle diffusion	**(mg L^{-1})**		**Kp (mg^{-1} g^{-1} $min^{1/2}$)**	**C**	**R^2**	**R^2adj**	**RMSE**	**AIC**
	5		2.048	8.736	0.579	0.438	5.232	1.211
	10		4.147	16.375	0.616	0.488	7.977	3.211
	20		8.099	33.913	0.588	0.45	9.949	4.493
	30		11.883	51.503	0.571	0.428	11.961	5.773
	40		15.837	65.345	0.59	0.453	14.848	9.951
	50		19.363	86.049	0.57	0.458	15.016	11.122

3.3. Dicamba Adsorption Isotherms

The equilibrium data of the adsorption process was validated by the Langmuir, Freundlich and Temkin isotherm models to study the surface properties and interaction mechanism between the MOF and the adsorbate molecule. From the calculated results in Table S1 and Figure 3, the Freundlich isotherm model best fits the adsorption process based on the regression analysis with the highest R^2 = 0.998, R^2adj (0.997); lowest RMSE (0.023) and the least AIC (−43.773) values. The Freundlich model shows a more linear curve that implies an adsorption process with multilayer interaction on heterogeneous surfaces with binding sites that are not equivalent [36].

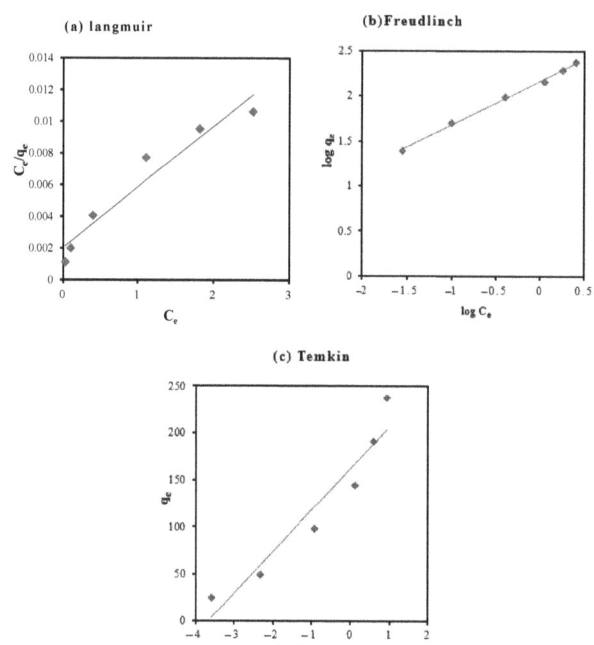

Figure 3. Isotherm adsorption models of dicamba. (**a**) Langmuir, (**b**) Freundlich, and (**c**) Temkin. Dosage: 20 mg; concentration of dicamba, 20 mg L^{-1}; temperature, 40 °C; equilibration time, 25 min and rpm, 150.

3.4. Effect of Temperature and Thermodynamic Studies

The thermodynamic studies were conducted by varying the temperature from 25 to 50 °C to understand the spontaneity of the adsorption process (Figure 4). An increase in temperature leads to an increase in the adsorption of dicamba. As temperature rises, the viscosity of the solution decreases, which accelerates the mobility of the adsorbate molecules that facilitate the adsorption process [36]. The thermodynamic parameters are described in Table 3. The continuous decrease in the values of the Gibbs free energy (ΔG°) due to an increase in temperature indicates a spontaneous process in the adsorption of dicamba. The positive enthalpy change (ΔH° = 27.920 kJ mol^{-1}) shows that the adsorption of dicamba unto MIL-101(Cr) is endothermic. Also, the positive values of the standard entropy change denotes the affinity and increased randomness at the liquid–solid interface between the MOF and dicamba during the adsorption process [31].

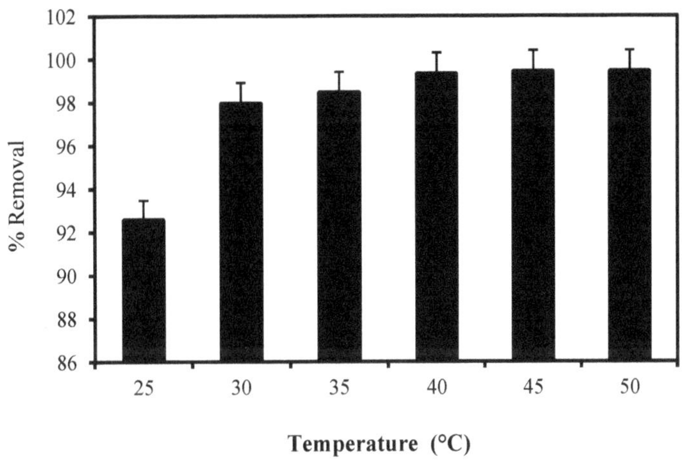

Figure 4. Effect of temperature on the dicamba adsorption (concentration of dicamba, 20 mg L^{-1}; equilibration time, 25 min; rpm, 150).

Table 3. Thermodynamic parameters for the adsorption of dicamba onto MIL-101(Cr).

Temp (°C)	ΔG° (kJ mol^{-1})	ΔH° (kJ mol^{-1})	ΔS° (kJ mol^{-1} K^{-1})
25	−155.781	27.920	522.850
30	−158.395		
35	−161.009		
40	−163.624		
45	−166.238		
50	−168.852		

3.5. Optimization of Process Parameters by Response Surface Methodology (RSM)

To study the interaction effect of the independent variables on the dicamba adsorption capacity (q_e mg g^{-1}), the central composite design (CCD) was selected for the experimental design data matrix for the statistical analysis. Thus, the significance of the data was ascertained by the analysis of variance (ANOVA) in Table 4, containing the Model *F*-value of 103.03 and *p*-values less than 0.05. The model signifies a minimum chance of 0.01% that an *F*-value of this magnitude could exist by noise. The less *p*-values represent a statistically significant model that can be used to predict the dicamba adsorption capacity. The second order polynomial equation was developed using the data based on the coded factors as shown in Equation (18). The coded levels and experimental input design are shown in Table S2. The result obtained from the CCD-RSM multiple regression analysis gave a significant prediction with an $R^2 = 0.990$, $R^2adj = 0.979$ and R^2 predicted = 0.955, which indicate a positive relationship between the experimental and predicted response values. Also, an adequate precision (AP) value of 37.738 that represents the ratio of response to noise, further describes the significance of the model used. Using the RSM model, the equation is represented as contact time (A), initial concentration (B), adsorbent dosage (C), pH (D), and temperature (E).

$$\text{Adsorption capacity of Dicamba (mg g}^{-1}) = 9.59 + 0.2910A + 2.58B + 0.0005C - 0.0093D + 0.0144E - 0.0135AB + 0.0991AC + 0.0984AD - 0.0576AE + 0.0689BC + 0.0951BD - 0.1031BE - 0.0288CD + 0.0201CE + 0.0372DE - 0.0713A^2 - 0.2494B^2 + 0.0178C^2 + 0.0061D^2 + 0.0233E^2 \quad (18)$$

Table 4. Analysis of variance (ANOVA) for dicamba removal.

Source	Sum of Square	df	Mean Square	*F*-Value	*p*-Value
Model	271.165	20	13.558	103.031	<0.0001
A-Contact time	2.800	1	2.800	21.279	<0.0001
B-Initial concentration	220.691	1	220.691	1677.078	<0.0001
C-Adsorbent dosage	9.130	1	9.126	6.940	0.993
D-pH	0.002	1	0.00284	0.021	0.0431
E-Temperature	0.006	1	0.006	0.052	0.0235
AB	0.005	1	0.005	0.044	<0.0001
AC	0.298	1	0.298	2.271	0.0354
AD	0.294	1	0.294	2.238	<0.0001
AE	0.111	1	0.111	0.849	0.3661
BC	0.151	1	0.151	1.148	0.2949
BD	0.287	1	0.287	2.188	0.1526
BE	0.338	1	0.338	2.569	0.1225
CD	0.026	1	0.026	0.200	0.6585
CE	0.012	1	0.012	0.097	0.7576
DE	0.044	1	0.044	0.335	0.5683
AÂ^2	0.257	1	0.257	1.957	0.1750
BÂ^2	2.744	1	2.744	20.858	0.0001
CÂ^2	0.015	1	0.015	0.121	0.7310
DÂ^2	0.001	1	0.001	0.014	0.9058
EÂ^2	0.027	1	0.027	0.209	0.6516
Residual	3.026	23	0.131		
Lack of Fit	3.026	22	0.137		
Pure Error	6.121	1	6.124		
Cor Total	274.192	43			
R^2	0.990				
R^2adj	0.979				
R^2pred	0.955				

The multivariate interaction between the independent variables that determine the dicamba adsorption capacity onto MIL-101(Cr) is depicted by the contour and 3D graph of the RSM plots in Figure 5. The optimum adsorption condition is given as contact time is 25 min, initial concentration 50 mg L^{-1}, adsorbent dosage 20 mg, pH 4 and temperature 40 °C. Hence, Figure 5a describes the shared interaction between initial concentration (5-50 mg L^{-1}) and contact time (5 to 60 min) with other parameters held at optimum conditions. It can be seen that the adsorption capacity increases with increase in the concentration of dicamba within a short time. The equilibration time of the adsorption is reached in ~25 min and remain static with no further changes as the time extends to 1 h. As the concentration increases, the force on the active and vacant pores of the adsorbent will be intensified. These values are closely correlated with the experimental (q_e) values and calculated (q_e) values of the kinetics model. The interaction between pH and time was also studied by varying the pH from 2 to 12, as shown in Figure 5b. Hence, when the pH is low (2 to 6), the solution of the herbicide will move to the anionic form, causing it to be negatively charged due to deprotonation, resulting to a positively charged MIL-101(Cr) surface [30]. This causes an electrostatic interaction to take place, resulting in a high removal capacity due to the attraction of the negatively charged molecule with a positive surface of the adsorbent. An increase in the solution pH value by varying the range from 7 to 12 results in a negative charge surface of the MIL-101(Cr) thus, hindering the electrostatic interactions to take place that lead to reduction in the adsorption capacity.

This can be caused by the strong competition for active vacant sites between the –OH and the herbicide molecules [37]. Also, when the pH of the solution is higher, some functional groups comprising of carbonyl and hydroxyl will be in their protonated cationic form, which retard efficient adoption. The influence of adsorbent dose and contact time on adsorption capacity is described in Figure 5c. The adsorption increases as the adsorbent dose increase from 5 to 20 mg. Further increase in the dosage above 20 mg did not result in a significant change in the adsorption capacity. As such, 20 mg is selected as the optimum dose for the effective removal of dicamba.

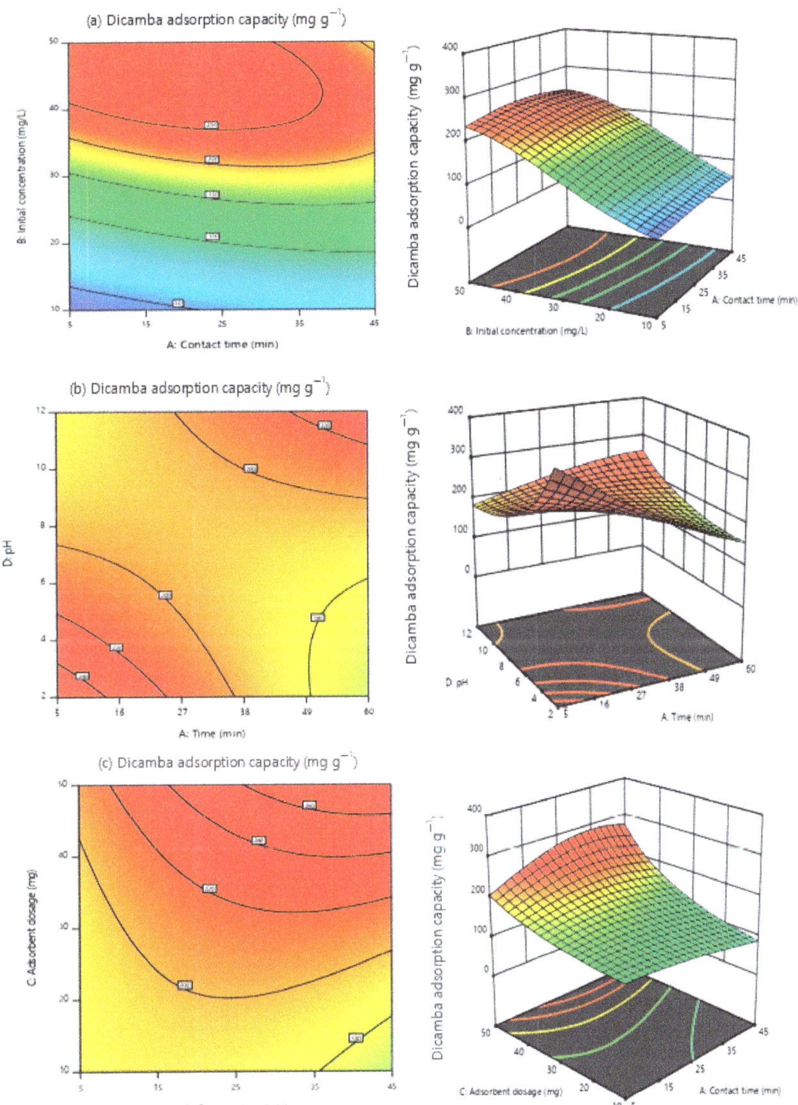

Figure 5. Multivariate interaction for adsorption capacity of dicamba (mg g^{-1}), (**a**) initial dicamba concentration and contact time, (**b**) solution pH and contact time (**c**) adsorbent dosage and contact time.

3.6. Prediction Modeling by ANN

The ANN architecture for this study consists of five predictor variables (contact time, initial concentration, adsorbent dosage, pH, and temperature), eight hidden neurons and one output (dicamba adsorption capacity, q_e (mg g^{-1})). Several topologies were trained, tested, and validated based on the trial and error approach to learn the pattern of the data for accurate prediction. The 5-8-1 topology developed after series of trial (Figure 6) gave the best prediction with good correlation with the experimental values, and $R^2 = 0.999$, $R^2adj = 0.992$ and RMSE = 0.053 as described in Table 5.

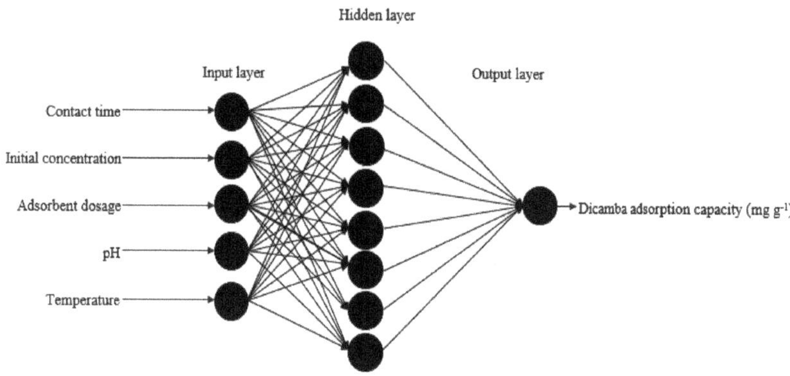

Figure 6. Artificial neural network architecture.

Table 5. Optimum conditions for designing the ANN prediction architecture.

		Training			Testing			Validation		
Numbers	Neurons	R^2	R^2adj	RMSE	R^2	R^2adj	RMSE	R^2	R^2adj	RMSE
1	[3]	0.992	0.988	4.774	0.966	0.951	3.012	0.988	0.901	2.930
2	[4]	0.994	0.991	1.512	0.984	0.975	0.431	0.984	0.954	2.270
3	[5]	0.987	0.975	0.958	0.990	0.983	1.011	0.991	0.986	0.824
4	[6]	0.991	0.977	0.973	0.987	0.981	0.621	0.986	0.911	0.716
5	[7]	0.993	0.991	0.403	0.991	0.980	1.006	0.984	0.972	0.531
6	[8]	0.999	0.992	0.053	0.998	0.996	0.033	0.994	0.988	0.043
7	[9]	0.995	0.991	0.061	0.993	0.988	1.210	0.996	0.941	0.094
8	[10]	0.998	0.981	0.166	0.995	0.992	0.922	0.992	0.987	0.428
9	[5 5]	0.988	0.985	0.975	0.977	0.969	0.221	0.988	0.930	0.807
10	[5 7]	0.983	0.980	0.392	0.981	0.980	1.861	0.990	0.906	0.278
11	[6 7]	0.982	0.972	0.866	0.980	0.971	1.901	0.983	0.966	0.081

3.7. Evaluation of the Prediction Performance of RSM and ANN Model

The RSM and ANN were used to model and predict the dicamba adoption capacity unto MIL-101(Cr). The results obtained from both models are in good agreement with the experimental findings in Table 6, but the ANN model performs better in comparison with the RSM. In every experimental condition selected in studying the adsorption process, the ANN model showed a better prediction with a high level of significance as well as validated the experimental results. The ANN has $R^2 = 0.999$, $R^2adj = 0.992$ and RMSE = 0.053, while for RSM, $R^2 = 0.990$ and $R^2adj = 0.979$. Less error is observed in the ANN model than the RSM. This is due to the fact that the ANN mimics the nervous system of the human by understanding the data combination, as well as generalizes the multivariate correlation between the experimental and the predicted variables.

Table 6. Comparison between RSM and ANN model for predicting dicamba adsorption capacity.

Contact Time (min)	Initial Concentration (mg L^{-1})	Dosage (mg)	pH	Temperature (°C)	Experimental (mg g^{-1})	Predicted RSM (mg g^{-1})	Error	Predicted ANN (mg g^{-1})	Error
15	20	20	4	30	94.861	91.994	2.867	94.169	0.692
5	10	30	2	35	43.040	44.319	1.279	44.109	1.069
15	20	40	4	30	97.956	93.388	4.568	97.325	0.631
15	20	20	8	30	94.780	92.110	2.670	95.551	0.771
25	40	20	4	30	190.904	194.411	3.507	190.680	0.224
25	10	10	2	35	48.874	48.904	0.030	49.317	0.443
15	20	20	4	45	97.956	96.903	1.053	97.973	0.017
5	30	10	2	35	132.660	133.261	0.601	133.230	0.570
5	30	10	6	25	136.617	136.390	0.227	135.918	0.699
5	30	30	2	25	132.660	136.389	3.729	135.871	3.211
5	10	30	6	35	45.299	39.294	6.005	43.782	1.517
25	10	10	6	35	48.874	50.598	1.724	49.056	0.182
25	30	30	6	35	144.423	153.978	9.555	143.836	0.587
25	30	10	2	35	144.150	143.894	0.256	144.506	0.356
5	10	10	2	25	43.040	44.340	1.300	42.542	0.498
5	30	10	6	35	132.660	135.743	3.083	133.234	0.574
35	20	20	4	30	98.011	97.774	0.237	96.554	1.457
5	30	30	6	35	132.660	133.959	1.299	134.128	1.168
5	30	30	6	25	136.617	132.745	3.872	136.245	0.372
25	10	30	2	25	49.504	49.940	0.436	51.683	2.179
25	30	10	6	35	144.423	146.144	1.721	144.190	0.233
15	20	20	10	30	94.861	92.521	2.340	95.741	0.880
25	10	30	6	25	49.504	47.926	1.578	50.537	1.033
25	10	30	6	35	49.504	51.227	1.723	49.754	0.250
45	20	20	4	30	97.956	96.483	1.473	99.382	1.426
25	20	10	4	40	98.011	94.005	4.006	96.337	1.674
5	10	10	6	25	43.040	38.892	4.148	41.626	1.414
5	10	30	6	25	40.710	33.681	7.029	41.380	0.670
25	30	30	2	35	144.423	144.560	0.137	143.657	0.766
5	30	10	2	25	133.890	137.371	3.481	135.820	1.930
25	30	30	2	25	144.150	152.528	8.378	144.157	0.007
5	10	10	2	35	43.040	47.535	4.495	42.640	0.400
5	30	30	2	35	132.660	134.149	1.489	133.280	0.620
25	30	10	6	25	144.423	142.457	1.966	144.494	0.071
5	10	30	2	25	42.710	40.210	2.500	42.755	0.045
25	10	10	6	25	49.504	48.430	1.074	49.506	0.002
15	20	50	4	30	94.861	95.130	0.269	94.935	0.074
15	20	20	4	30	94.861	91.994	2.867	94.930	0.069
25	50	10	4	40	237.384	230.865	6.519	237.368	0.016
5	10	10	6	35	45.089	46.618	1.529	42.385	2.704
15	10	10	2	25	47.710	47.537	0.173	47.807	0.097
25	30	30	6	25	144.423	158.427	14.004	144.556	0.133
25	30	10	2	25	140.103	143.894	3.791	140.072	0.031
25	20	10	4	35	49.504	43.880	5.624	49.491	0.013

3.8. Reusability Studies

The feasibility for the repeated removal of dicamba in aqueous medium by MIL-101(Cr) was evaluated to determine the possibility of regeneration and reuse (Figure 7). High removal percentage was maintained by the adsorbent after the third cycle (~99.4%). A small decline in the removal (2, 5, and 6%) is noticed after the fourth, fifth and sixth cycles, respectively. Nevertheless, the MOF retain > 90% removal efficiency even after the sixth cycle.

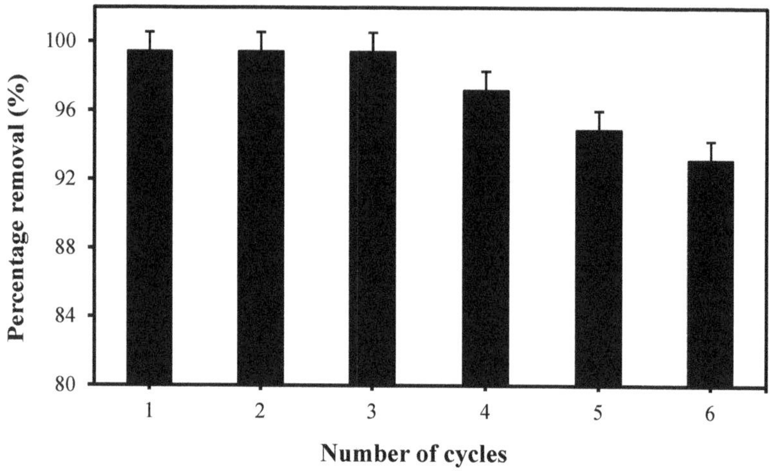

Figure 7. Reusability of MIL-101(Cr) adsorbent.

3.9. Comparison with Different Adsorbents

The adsorption characteristics of several adsorbent materials that were previously reported for the remediation of dicamba from aqueous medium are summarised in Table 7. MIL-10(Cr) adsorbent shows more superiority in terms of the surface area that is higher (1439 $m^2\ g^{-1}$), adsorption capacity (237.384 mg g^{-1}), % removal efficiency (99.432%), as well as fast equilibration time (~25 min). Comparison of reusability is not possible for the other adsorbents as it is not mentioned in all the earlier studies (Table 7).

Table 7. Comparison of different materials reported for the remediation of dicamba from water.

Adsorbent	Surface Area, ($m^2\ g^{-1}$)	Concentrations (mg L^{-1})	(%) R	Q_e (mg g^{-1})	Equilibrium Time (min)	Reuse	Ref.
Carbon nanotubes	600	50	86	21	Not reported	Not reported	[38]
Clay material	204	50	80		30	Not reported	[39]
Mesoporous carbon	876	50	NIL	222	60	Not reported	[40]
Vinyl and NH_2@COF	336		92	13	Not reported	Not reported	[41]
MIL-10(Cr)	1439	50	99	237	25	6	This work

4. Conclusions

A detailed evaluation of the optimization and adsorption of dicamba from aqueous solution was successfully demonstrated by using MIL-101(Cr). The adsorption best fitted the pseudo-second order kinetics and the Freundlich isotherm. The removal of dicamba was spontaneous and was endothermic in nature. The RSM and ANN models were used to optimize and model the adsorption process with a high level of significance. The shared interaction of the adsorption parameters were studied to understand the multivariate impact on the removal process. ANN gave better prediction with the highest coefficient of determination and minimum error for each studied experimental condition when compared with RSM. The adsorption capacity of dicamba (q_e mg g^{-1}) is in good agreement with the experimental and calculated q_e kinetics values. The MIL-101(Cr) displayed numerous advantageous features such as fast equilibration (~25 min), high adsorption capacity (237.384 mg g^{-1}), excellent percentage removal (99.432%) and high surface area (1439 m^2 g^{-1}) when compared to other reported adsorbents. Furthermore, prospects for reusability were good as the adsorbent retained removal efficiency of 93% even after the sixth cycle. Commercial exploitation of this adsorbent must focus on production routes that are not only cost-effective but also environmentally benign.

Supplementary Materials: The following are available online at https://www.mdpi.com/2227-9717/9/3/419/s1, Figure S1: (a) Pseudo-second-order kinetics and (b) intraparticle diffusion model kinetics for dicamba adsorption (Dosage: 20 mg; 40 °C; equilibrium time: 25 min, rpm: 150); Table S1: Isotherm parameters for adsorption of dicamba onto MIL-101(Cr); Table S2: Coded range for independent variables for the CCD-RSM design matrix.

Author Contributions: Conceptualization, methodology, software, writing—original draft preparation, H.A.I.; validation, Z.U.Z.; resources, supervision, project administration and funding acquisition, K.J. and N.S.S.; formal analysis and visualization, J.W.L. and A.R.; writing—review and editing, H.A.I. and B.S. All authors have read and agreed to the published version of the manuscript.

Funding: This research was funded by the Fundamental Research Grant Scheme (FRGS-015MA0-127), Ministry of Higher Education (MOHE), Malaysia and Universiti Teknologi PETRONAS under the YUTP research grant cost center (015LCO-211) and UTP-UIR (015-MEO-166), FRGS/1/2020/STG04/UTP/02/3 Ministry of Higher Education (MOHE) grant.

Institutional Review Board Statement: Not applicable.

Informed Consent Statement: Not applicable.

Data Availability Statement: Not applicable.

Conflicts of Interest: The authors declare no conflict of interest.

References

1. Mojiri, A.; Zhou, J.L.; Robinson, B.; Ohashi, A.; Ozaki, N.; Kindaichi, T.; Farraji, H.; Vakili, M. Pesticides in aquatic environments and their removal by adsorption methods. *Chemosphere* **2020**, *253*, 126646. [CrossRef]
2. Yao, L.; Jia, X.; Zhao, J.; Cao, Q.; Xie, X.; Yu, L.; He, J.; Tao, Q. Degradation of the herbicide dicamba by two sphingomonads via different O-demethylation mechanisms. *Int. Biodeterior. Biodegrad.* **2015**, *104*, 324–332. [CrossRef]
3. Meftaul, I.M., Venkateswarlu, K.; Dharmarajan, R.; Annamalai, P.; Megharaj, M. Pesticides in the urban environment: A potential threat that knocks at the door. *Sci. Total Environ.* **2020**, *711*, 134612. [CrossRef]
4. Huang, Y.; Yuan, L.; Reddy, K.N.; Zhang, J. In-situ plant hyperspectral sensing for early detection of soybean injury from dicamba. *Biosyst. Eng.* **2016**, *149*, 51–59. [CrossRef]
5. Gupta, P.K. *Toxicity of Herbicides*, 3rd ed.; Elsevier: Amsterdam, The Netherlands, 2018; pp. 553–567. [CrossRef]
6. De Arcaute, C.R.; Larramendy, M.L.; Soloneski, S. Genotoxicity by long-term exposure to the auxinic herbicides 2,4-dichlorophenoxyacetic acid and dicamba on Cnesterodon decemmaculatus (Pisces: Poeciliidae). *Environ. Pollut.* **2018**, *243*, 670–678. [CrossRef] [PubMed]
7. Carmalin, S.A.; Lima, E.C. Removal of emerging contaminants from the environment by adsorption. *Ecotoxicol. Environ. Saf.* **2018**, *150*, 1–17. [CrossRef]
8. Ighalo, J.O.; Ajala, O.J.; Umenweke, G.; Ogunniyi, S.; Adeyanju, C.A.; Igwegbe, C.A.; Adeniyi, A.G. Mitigation of clofibric acid pollution by adsorption: A review of recent developments. *J. Environ. Chem. Eng.* **2020**, *8*, 104264. [CrossRef]
9. Singh, N.; Nagpal, G.; Agrawal, S. Rachna Water purification by using Adsorbents: A Review. *Environ. Technol. Innov.* **2018**, *11*, 187–240. [CrossRef]
10. Ghanizadeh, H.; Harrington, K.C.; James, T.K. A comparison of dicamba absorption, translocation and metabolism in Chenopodium album populations resistant and susceptible to dicamba. *Crop. Prot.* **2018**, *110*, 112–116. [CrossRef]
11. Beyki, T.; Asadollahzadeh, M.J. Selective removal of dicamba from aqueous samples using molecularly imprinted polymer nanospheres. *J. Water Environ. Nanotechol.* **2016**, *1*, 19–25. [CrossRef]
12. Azejjel, H.; Del Hoyo, C.; Draoui, K.; Rodríguez-Cruz, S.; Sánchez-Martín, M.J. Natural and modified clays from Morocco as sorbents of ionizable herbicides in aqueous medium. *Desalination* **2009**, *249*, 1151–1158. [CrossRef]

13. Attari, M.; Bukhari, S.S.; Kazemian, H.; Rohani, S. A low-cost adsorbent from coal fly ash for mercury removal from industrial wastewater. *J. Environ. Chem. Eng.* **2017**, *5*, 391–399. [CrossRef]
14. Dhaka, S.; Kumar, R.; Deep, A.; Kurade, M.B.; Ji, S.-W.; Jeon, B.-H. Metal–organic frameworks (MOFs) for the removal of emerging contaminants from aquatic environments. *Coord. Chem. Rev.* **2019**, *380*, 330–352. [CrossRef]
15. Roy, D.; Neogi, S.; De, S. Adsorptive removal of heavy metals from battery industry effluent using MOF incorporated polymeric beads: A combined experimental and modeling approach. *J. Hazard. Mater.* **2020**, *403*, 123624. [CrossRef] [PubMed]
16. Rasheed, T.; Hassan, A.A.; Bilal, M.; Hussain, T.; Rizwan, K. Metal-organic frameworks based adsorbents: A review from removal perspective of various environmental contaminants from wastewater. *Chemosphere* **2020**, *259*, 127369. [CrossRef]
17. Yoo, D.K.; Bhadra, B.N.; Jhung, S.H. Adsorptive removal of hazardous organics from water and fuel with functionalized metal-organic frameworks: Contribution of functional groups. *J. Hazard. Mater.* **2020**, *403*, 123655. [CrossRef]
18. Zhao, H.Z.; Li, Q.; Wang, Z.; Wu, T.; Zhang, M. Synthesis of MIL-101(Cr) and its water adsorption performance. *Microporous Mesoporous Mater.* **2020**, *297*, 110044. [CrossRef]
19. Maksimchuk, N.V.; Zalomaeva, O.V.; Skobelev, I.Y.; Kovalenko, K.A.; Fedin, V.P.; Kholdeeva, O.A. Metal–organic frameworks of the MIL-101 family as heterogeneous single-site catalysts. *Proc. R. Soc. A: Math. Phys. Eng. Sci.* **2012**, *468*, 2017–2034. [CrossRef]
20. Férey, S.S.G.; Mellot-Draznieks, C.; Serre, C.; Millange, F.; Dutour, J.; Margiolaki, I. A chromium terephthalate-based solid with unusually large pore volumes and surface area. *Science* **2005**, *309*, 2040–2042. [CrossRef]
21. Pinheiro, D.; Pai, S.D.K.R.; Jose, A.; Bharadwaj, N.R.; Thomas, K. Effect of surface charge and other critical parameters on the adsorption of dyes on SLS coated ZnO nanoparticles and optimization using response surface methodology. *J. Environ. Chem. Eng.* **2020**, *8*, 103987. [CrossRef]
22. Isiyaka, H.A.; Jumbri, K.; Sambudi, N.S.; Zango, Z.U.; Ain, N.; Abdullah, F.; Saad, B.; Mustapha, A. Adsorption of dicamba and MCPA onto MIL-53(Al) metal–organic framework: Response surface methodology and artificial neural network model studies. *RSC Adv.* **2020**, *10*, 43213–43224. [CrossRef]
23. Zhu, S.; Khan, M.A.; Wang, F.; Bano, Z.; Xia, M. Exploration of adsorption mechanism of 2-phosphonobutane-1,2,4-tricarboxylic acid onto kaolinite and montmorillonite via batch experiment and theoretical studies. *J. Hazard. Mater.* **2020**, *403*, 123810. [CrossRef] [PubMed]
24. Chaturvedi, G.; Kaur, A.; Umar, A.; Khan, M.A.; Algarni, H.; Kansal, S.K. Removal of fluoroquinolone drug, levofloxacin, from aqueous phase over iron based MOFs, MIL-100(Fe). *J. Solid State Chem.* **2020**, *281*, 121029. [CrossRef]
25. Bahrami, M.; Amiri, M.J.; Bagheri, F. Optimization of the lead removal from aqueous solution using two starch based adsorbents: Design of experiments using response surface methodology (RSM). *J. Environ. Chem. Eng.* **2019**, *7*, 102793. [CrossRef]
26. Soleimanzadeh, H.; Niaei, A.; Salari, D.; Tarjomannejad, A.; Penner, S.; Grünbacher, M.; Hosseini, S.A.; Mousavi, S.M. Modeling and optimization of V2O5/TiO2 nanocatalysts for NH_3-Selective catalytic reduction (SCR) of NOx by RSM and ANN techniques. *J. Environ. Manag.* **2019**, *238*, 360–367. [CrossRef]
27. Banerjee, P.; Sau, S.; Das, P.; Mukhopadhayay, A. Optimization and modelling of synthetic azo dye wastewater treatment using Graphene oxide nanoplatelets: Characterization toxicity evaluation and optimization using Artificial Neural Network. *Ecotoxicol. Environ. Saf.* **2015**, *119*, 47–57. [CrossRef]
28. Altowayti, W.A.H.; Algaifi, H.A.; Abu Bakar, S.; Shahir, S. The adsorptive removal of As (III) using biomass of arsenic resistant Bacillus thuringiensis strain WS3: Characteristics and modelling studies. *Ecotoxicol. Environ. Saf.* **2019**, *172*, 176–185. [CrossRef]
29. Yusuf, M.; Song, K.; Li, L. Fixed bed column and artificial neural network model to predict heavy metals adsorption dynamic on surfactant decorated graphene. *Coll. Surf. A* **2020**, *585*, 124076. [CrossRef]
30. Shadmehr, J.; Zeinali, S.; Tohidi, M. Synthesis of a chromium terephthalate metal organic framework and use as nanoporous adsorbent for removal of diazinon organophosphorus insecticide from aqueous media. *J. Dispers. Sci. Technol.* **2019**, *40*, 1423–1440. [CrossRef]
31. Karmakar, S.; Roy, D.; Janiak, C.; De, S. Insights into multi-component adsorption of reactive dyes on MIL-101-Cr metal organic framework: Experimental and modeling approach. *Sep. Purif. Technol.* **2019**, *215*, 259–275. [CrossRef]
32. Alivand, M.S.; Shafiei-Alavijeh, M.; Tehrani, N.H.M.H.; Ghasemy, E.; Rashidi, A.; Fakhraie, S. Facile and high-yield synthesis of improved MIL-101(Cr) metal-organic framework with exceptional CO_2 and H_2S uptake; the impact of excess ligand-cluster. *Microporous Mesoporous Mater.* **2019**, *279*, 153–164. [CrossRef]
33. Niknam, E.; Panahi, F.; Daneshgar, F.; Bahrami, F.; Khalafi-Nezhad, A. Metal–Organic Framework MIL-101(Cr) as an Efficient Heterogeneous Catalyst for Clean Synthesis of Benzoazoles. *ACS Omega* **2018**, *3*, 17135–17144. [CrossRef] [PubMed]
34. Liu, Q.; Ning, L.; Zheng, S.; Tao, M.; Shi, Y.; He, Y. Adsorption of Carbon Dioxide by MIL-101(Cr): Regeneration Conditions and Influence of Flue Gas Contaminants. *Sci. Rep.* **2013**, *3*, 1–6. [CrossRef] [PubMed]
35. Gao, Y.; Liu, K.; Kang, R.; Xia, J.; Yu, G.; Deng, S. A comparative study of rigid and flexible MOFs for the adsorption of pharmaceuticals: Kinetics, isotherms and mechanisms. *J. Hazard. Mater.* **2018**, *359*, 248–257. [CrossRef]
36. Shahnaz, T.; Sharma, V.; Subbiah, S.; Narayanasamy, S. Multivariate optimisation of Cr (VI), Co (III) and Cu (II) adsorption onto nanobentonite incorporated nanocellulose/chitosan aerogel using response surface methodology. *J. Water Process. Eng.* **2020**, *36*, 101283. [CrossRef]
37. Zhou, M.; Wu, Y.-N.; Qiao, J.; Zhang, J.; McDonald, A.; Li, G.; Li, F. The removal of bisphenol A from aqueous solutions by MIL-53(Al) and mesostructured MIL-53(Al). *J. Coll. Interface Sci.* **2013**, *405*, 157–163. [CrossRef]

38. Pyrzyńska, K.; Stafiej, A.; Biesaga, M. Sorption behavior of acidic herbicides on carbon nanotubes. *Microchim. Acta* **2007**, *159*, 293–298. [CrossRef]
39. You, Y.; Zhao, H.; Vance, G.F. Adsorption of dicamba (3,6-dichloro-2-methoxy benzoic acid) in aqueous solution by calcined–layered double hydroxide. *Appl. Clay Sci.* **2002**, *21*, 217–226. [CrossRef]
40. Pinto, M.D.C.E.; Gonçalves, R.G.L.; Dos Santos, R.M.M.; Araújo, E.A.; Perotti, G.F.; Macedo, R.D.S.; Bizeto, M.A.; Constantino, V.R.L.; Pinto, F.G.; Tronto, J. Mesoporous carbon derived from a biopolymer and a clay: Preparation, characterization and application for an organochlorine pesticide adsorption. *Microporous Mesoporous Mater.* **2016**, *225*, 342–354. [CrossRef]
41. Ji, W.-H.; Guo, Y.-S.; Wang, X.; Lu, X.-F.; Guo, D.-S. Amino-modified covalent organic framework as solid phase extraction absorbent for determination of carboxylic acid pesticides in environmental water samples. *J. Chromatogr. A* **2019**, *1595*, 11–18. [CrossRef]

processes

MDPI

Article

Low-Viscosity Ether-Functionalized Ionic Liquids as Solvents for the Enhancement of Lignocellulosic Biomass Dissolution

Asyraf Hanim Ab Rahim [1], Normawati M. Yunus [1,*], Wan Suzaini Wan Hamzah [1], Ariyanti Sarwono [2] and Nawshad Muhammad [3]

1 Centre of Research in Ionic Liquids (CORIL), Department of Fundamental and Applied Sciences, Universiti Teknologi PETRONAS, Seri Iskandar 32610, Perak, Malaysia; asyrafhanim92@gmail.com (A.H.A.R.); wansuzaini@gmail.com (W.S.W.H.)
2 Department of Environmental Engineering, Pertamina University, Jalan Teuku Nyak Arief, Simprug, Kebayoran Lama, Jakarta 12220, Indonesia; arisarwono3@gmail.com
3 Institute of Basic Medical Sciences, Khyber Medical University, Peshawar, Khyber Pakhtunkhwa 25100, Pakistan; nawshadchemist@yahoo.com
* Correspondence: normaw@utp.edu.my

Abstract: Due to the substantial usage of fossil fuels, the utilization of lignocellulosic biomass as renewable sources for fuels and chemical production has been widely explored. The dissolution of lignocellulosic biomass in proper solvents is vital prior to the extraction of its important constituents, and ionic liquids (ILs) have been found to be efficient solvents for biomass dissolution. However, the high viscosity of ILs limits the dissolution process. Therefore, with the aim to enhance the dissolution of lignocellulosic biomass, a series of new ether-functionalized ILs with low viscosity values were synthesized and characterized. Their properties, such as density, viscosity and thermal stability, were analyzed and discussed in comparison with a common commercial IL, namely 1-butyl-3-methylimidazolium chloride (BMIMCl). The presence of the ether group in the new ILs reduces the viscosity of the ILs to some appreciable extent in comparison to BMIMCl. 1-2(methoxyethyl)-3-methylimidazolium chloride (MOE-MImCl), which possesses the lowest viscosity value among the other ether-functionalized ILs, demonstrates an ability to be a powerful solvent in the application of biomass dissolution via the sonication method. In addition, an optimization study employing response surface methodology (RSM) was carried out in order to obtain the optimum conditions for maximum dissolution of biomass in the solvents. Results suggested that the maximum biomass dissolution can be achieved by using 3 weight% of initial biomass loading with 40% amplitude of sonication at 32.23 min of sonication period.

Keywords: biomass; dissolution; ionic liquids; ultrasonic; optimization; RSM

Citation: Rahim, A.H.A.; Yunus, N.M.; Hamzah, W.S.W.; Sarwono, A.; Muhammad, N. Low-Viscosity Ether-Functionalized Ionic Liquids as Solvents for the Enhancement of Lignocellulosic Biomass Dissolution. *Processes* **2021**, *9*, 261. https://doi.org/10.3390/pr9020261

Received: 31 December 2020
Accepted: 27 January 2021
Published: 29 January 2021

1. Introduction

A rapid growth in world population has led to high demand in fuel and chemicals supply [1]. The annual consumption of petroleum in 2007 was at 4.4×10^6 tons per year with 10% utilization for petrochemical feedstocks [2,3]. High demand for raw materials supplies in the chemical production industry has led to elevated prices of petroleum due to its limited source. A statement provided by BP on World Energy Day in 2014 suggested that the raw crude oil supply will only be able to last for 53.3 years [4]. Therefore, due to this alarming situation, the search for new renewable feedstock has become the main target for researchers.

Lignocellulosic biomass, an inedible portion of dry plant materials, contains cellulose, hemicellulose and lignin, and this biomass is typically disposed of by combustion [5]. Since decades ago, lignocellulosic biomass has gain researcher interest as a new alternative chemical feedstock as it is abundantly available and easily obtained through agriculture activities [6,7]. This signifies a large supply with affordable cost for biofuel and chemical production [6,8]. Numerous studies have shown that all biomass constituents have the potential to be converted into various simple chemicals such as fuel ethanol, furfural, lactic acid, phenolic aldehydes and cyclohexane [9,10]. However, due to the high recalcitrance of lignocellulosic biomass, mainly towards organic solvents, which is due to amphiphilic properties of cellulose, the search for the suitable solvents for dissolution is a crucial step prior to any conversion processes in order to extract all constituents and reduce the biomass crystallinity thus, in turn, helps to improve cellulose accessibility [7,11]. Several pretreatment methods have been developed, such as ammonia explosion, acid and alkali pretreatment [12]. However, these processes suffer several drawbacks, such as the generation of a huge amount of wastes into the environment, the requirement of expensive catalysts and the formation of undesirable compounds [13,14].

Ionic liquids (ILs) have been recognized as potential solvents for dissolving biomass and extracting cellulose, hemicellulose and lignin. Generally, an IL is defined as a molten salt, composed entirely of ions and has a melting point lower than 100 °C [15]. The availability of numerous anions and cations combinations allows ILs to be synthesized based on applications of interest with tunable properties. In addition, ILs possess exceptional physical properties such as broad liquidus range, negligible vapor pressure and high thermal stability [16,17]. As a large portion of lignocellulosic biomass consists of cellulose with the amphiphilic property [18], ILs have been recognized as a potential pretreatment solvent to dissolve lignocellulosic biomass. Nevertheless, ILs normally possess high viscosity values, and this provides the main challenge in the application of ILs as solvents for lignocellulosic biomass dissolution. Highly viscous ILs such as 1-butyl-3-methylimidazolium chloride (BMIMCl) and 1-ethyl-3-methylimidazolium chloride (EMIMCl) cause a reduction in the mass transfer of solute during dissolution, and this eventually leads to the decline of the percentage of biomass dissolved in the ILs. The studies conducted by Fukaya et al. and Zhang et al. have shown that ally-based ILs with lower viscosity values as compared to alkyl-based ILs demonstrated better dissolution ability [19,20].

Therefore, the introduction of various functionalized groups has been explored to alter the viscosity of ILs. In order to reduce the viscosity of ILs, an ether functional group has been proposed to replace one of the methylene groups, CH_2, in the alkyl chain structure of the ILs. Unlike the carbon atom, which forms bonds with four atoms in the alkyl chain, the oxygen atom in the ether functional group may only bond to another two atoms, and this provides rotational flexibility to the structure of the ILs and consequently reduces the viscosity of the ILs. However, this theory is vaguely discussed. The molecular dynamic study done by Siqueira and Ribeiro has shown that flexible alkyl chain has resulted in less effective structure assembly, which then caused a reduction in ILs viscosity [21]. Although the work related to low viscosity ILs as solvents for biomass dissolution has extensively been done, the combination of ether-functionalized ILs with the implementation of sonication has not yet been reported in detail. Previously, our group has reported a comparison study between ether-functionalized and non-functionalized ILs for bamboo biomass dissolution [22]. Therefore, to further explore the effect of ether-functionalized ILs towards biomass dissolution, another three ether-functionalized, which include alkoxyethyl and alkoxymethyl groups, ILs were synthesized. Additionally, the effect of two different ether groups, which are methoxyethyl and methoxymethyl, towards melting point and thermal stability, was also discussed.

In addition, the present work was carried out to determine the optimization condition for biomass dissolution in newly synthesized ether-functionalized ILs assisted by sonication using response surface methodology (RSM). RSM is an empirical model involving mathematical and statistical methods to analyze the proposed experimental condition with an objective to obtain an alternative strategy for various processes and optimize selected variables to find the best performance for the selected response [23]. The RSM study normally involves multiple steps, which are (1) preliminary study for identifying independent variables that affect the response, (2) the selection of suitable experimental design and experimental works, (3) statistical analyses, (4) the evaluation of model fitness,

(5) model verification, and (6) getting an optimum value for each variable [24]. In this work, the preliminary study was performed at conditions that were predetermined during the screening process of selecting the best IL for the optimization study. Further work focusing on the optimization process using selected ether-functionalized IL and model validation was carried out by using RSM.

2. Materials and Methods

2.1. Materials

Chemicals of analytical grade were used without further purification process for the synthesis of ether-functionalized ILs and biomass dissolution. The CAS number, source and grades of the chemicals used are as follows: 1-methylimidazole (616-47-7, Merck, 98%), 1-butylimidazole (4316-42-1, Merck, 98%), 2-chloroethyl methyl ether (627-42-9, Merck, 98%), chloromethyl ethyl ether (3188-13-44, Sigma-Aldrich, 95%), 2-chloroethyl ethyl ether (628-34-2, Sigma-Aldrich, 99%), dimethyl sulfoxide (67-68-5, Merck, 99.9%), dichloromethane (75-09-2, Merck, 99.9%), diethyl ether (60-29-7, Merck, 99.9%), ethyl acetate (141-78-6, Merck, 99.9%), 1-butyl-3-methylimidazolium chloride (79917-90-1, Merck, 99%).

The bamboo biomass from species of *Gigantochloa scortcheninii*, a native plant of Malaysia commonly known as "buluh semantan", was obtained from the bamboo processing industry in Seri Iskandar, Malaysia. The bamboo was ground into a powder and sieved into particle sizes of <500 μm by Retsch Test Sieve, AS 200.

2.2. Synthesis of Ether-Functionalized ILs

The synthesis procedure for each ether-functionalized ILs is described in this section. The structural confirmation of the synthesized ether-functionalized ILs was done by ^{1}H and ^{13}C NMR (Bruker Advance III, 500 MHz) and DMSO-d_6 was used as a solvent. The ^{1}H and ^{13}C chemical shifts are reported in part per million (ppm). The multiplicities are abbreviated as singlet (s), duplet (d), triplet (t) and multiplet (m). The structures of the ILs are shown in Figure 1, and the NMR results are also given in this section.

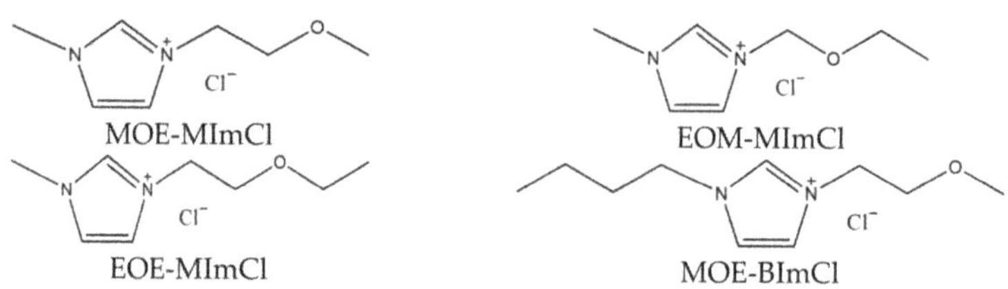

Figure 1. Structures of ether-functionalized ionic liquids (ILs).

2.2.1. 1-(2-Methoxyethyl)-3-Methylimidazolium Chloride (MOE-MImCl)

An equal molar of 2-chloroethyl methyl ether was added into a 50 mL round-bottomed flask containing 0.05 mol of 1-methylimidazole. The mixture was refluxed at a temperature of 80 °C for 48 h with continuous stirring resulted in an amber color liquid. Then, 15 mL of ethyl acetate was added into the IL, and the mixture was shaken several times to remove excess reactants. Ethyl acetate was then removed under reduced pressure using a rotary evaporator. This step was repeated five times to ensure complete removal of excess reactant.

^{1}H NMR (500 MHz, DMSO-d_6): δ 9.536 (s, 1H, (imi)), 7.887 (d, 1H, (imi)), 7.833 (d, 1H, (imi)), 4.402–4.422 (t, 2H, CH_2), 3.3661–3.681, (t, 2H, CH_2), 3.902 (s, 3H, CH_3), 3.240 (s. 3H, CH_3), ^{13}C NMR (500 MHz, DMSO-d_6): δ 137.43123.87, 123.05, 70.05, 58.48, 48.92, 36.16.

2.2.2. 1-Ethyoxymethyl-3-Methylimidazolium Chloride (EOM-MImCl)

0.05 mol of 1-methylimidazole was dissolved in 30 mL dichloromethane (DCM). The reaction was stirred continuously at room temperature for a few hours. Then, the flask was immersed in an ice bath. As the temperature reached 0 °C, 0.05 mol of chloromethyl ethyl ether was added dropwise. The mixture was continuously stirred at room temperature for 6 h. DCM was then removed using a rotary evaporator. The resulting white solid was washed with 20 mL diethyl ether three times to remove the unreacted reactant. Diethyl ether was removed under vacuum.

^{1}H NMR (500 MHz, DMSO-d_6): δ 9.660 (s, 1H, (imi)), 7.943 (d, 1H, (imi)), 7.862 (d, 1H, (imi)), 5.624 (s, 2H, CH_2) 3.919 (s, 3H, CH_3) 3.540–3.554 (m, 2H, CH_2), 1.091–1.119 (t, 3H, CH_3), ^{13}C NMR (500 MHz, DMSO-d_6): δ 137.84, 124.41, 122.36, 78.14, 65.00, 36.37, 15.06.

2.2.3. 1-(2-Ethoxyethyl)-3-Methylimidazolium Chloride (EOE-MImCl)

0.05 mol of 2-chloroethyl ethyl ether was added into a 50 mL round-bottomed flask containing 0.05 mol 1-methylimidazole. The reaction was stirred continuously for 96 h at 80 °C resulting in a yellowish color liquid. The IL was washed with 15 mL of ethyl acetate five times to remove excess reactants. Ethyl acetate was then dried under vacuum pressure.

^{1}H NMR (500 MHz, DMSO-d_6): δ 9.530 (s, 1H, (imi)), 7.887 (d, 1H, (imi)), 7.833 (d, 1H, (imi)), 4.402–4.422 (t, 2H, CH_2), 3.3661–3.681, (t, 2H, CH_2), 3.902 (s, 3H, CH_3), 3.240 (s. 3H, CH_3), ^{13}C NMR (500 MHz, DMSO-d_6): δ 137.40, 123.86, 123.08, 68.01, 65.92, 49.18, 36.18, 15.33.

2.2.4. 1-(2-Methoxyethyl)-3-Butylimidazolium Chloride (MOE-BImCl)

An equal molar of 2-chloroethyl methyl ether was added into a 50 mL round-bottomed flask containing 0.05 mol of 1-butylimidazole. The mixture was refluxed with continuous stirring at temperature 80 °C for a week resulting in a viscous, yellowish liquid. Ethyl acetate was added, and the mixture was shaken several times to remove an unreacted reactant. Ethyl acetate was then discarded while the excess solvent was removed using a rotary evaporator.

^{1}H NMR (500 MHz, DMSO-d_6): δ 9.362 (s, 1H, (imi)), 7.843 (d, 1H, (imi)), 7.812 (d, 1H, (imi)), 4.367–4.385 (t, 2H, CH_2), 4.192–4.221 (t, 2H, CH_2), 3.682–3.702 (m, 2H, CH_2), 3.261 (s, 3H, CH_3), 1.1742–1.801 (m, 2H, CH_2), 1.223–1.268 (m, 2H, CH_2), 0.895–0.912 (m. 3H, CH_3), ^{13}C NMR (500 MHz, DMSO-d_6): δ 137.02, 123.20, 122.74, 70.03, 58.43, 49.00, 48.90, 31.84, 19.21, 13.00.

2.3. Characterization of Ether-Functionalized ILs

The structural confirmation of the synthesized ether-functionalized ILs was done by ^{1}H and ^{13}C NMR (Bruker Advance III, 500 MHz) and DMSO-d_6 was used as a solvent. The ^{1}H and ^{13}C chemical shifts are reported in part per million (ppm). The multiplicities are abbreviated as singlet (s), duplet (d), triplet (t) and multiplet (m).

The thermal stability measurement of ether-functionalized ILs was done by using TGA (PerkinElmer, STA 6000). An approximate 5.00 mg sample was loaded in a crucible pan, and the measurement was carried out with a heating rate at 10 °C/min in the temperature range at 50–650 °C under 20 mL/min nitrogen flow.

The calorimetric measurements were done by employing DSC (Mettler Toledo, DSC-1) in the temperature range of −150 to 150 °C at a heating rate of 10 °C/min. Each sample was weighed to approximately 10.00 mg and sealed in aluminum pans. The samples were cooled to the temperature of −150 °C at the rate of 10 °C/min before heating to 150 °C at the same rate. The data were collected at the second cooling heating scan.

The measurements of viscosity and density of ether-functionalized ILs were done simultaneously by using SVM 3000 from Anton Parr. Then, 4–5 mL of samples were injected by using a 5 mL syringe into the instruments, and the measurement was carried out at atmospheric pressure in the temperature range of 293.15–363.15 K.

2.4. Dissolution of Bamboo in Ether-Functionalized ILs

Bamboo biomass (5 wt %) was added into a 16 mL vial containing 2.0 g of IL. The dissolution of biomass was done using a direct 6 mm sonication probe couples with Sonic Vibra Cell processor with power and frequency of 500 W and 20 kHz, respectively. The probe sonicator was inserted into the mixture IL/biomass to a point about 1/3 of the total height from the surface to ensure optimum acoustic power supply [25]. The performance of biomass dissolution was quantitatively measured based on the method suggested by Sun et al. [26]. The mixture was added into a beaker containing DMSO and stirred for 30 min. The precipitate was separated under centrifugation (4000 rpm, 15 min) and washed with distilled water a few times to remove excess ILs. The precipitate was dried in an oven at 80 °C for 6 h. The calculation for the percentage of biomass dissolution in ILs is shown in Equation (1):

$$\text{Dissolution percentage} = [(Wi - Wp) / Wi] \times 100 \quad (1)$$

Wi is the initial mass of sample loading, and Wp is the undissolved residue left after the addition of DMSO. It must be noted that the addition of DMSO did not cause the dissolution of biomass as biomass has a more complicated structure, which prevents it from dissolving in DMSO. This was proven by functional group analysis done by Sun and coworkers in which the collected undissolved residue spectra were in line with spectra of biomass before undergoing dissolution [26].

Optimization of Bamboo Dissolution in Ether-Functionalized ILs

This part of the study investigates the effect of three independent variables towards bamboo biomass dissolution in the selected ether–functionalized IL by employing Box–Behnken (BB) design using Design Expert. The objective of the experimental design is to optimize the response variables, which is the percentage of biomass dissolved. The IL chosen is MOE-MImCl because it exhibits the best biomass dissolution performance during the initial stage of the study. Three independent variables, which are ultrasonic amplitude (A, 20 to 40%), the period of sonication (B, 10 to 60 min) and initial biomass loading (C, 2.0 to 8.0 wt %), were studied at three levels for a total of 17 runs with five repetitions at the central point as shown in Table 1, and the percentage of bamboo dissolved in ILs reported as the response.

Table 1. Experimental design for response surface methodology (RSM).

Variables	Coding	Unit	Coded Level Experiments Factor		
			−1	0	1
Ultrasonic amplitude	A	%	20	30	40
Period of sonication	B	minutes	10	35	60
Initial biomass loading	C	wt %	2	5	8

The lowest and highest points for each variable were selected based on the literature review and preliminary studies. The experimental data obtained were analyzed for the determination of significance level using one-way analysis of variance (ANOVA) [27]. Meanwhile, the model equation proposed by design was validated by carrying out experiments based on the suggested solution for optimum conditions.

3. Results

3.1. Thermal Properties of Ether-Functionalized ILs

3.1.1. Thermal Stability

Various studies that have been carried out suggested that the introduction of ether will alter the thermal properties of ILs by reducing their thermal stability and melting point in comparison to imidazolium ILs without ether group [28–30]. Table 2 lists the onset temperature (T_o) for synthesized ether-functionalized ILs, and the result was compared with BMIMCl from a previous publication [31]. The incorporating of ether should have reduced the T_o of ether-functionalized ILs as compared to BMIMCl due to the weak electrostatic interactions between cation and anion caused by the electron-donating effect, as can be seen in EOM-MImCl [32,33]. However, the opposite result was observed for MOE-MImCl, EOE-MImCl and MOE-BImCl in which their T_o do not significantly differ from BMIMCl at the same reaction rate. Besides this, EOM-MImCl with the substituent group of alkoxymethyl showed a noticeable reduction in T_o. This may be due to the short space that exists between the alkoxy oxygen atom and quaternary nitrogen atom as it is only separated by one methylene group in the cation's structure as compared to ILs with alkoxyethyl group [34].

Table 2. Onset, glass transition and melting temperature for ILs.

Ionic Liquids	T_o (°C)	T_g (°C)	T_m (°C)
MOE-MImCl	273.13	−86.45	-
EOM-MImCl	178.00	−83.84	-
EOE-MImCl	267.88	−91.43	-
MOE-BImCl	264.76	−69.09	-
BMIMCl	268.89	-	74.1

3.1.2. Phase Transition

The study on ILs phase transition behavior was done by DSC in the temperature range of −150 to 150 °C. Table 2 shows the glass transition (T_g) and melting temperature (T_m) of all synthesized ether-functionalized ILs. Based on the DSC measurement, all ether-functionalized ILs displayed glass transition behavior instead of melting. These phase transition behavior of ether-functionalized ILs were different as being compared with IL without the presence of ether group, and this is supported by the presence of T_m in the phase transition of BMIMCl as shown in Table 2 [31]. In general, this type of behavior observed in ether-functionalized ILs indicates the formation of amorphous glass compounds during the cooling process, and reformation of the liquid phase on the heating stage without the crystallization process occurred [35]. The structural flexibility of ether in the alkyl side chain of ether-functionalized ILs had increased the rotational freedom, thus suppressing the ILs crystallization ability.

3.1.3. Density

The density and viscosity measurements were carried out for all ether-functionalized ILs except for EOM-MImCl due to its solid appearance, which is not suitable for viscosity measurement. The experimental density values for ether-functionalized ILs and BMIMCl are given in Figure 2. In the ether-functionalized ILs family, the density of ILs reduced as the alky chain increased from methyl to butyl group. Similar cases were observed in tetraalkylammonium, trialkyl sulfonium and 1,2,3-dialkyimidazolium-based ILs [36,37]. The cation size becomes larger as the alkyl chain gets longer, thus leads to unfavorable packing in IL structure, which in turn contributes to the high-density of ether ILs [37]. To highlight the effect of ether in ILs density, a similar measurement was done for BMIMCl. According to Zhang and coworkers, the presence of ether in ILs alkyl chain had increased the density of ILs due to the better entanglement of the ether group caused by its flexibility [37].

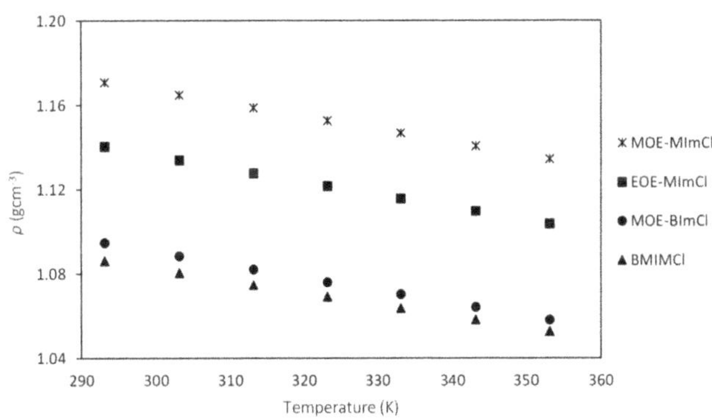

Figure 2. Density of ILs.

Additionally, the experimental density values can be plotted to form a straight line by inserting the values according to empirical Equation (2) as follows:

$$\ln\rho/\text{gcm}^{-3} = b - \alpha T/K \quad (2)$$

where b is empirical constant, and α is a thermal expansion coefficient. Figure 3 shows the empirical plot of ILs, while Table 3 lists the correlation coefficient (R^2) and thermal expansion coefficient (α) values of ILs.

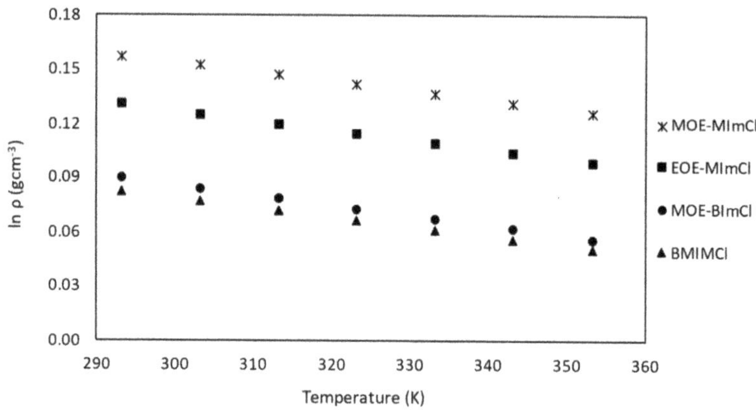

Figure 3. Empirical plot of ILs density.

Table 3. The value for linear fitting parameters (R^2), thermal expansion coefficient (α), molar volume (V), lattice energy (U_{POT}) and standard molar entropy (S°) for ILs.

Ionic Liquids	R^2	α	V (nm^3)	U_{POT} (kJ.mol)	S° (J.K. mol^{-1})
MOE-MImCl	0.9999	5.3×10^{-4}	0.2512	475.6455	342.5681
EOE-MImCl	0.9999	5.4×10^{-4}	0.2785	463.0592	376.6383
MOE-BImCl	1.0000	5.7×10^{-4}	0.3327	442.3739	444.2305
[BMIM][Cl]	0.9999	5.2×10^{-4}	0.26782	467.7663	363.3433

As the R^2 value is more than 0.99, it was concluded that the experimental density values of ILs were well fitted with the proposed empirical linear equation. The values of α for all ILs were in the range of 5.2×10^{-4} to 5.7×10^{-4}, which is lower than common solvents.

Besides this, the standard molar volume (V), lattice energy (U_{POT}) and standard molar entropy (S°) at T = 298.15 K can be calculated based on experimental density values using the following equations:

$$V = M/[(N\cdot\rho)] \quad (3)$$

$$S^\circ = 1246.5\,(V) + 29.5 \quad (4)$$

$$U_{POT} = 1981.2\,[(\rho/M)\hat{}(1/3)] + 103.8 \quad (5)$$

where M is molar mass, N is Avogadro's constant (6.0221×10^{23}) and ρ is the density. All the data gained from those equations were also listed in Table 3.

3.1.4. Viscosity

The dynamic viscosity of the synthesized ether-functionalized ILs and commercial BMIMCl as a comparison was carried out simultaneously with density measurement in the temperature range of 293.15 K to 353.15 K by using Anton Parr SVM 3000. The dynamic viscosities of ILs decreased exponentially with increasing temperature, as shown in Figure 4. The viscosity values decreased in the order of MOE-BImCl > BMIMCl > EOE-MImCl >

MOE-MImCl. Generally, the viscosity of ether-functionalized ILs increased as the alky chain increased. To highlight the effect of the ether-functionalized group on the viscosity of ILs, a comparison was made between MOE-MImCl and BMIMCl, which contain a common ethyl chain in their structures. MOE-MImCl, which has an ether group, exhibited lower viscosity at all temperatures studied than that of BMIMCl. The existence of an oxygen atom with two lone pairs in the ILs alkyl chain causes an electron-donating effect that weakens the electrostatic interaction that exists between cation and anion and thus contributing to a reduction of MOE-MImCl viscosity [37].

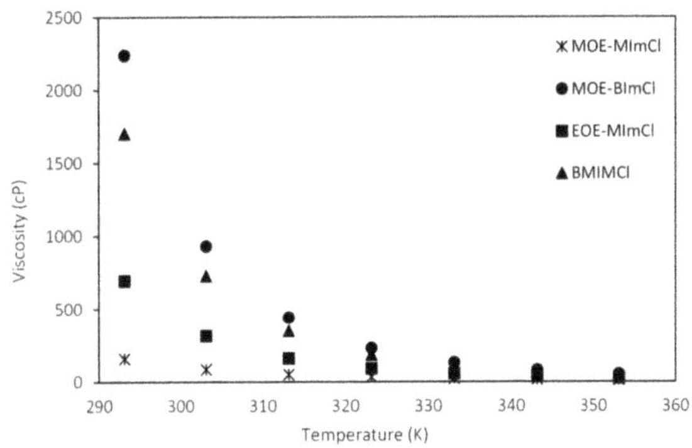

Figure 4. Viscosity for ILs.

Figure 5 presents the Arrhenius plot of ILs dynamic viscosity, which were done according to Equation (6):

$$\ln \eta = \ln A + E_n/RT \tag{6}$$

where η is the dynamic viscosity, E_n is the activation energy, and R is the universal gas constant (8.314 J/mol.K).

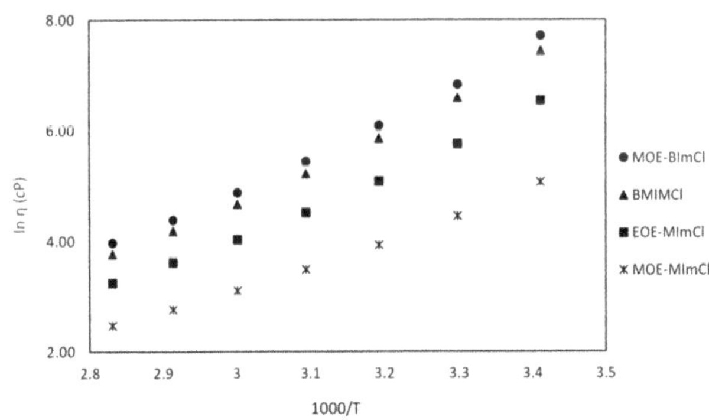

Figure 5. Arrhenius plot for ILs.

The values of E_n, A and linear fitting parameters (R^2) were calculated and tabulated in Table 4. The high R^2 values for all ILs, which are >0.90, indicate all ILs were well fitted with the proposed Arrhenius model. The E_n values generated based on constructed Arrhenius plot represent the energy barrier that needs to be overcome by ion or mass transport. Generally, the viscosity flow is influenced by the value of E_n owned by the fluid. The changes in E_n will be constant with the changes in viscosity. Based on Table 4, by comparing two types of ILs with the same number of alkyl length, namely BMIMCl and MOE-MImCl, the incorporating of ether group in MOE-MImCl has reduced its E_n value as compared to BMIMCl. This indicates that the ether-functionalized group lowers the energy barrier and thus causes a reduction in the viscosity of the IL [34].

Table 4. The value for R^2 and activation energy (E_n) for ILs.

Ionic Liquids	R^2	E_n ($kJ \cdot mol^{-1}$)
MOE-MImCl	0.9945	37.08
EOE-MImCl	0.9926	47.18
MOE-BImCl	0.9949	53.58
BMIMCl	0.9955	52.60

3.2. *Dissolution of Bamboo Biomass in Ether-Functionalized ILs*

Generally, lignocellulosic biomass dissolution is a crucial step in the pretreatment process in order to extract its important constituents, namely cellulose, hemicellulose and lignin, which later can be converted into value-added chemicals. After a successful dissolution process, each constituent can be recovered by adding various solvents into the mixture containing dissolved lignocellulosic biomass and IL [38,39]. Nevertheless, this work is focusing on the improvement of the lignocellulosic biomass dissolution process utilizing ILs. In this work, the dissolution of bamboo was carried out in the synthesized ether-functionalized ILs to identify the best ether ILs for lignocellulosic biomass pretreatment. All parameters for the dissolution process were predetermined during previous studies [22]. Figure 6 shows the percentage of bamboo dissolved in all ILs. MOE-MImCl recorded the highest dissolution percentage, which is more than 90%, as compared to other ether-functionalized ILs. As shown in the viscosity data previously, MOE-MImCl recorded the lowest viscosity values. Besides promoting the high mass transfer of solute during dissolution, a low viscosity environment is also preferable for the sonication technique [40]. In a low viscous solution, the cavitation occurs rapidly to generate microbubbles that later burst and provide higher energy for breaking down the cell wall of lignocellulosic biomass [41]. This process enhances the amount of biomass dissolved in MOE-MImCl within a shorter period. The lowest dissolution percentage of bamboo was recorded in EOM-MImCl, which is less than 10%. EOM-MImCl exists in a solid state in which the ultrasonic waves were unable to be properly transmitted, thus causing an immediate reduction in power generated by ultrasonic.

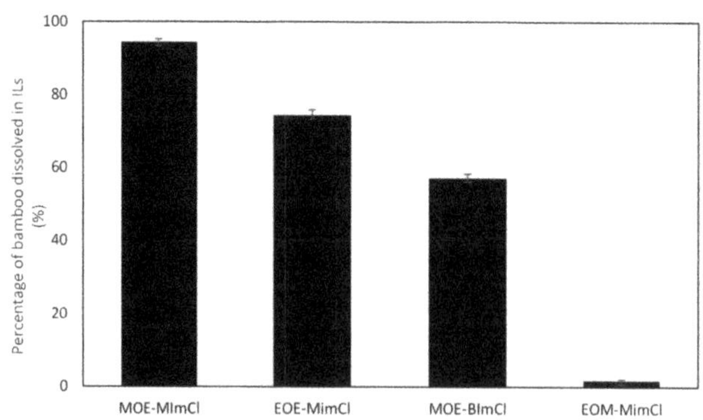

Figure 6. Percentage of bamboo dissolved in ether-functionalized ILs.

3.2.1. RSM Study in Biomass Dissolution in ILs by Analysis on ANOVA

RSM is applied for the optimization condition determination to improve the response to maximum or minimum [42,43]. In this case, RSM was employed to determine the optimum condition for bamboo dissolution in MOE-MImCl. The model proposed empirical relationship as shown in Equation (7) as follows:

$$Y = 12.0658 + 2.58055A - 0.91528B + 3.20842C + 0.27320AB - 0.34650AC + 0.018367BC \quad (7)$$

in which y is the percentage of bamboo, A is ultrasonic amplitude, B is period of dissolution, and C is initial biomass loading. Table 5 shows the actual and predicted value for optimization experiments.

Table 5. Predicted vs. actual value for bamboo dissolution in 1-2(methoxyethyl)-3-methylimidazolium chloride (MOE-MImCl).

Run	Amplitude (%)	Period of Sonication (min)	Initial Biomass Loading (wt %)	Dissolution Percentage (Predicted)	Dissolution Percentage (Actual)
1	40	10	5.0	64.72	69.7
2	20	35	8.0	26.14	23.11
3	30	60	2.0	71.57	70.54
4	30	35	5.0	53.42	59.89
5	30	35	5.0	53.42	57.78
6	30	60	8.0	35.07	36.64
7	40	35	2.0	101.48	95.63
8	30	35	5.0	53.42	53.78
9	40	60	5.0	78.19	84.11
10	30	10	2.0	74.52	74.43
11	20	10	5.0	42.30	41.84
12	30	10	8.0	32.50	35.02
13	30	35	5.0	53.42	48.62
14	30	35	5.0	53.42	50.87
15	40	35	8.0	41.43	32.97
16	20	35	2.0	44.61	44.19
17	20	60	5.0	28.45	28.93

Table 6 shows the analysis of variance (ANOVA) as the software has suggested a two-factor interaction (2FI) to describe the desired response, which is the dissolution percentage of bamboo. Generally, ANOVA provides information such as *p*-value, F-value and lack of fit to evaluate the adequacy of the model for ensuring the tolerable fit is achieved. The *p*-value is used to evaluate the significance of each variable while simultaneously determine the effect of each factor in which as the *p*-value is <0.05 (also known as Prob > F-value), the model can be considered as significantly fitted [44]. In this study, the model *p*-value was <0.0001 with model F-value of 38.14 for dissolution percentage of bamboo indicates the significance of the suggested model. Other than this, the *p*-value also determines which variables provide a significant effect on the response. As shown in Table 6, the parameters A, C, and interactive parameters, which are AB and AC, have *p*-values <0.05 suggesting a significant effect towards response. The values of B and BC are >0.05, implying no significant effect towards bamboo dissolution in IL. However, the insignificant variable and interactive effect cannot be omitted in order to retain the model hierarchy as the other variables are highly significant.

Meanwhile, lack of fit that is used for verifying whether systematic or random error responsible for deviation of expected values from the measured one had a *p*-value of 0.3792, which indicates the model also fits well with the experiments [45]. An adequate precision is a signal-to-noise ratio that measures the range of predicted response relative to an associate error with a value greater than four is desired. Our work displays the value of 22.354 confirmed an adequate signal. The coefficient of determination (R^2) is one of the main criteria to check the adequacy of the model as it indicates the goodness of the data fit in the model. The closer the value of predicted data by the model to experimental data, the closer the value of R^2 into 1. In this study, the value of R^2 is 0.9581, which signage that 95.81% of the percentage of bamboo dissolution is attributed to the considered variables. A high R^2 value also implies a strong correlation between actual and predicted value, as shown in Figure 7. Moreover, the model fitting can be discussed based on the values of adjusted coefficient determination (adj-R^2) and predicted coefficient determination (pred-R^2), in which the difference between both values should be within 0.2 for the model to be reasonable [45]. The values of adj-R^2 and pred-R^2, which are 0.9330 and 0.84, also supported the previous R^2 value.

Table 6. ANOVA table.

Source of Variations	Sum of Squares	Degree of Freedom	Mean Square	F-Value	*p*-Value	
Model	6313.83	6	1052.30	38.14	<0.0001	significant
A—ultrasonic amplitude	2604.25	1	2604.25	94.39	<0.0001	
B—period of dissolution	0.074	1	0.074	2.686×10^{-3}	0.9597	
C—initial sample-loading	3083.09	1	3083.09	111.74	<0.0001	
AB	186.60	1	186.60	6.76	0.0265	
AC	432.22	1	432.22	15.67	0.0027	
BC	7.59	1	7.59	0.28	0.6114	
Residual	275.91	10	27.59			
Lack of fit	188.31	6	31.39	1.43	0.3792	Not significant

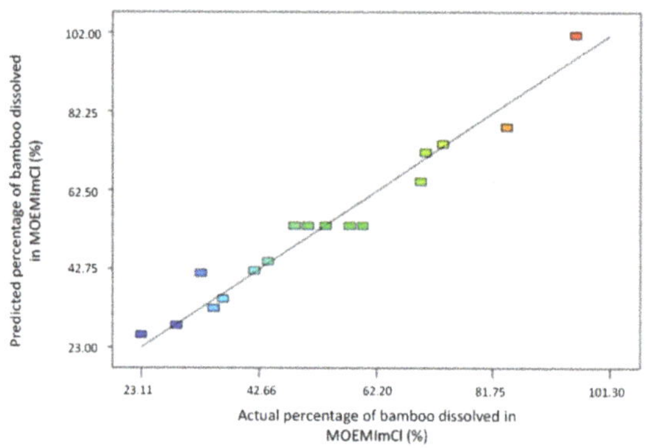

Figure 7. Actual vs. predicted plot for response surface methodology (RSM) of bamboo dissolution in MOE-MImCl.

3.2.2. Effect of Ultrasonic Amplitude, Period of Dissolution and Initial Sample-Loading towards Bamboo Dissolution in MOE-MImCl

The highest percentage for bamboo dissolved, which is 95.63%, was recorded at 40% amplitude with 2 wt % initial bamboo-loading at 35 min period of sonication. Meanwhile, increasing the initial bamboo-loading to 8 wt % while sonicated at a minimum amplitude of 20% reduced the percentage of dissolution to 23.11 within the same reaction time. The contour plot in Figure 8 describes the relationship between each variable towards lignocellulosic biomass dissolution.

In our work, the initial reaction temperature was recorded at 27 °C. As the mixture of bamboo/ILs was exposed into sonication, the final reaction temperature has reached about 80 °C and 120 °C at 20% and 40% of ultrasonic amplitude, respectively. Generally, ultrasonic amplitude provides a significant effect on the reaction temperature. The reaction temperature was found to be proportional to the ultrasonic amplitude, and this relationship has been further elaborated in our previous publication [46]. As the temperature is highly dependent on ultrasonic amplitude, thus in this work, we focus on varying the ultrasonic amplitude. Table 5 indicates that the ultrasonic amplitude used in each reaction has a significant influence on the dissolution percentage of bamboo in ILs. The maximum amplitude of 40% with 2 wt % of bamboo-loading at 35 min period of dissolution had resulted in a high percentage of dissolved bamboo, which is 95.63%. However, the percentage of biomass dissolved had been reduced to 44.19% as the ultrasonic amplitude decreased to 20% at the same initial loading and period of dissolution.

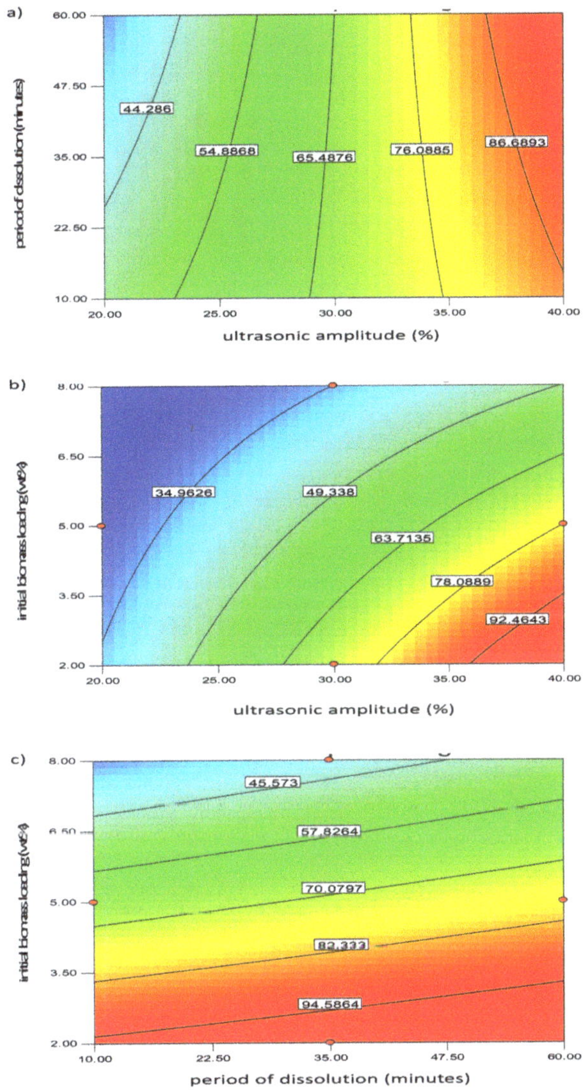

Figure 8. Contour plot for the effect of each parameter for bamboo dissolution in MOE-MImCl; (**a**) plot of period of dissolution against ultrasonic amplitude, (**b**) plot of initial biomass loading against ultrasonic amplitude and (**c**) plot of initial biomass loading against period of dissolution.

The relationship between the period of dissolution and the percentage of bamboo dissolved in MOE-MImCl not showing a constant trend as different initial loading shown different results. For example, at initial loading of 2.0 wt % with 30% amplitude and 10 min of reaction time, the dissolution percentage of bamboo was 74.43%. As the reaction time increased to 60 min, the percentage of bamboo dissolved slightly decreased to 70.54%. In the meantime, this trend was reversed for maximum initial bamboo-loading of 8.0 wt % within the same ultrasonic amplitude. At 10 min of reaction time, 35.02% of bamboo had been dissolved. As the reaction was extended to 60 min, the percentage of bamboo dissolved had increased to 36.64%. Based on this study, it can be concluded that a longer reaction time does not assure a higher dissolution percentage as amplitude and initial sample-loading provide a more significant effect.

However, a negative correlation was observed between the initial biomass-loading and the percentage of biomass dissolved. For example, 8 wt % of initial biomass-loading had resulted in less than 35% of biomass had dissolved in ILs. Meanwhile, the minimum initial biomass-loading of 2 wt % resulted in more than 90% of biomass dissolved in the IL within the same amplitude of 40% and period of sonication at 35 min. The same result was obtained by Muhammad et al. in which the lignin recovery decreased as the loading of bamboo increased [47]. Cruz et al., indicated that an increase in initial biomass-loading could cause an increase in the IL/biomass mixture's viscosity [48]. This, in turn, contributes to a low dissolution percentage due to the reduction of the ILs ion mobility, which restricts the interaction of IL with biomass.

For the optimization process, the response was set at maximum while ultrasonic amplitude was to be in the range while a minimum period of dissolution was chosen since, in any dissolution process, a shorter period of dissolution with high biomass loading is preferable. Although the maximum dissolution was achieved at the initial biomass-loading at 2 wt %, 3 wt % of initial biomass-loading has been chosen due to a higher chance of extracting biomass constituents. In order to further verify the model, another experiment was conducted based on the optimization condition suggested by the software to calculate the relative error. The relative error was 5.83, which is acceptable for model verification.

3.2.3. Comparison Study on Bamboo Dissolution in ILs

For comparison study, the bamboo was dissolved in BMIMCl based on optimum condition obtained in the previous RSM study. The result was then compared with MOE-MImCl. Figure 9 shows the percentage of bamboo dissolved in both ILs. It was identified that ether-functionalized ILs successfully dissolved 96.45% of bamboo compared to BMIMCl, which recorded 78.35% of dissolution. Previous studies suggested that the lignocellulosic biomass dissolution in ILs was caused by several factors such as temperature, the particle size of biomass, time and viscosity of ILs [49,50]. The low viscosity property of MOE-MImCl compared to BMIMCl was discussed in the previous section. Apparently, the low viscosity of MOE-MImCl becomes a major contribution in improving the bamboo dissolution in ILs. Low viscosity MOE-MImCl boosts the mass transfer of solute, thus leads to a higher percentage of dissolution. The same results were obtained by Fukaya et al. in which low viscosity of ally-based ILs displayed a better dissolution ability compared to alkyl-based ILs [19]. Other than this, the dissolution of bamboo in ILs was done by implementing a direct probe sonication technique. Generally, sonication is preferably done in a low viscous medium in order to maximize cavitation during the process. In a highly viscous medium such as BMIMCl, the shockwave will not be able to transfer properly, which leads to a reduction in cavitation, thus causes low dissolution of bamboo.

Besides this, the lignocellulosic bamboo dissolution involves the disruption of the hydrogen bonds network in cellulose. Numerous studies have shown that anions do play a huge role in determining the ability of ILs to break chemical bonds in lignocellulosic biomass. However, current works by Okushita et al. and Chang et al. have confirmed the involvement of cation in the disruption of hydrogen bonds in lignocellulosic biomass based on FTIR and solid-state NMR studies [51,52]. Apart from this, a simulation study by Liu and coworkers had revealed the existence of hydrophobic interaction between imidazolium cation with D-glucose in cellulose. This interaction also plays an important role in ensuring lignocellulosic biomass dissolution occurred in ILs [53]. In the case of MOE-MImCl, the oxygen atom of the methoxy group with two lone pairs acts as an electron-withdrawing group, thus becomes an extra point for the formation of hydrogen bond compared with

BMIMCl. However, the presence of ether group in MOE-MImCl also causes the IL to become more sensitive towards the presence of moisture and consequently owns a higher moisture content compared to BMIMCl. Besides this, the reaction was done in an open system, which may increase the probability of ILs attracting moisture. The presence of extra moisture is well known to reduce the dissolution ability of ILs [54]. Therefore, the moisture in MOE-MImCl was dried under vacuum to ensure minimum content of moisture to avoid any disturbance to the dissolution process.

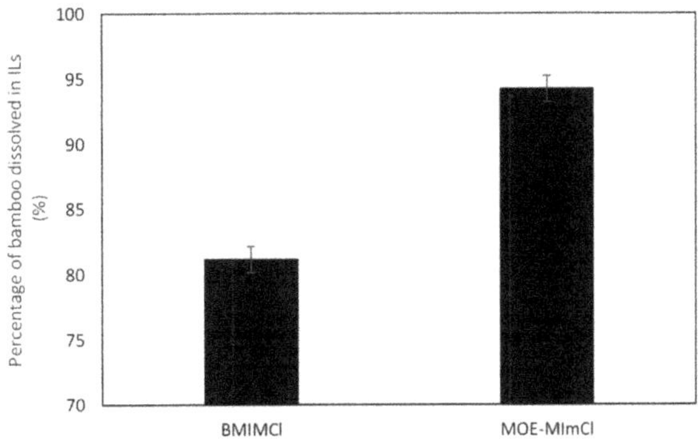

Figure 9. Dissolution of bamboo in MOE-MImCl and [BMIM][Cl].

4. Conclusions

The presence of ether—the functionalized group—was proven to alter the thermal and physicochemical properties of ILs. Further studies on the dissolution of bamboo biomass have shown that the highest amount of bamboo dissolved was achieved in MOE-MImCl, which has the lowest viscosity value as compared to other ILs. Furthermore, the optimization study by using RSM has proven that the amplitude and the initial loading of bamboo biomass play significant roles in the dissolution process. However, more research works need to be carried out to extract constituents in lignocellulosic biomass with a more effective technique to improve the dissolution.

Author Contributions: Conceptualization, A.H.A.R. and N.M.Y.; methodology, A.H.A.R., W.S.W.H., A.S. and N.M., validation, N.M.Y.; writing—original draft preparation, A.H.A.R.; writing—review and editing, N.M.Y.; supervision, N.M.Y. All authors have read and agreed to the published version of the manuscript.

Funding: This research was funded by the Petroleum Research Fund, grant number 0153AB-A30.

Institutional Review Board Statement: Not applicable.

Informed Consent Statement: Not applicable.

Data Availability Statement: The data that support the findings in the present study are available from the corresponding author upon request.

Acknowledgments: Financial assistance and support from Universiti Teknologi PETRONAS and Centre of Research in Ionic Liquids (CORIL), UTP, are greatly acknowledged.

Conflicts of Interest: The authors declare no conflict of interest.

References

1. Owusu, P.A.; Asumadu-Sarkodie, S. A review of renewable energy sources, sustainability issues and climate change mitigation. *Cogent. Eng.* **2016**, *3*, 1–14. [CrossRef]
2. Carlsson, A.S. Plant oils as feedstock alternatives to petroleum—A short survey of potential oil crop platforms. *Biochimie* **2009**, *91*, 665–670. [CrossRef] [PubMed]
3. Roddy, D.J. Biomass in a petrochemical world. *Interface Focus* **2013**, *3*, 1–8. [CrossRef] [PubMed]

4. Stastical Review of World Energy. June 2014. Available online: http://large.stanford.edu/courses/2014/ph240/milic1/docs/bpreview.pdf (accessed on 15 August 2019).
5. Cai, J.; He, Y.; Yu, X.; Banks, S.W.; Yang, Y.; Zhang, X.; Yu, Y.; Liu, R.; Bridgwater, A.V. Review of physicochemical properties and analytical characterization of lignocellulosic biomass. *Renew. Sustain. Energy Rev.* **2017**, *76*, 309–322. [CrossRef]
6. Dhyani, V.; Bhaskar, T. A comprehensive review on the pyrolysis of lignocellulosic biomass. *Renew. Energy* **2018**, *129*, 695–716. [CrossRef]
7. Arevalo-Gallegos, A.; Ahmad, Z.; Asgher, M.; Parra-Saldivar, R.; Iqbal, H.M. Lignocellulose: A sustainable material to produce value-added products with a zero waste approach—A review. *Int. J. Biol. Macromol.* **2017**, *99*, 308–318. [CrossRef]
8. Zhang, K.; Pei, Z.; Wang, D. Organic solvent pretreatment of lignocellulosic biomass for biofuels and biochemicals: A review. *Bioresour. Technol.* **2016**, *199*, 21–33. [CrossRef]
9. Sun, Z.; Fridrich, B.L.; de Santi, A.; Elangovan, S.; Barta, K. Bright side of lignin depolymerization: Toward new platform chemicals. *Chem. Rev.* **2018**, *118*, 614–678. [CrossRef]
10. Wendisch, V.F.; Kim, Y.; Lee, J.-H. Chemicals from lignin: Recent depolymerization techniques and upgrading extended pathways. *Curr. Opin. Green Sustain. Chem.* **2018**, *14*, 33–39. [CrossRef]
11. Zhang, J.; Xu, L.; Yu, J.; Wu, J.; Zhang, X.; He, J.; Zhang, J. Understanding cellulose dissolution: Effect of the cation and anion structure of ionic liquids on the solubility of cellulose. *Sci. China Chem.* **2016**, *59*, 1421–1429. [CrossRef]
12. Antunes, F.A.F.; Chandel, A.K.; Teran-Hilares, R.; Ingle, A.P.; Rai, M.; dos Santos Milessi, T.S.; da Silva, S.S.; dos Santos, J.C. Overcoming challenges in lignocellulosic biomass pretreatment for second-generation (2G) sugar production: Emerging role of nano, biotechnological and promising approaches. *3 Biotech* **2019**, *9*, 1–17. [CrossRef] [PubMed]
13. Brodeur, G.; Yau, E.; Badal, K.; Collier, J.; Ramachandran, K.B.; Ramakrishnan, S. Chemical and physicochemical pretreatment of lignocellulosic biomass: A review. *Enzyme Res.* **2011**, *2011*, 787532. [CrossRef] [PubMed]
14. Hassan, S.S.; Williams, G.A.; Jaiswal, A.K. Emerging technologies for the pretreatment of lignocellulosic biomass. *Bioresour. Technol.* **2018**, *262*, 310–318. [CrossRef] [PubMed]
15. Lei, Z.; Chen, B.; Koo, Y.-M.; MacFarlane, D.R. Introduction: Ionic liquids. *Chem. Rev.* **2017**, *117*, 6633–6635. [CrossRef] [PubMed]
16. Mahmood, H.; Moniruzzaman, M.; Yusup, S.; Akil, H.M. Pretreatment of oil palm biomass with ionic liquids: A new approach for fabrication of green composite board. *J. Clean. Prod.* **2016**, *126*, 677–685. [CrossRef]
17. Sivapragasam, M.; Moniruzzaman, M.; Goto, M. Recent advances in exploiting ionic liquids for biomolecules: Solubility, stability and applications. *Biotechnol. J.* **2016**, *11*, 1000–1013. [CrossRef]
18. Medronho, B.; Romano, A.; Miguel, M.G.; Stigsson, L.; Lindman, B. Rationalizing cellulose (in) solubility: Reviewing basic physicochemical aspects and role of hydrophobic interactions. *Cellulose* **2012**, *19*, 581–587. [CrossRef]
19. Fukaya, Y.; Sugimoto, A.; Ohno, H. Superior solubility of polysaccharides in low viscosity, polar, and halogen-free 1, 3-dialkylimidazolium formates. *Biomacromolecules* **2006**, *7*, 3295–3297. [CrossRef]
20. Zhang, H.; Wu, J.; Zhang, J.; He, J. 1-Allyl-3-methylimidazolium chloride room temperature ionic liquid: A new and powerful nonderivatizing solvent for cellulose. *Macromolecules* **2005**, *38*, 8272–8277. [CrossRef]
21. Siqueira, L.J.; Ribeiro, M.C. Alkoxy chain effect on the viscosity of a quaternary ammonium ionic liquid: Molecular dynamics simulations. *J. Phys. Chem. B* **2009**, *113*, 1074–1079. [CrossRef]
22. Rahim, A.H.A.; Yunus, N.M.; Man, Z.; Sarwono, A.; Hamzah, W.S.W.; Wilfred, C.D. Ultrasonic assisted dissolution of bamboo biomass using ether-functionalized ionic liquid. In Proceedings of the International Conference on Applied Science and Technology, Penang, Malaysia, 10–12 April 2018; Nifa, F.A.A., Chong, K.L., Hussain, A., Eds.; AIP Publishing: New York, NY, USA, 2018.
23. Tauler, R.; Walczak, B.; Brown, S.D. *Comprehensive Chemometrics: Chemical and Biochemical Data Analysis*, 2nd ed.; Elsevier: Amsterdam, The Netherlands, 2020; ISBN 9780444641656.
24. Bezerra, M.A.; Santelli, R.E.; Oliveira, E.P.; Villar, L.S.; Escaleira, L.A. Response surface methodology (RSM) as a tool for optimization in analytical chemistry. *Talanta* **2008**, *76*, 965–977. [CrossRef] [PubMed]
25. Sarwono, A.; Man, Z.; Muhammad, N.; Khan, A.S.; Hamzah, W.S.W.; Rahim, A.H.A.; Ullah, Z.; Wilfred, C.D. A new approach of probe sonication assisted ionic liquid conversion of glucose, cellulose and biomass into 5-hydroxymethylfurfural. *Ultrason. Sonochem.* **2017**, *37*, 310–319. [CrossRef] [PubMed]
26. Sun, N.; Rahman, M.; Qin, Y.; Maxim, M.L.; Rodríguez, H.; Rogers, R.D. Complete dissolution and partial delignification of wood in the ionic liquid 1-ethyl-3-methylimidazolium acetate. *Green Chem.* **2009**, *11*, 646–655. [CrossRef]
27. Behera, S.K.; Meena, H.; Chakraborty, S.; Meikap, B. Application of response surface methodology (RSM) for optimization of leaching parameters for ash reduction from low-grade coal. *Int. J. Min. Sci. Technol.* **2018**, *28*, 621–629. [CrossRef]
28. Chauhan, V.; Singh, S.; Kamboj, R. Self-Assembly and thermal stability of ether-functionalized imidazolium ionic liquids. *Ind. Eng. Chem. Res.* **2014**, *53*, 13247–13255. [CrossRef]
29. Chen, Z.; Huo, Y.; Cao, J.; Xu, L.; Zhang, S. Physicochemical properties of ether-functionalized ionic liquids: Understanding their irregular variations with the ether chain length. *Ind. Eng. Chem. Res.* **2016**, *55*, 11589–11596. [CrossRef]
30. Tang, S.; Baker, G.A.; Zhao, H. Ether- and alcohol-functionalized task-specific ionic liquids: Attractive properties and applications. *Chem. Soc. Rev.* **2012**, *41*, 4030–4066. [CrossRef]
31. Efimova, A.; Hubrig, G.; Schmidt, P. Thermal stability and crystallization behavior of imidazolium halide ionic liquids. *Thermochim. Acta* **2013**, *573*, 162–169. [CrossRef]

32. Chai, M.; Jin, Y.; Fang, S.; Yang, L.; Hirano, S.-I.; Tachibana, K. Ether-functionalized pyrazolium ionic liquids as new electrolytes for lithium battery. *Electrochim. Acta* **2012**, *66*, 67–74. [CrossRef]
33. Zhang, Z.; Salih, A.A.; Li, M.; Yang, B. Synthesis and characterization of functionalized ionic liquids for thermal storage. *Energy Fuels* **2014**, *28*, 2802–2810. [CrossRef]
34. Chen, Z.; Liu, S.; Li, Z.; Zhang, Q.; Deng, Y. Dialkoxy functionalized quaternary ammonium ionic liquids as potential electrolytes and cellulose solvents. *New J. Chem.* **2011**, *35*, 1596–1606. [CrossRef]
35. Fredlake, C.P.; Crosthwaite, J.M.; Hert, D.G.; Aki, S.N.; Brennecke, J.F. Thermophysical properties of imidazolium-based ionic liquids. *J. Chem. Eng. Data* **2004**, *49*, 954–964. [CrossRef]
36. Lee, C.-P.; Peng, J.-D.; Velayutham, D.; Chang, J.; Chen, P.-W.; Suryanarayanan, V.; Ho, K.-C. Trialkylsulfonium and tetraalkylammonium cations-based ionic liquid electrolytes for quasi-solid-state dye-sensitized solar cells. *Electrochim. Acta* **2013**, *114*, 303–308. [CrossRef]
37. Zhang, J.; Fang, S.; Qu, L.; Jin, Y.; Yang, L.; Hirano, S.-I. Synthesis, characterization, and properties of ether-functionalized 1, 3-dialkylimidazolium ionic liquids. *Ind. Eng. Chem. Res.* **2014**, *53*, 16633–16643. [CrossRef]
38. Lan, W.; Liu, C.F.; Sun, R.C. Fractionation of bagasse into cellulose, hemicelluloses, and lignin with ionic liquid treatment followed by alkaline extraction. *J. Agric. Food Chem.* **2011**, *59*, 8691–8701. [CrossRef] [PubMed]
39. Neubert, L.; Sunthornvarabhas, J.; Sakulsombat, M.; Sriroth, K. Delignification and fractionation of sugarcane baggase with ionic liquids. *Cell Chem. Technol.* **2020**, *54*, 301–318. [CrossRef]
40. Wu, T.Y.; Guo, N.; Teh, C.Y.; Hay, J.X.W. Theory and fundamentals of ultrasound. In *Advances in Ultrasound Technology for Environmental Remediation*; Springer: Berlin, Germany, 2013; pp. 5–12. ISBN 978-94-007-5533-8.
41. Kuna, E.; Behling, R.; Valange, S.; Chatel, G.; Colmenares, J.C. Sonocatalysis: A potential sustainable pathway for the valorization of lignocellulosic biomass and derivatives. *Top. Curr. Chem.* **2017**, *375*, 1–20. [CrossRef]
42. Khuri, A.I.; Mukhopadhyay, S. Response surface methodology. *Wiley Interdiscip. Rev. Comput. Stat.* **2010**, *2*, 128–149. [CrossRef]
43. Myers, R.H.; Montgomery, D.C.; Anderson-Cook, C.M. Response surface methodology. In *Response Surface Methodology: Process and Product Optimization Using Designed Experiments*, 4th ed.; John Wiley & Sons: Hoboken, NJ, USA, 2016; ISBN 978-1-118-91601-8.
44. Trinh, L.T.P.; Lee, Y.-J.; Lee, J.-W.; Lee, W.-H. Optimization of ionic liquid pretreatment of mixed softwood by response surface methodology and reutilization of ionic liquid from hydrolysate. *Biotechnol. Bioprocess Eng.* **2018**, 1–10. [CrossRef]
45. Houshmand, A.; Daud, W.M.A.W.; Shafeeyan, M.S. Tailoring the surface chemistry of activated carbon by nitric acid: Study using response surface method. *Bull. Chem. Soc. Jpn.* **2011**, *84*, 1251–1260. [CrossRef]
46. Rahim, A.H.A.; Man, Z.; Sarwono, A.; Muhammad, N.; Khan, A.S.; Hamzah, W.S.W.; Yunus, N.M.; Elsheikh, Y.A. Probe sonication assisted ionic liquid treatement for rapid dissolution of lignocellulosic biomass. *Cellulose* **2020**, *27*, 2135–2148. [CrossRef]
47. Muhammad, N.; Man, Z.; Bustam, M.A.; Mutalib, M.A.; Rafiq, S. Investigations of novel nitrile-based ionic liquids as pre-treatment solvent for extraction of lignin from bamboo biomass. *J. Ind. Eng. Chem.* **2013**, *19*, 207–214. [CrossRef]
48. Cruz, A.G.; Scullin, C.; Mu, C.; Cheng, G.; Stavila, V.; Varanasi, P.; Xu, D.; Mentel, J.; Chuang, Y.-D.; Simmons, B.A. Impact of high biomass loading on ionic liquid pretreatment. *Biotechnol. Biofuel.* **2013**, *6*, 1–59. [CrossRef] [PubMed]
49. Leskinen, T.; King, A.W.; Kilpeläinen, I.; Argyropoulos, D.S. Fractionation of lignocellulosic materials using ionic liquids: Part 2. Effect of particle size on the mechanisms of fractionation. *Ind. Eng. Chem. Res.* **2013**, *52*, 3958–3966. [CrossRef]
50. Li, Y.; Wang, J.; Liu, X.; Zhang, S. Towards a molecular understanding of cellulose dissolution in ionic liquids: Anion/cation effect, synergistic mechanism and physicochemical aspects. *Chem. Sci.* **2018**, *9*, 4027–4043. [CrossRef]
51. Okushita, K.; Chikayama, E.; Kikuchi, J. Solubilization mechanism and characterization of the structural change of bacterial cellulose in regenerated states through ionic liquid treatment. *Biomacromolecules* **2012**, *13*, 1323–1330. [CrossRef]
52. Chang, H.-C.; Zhang, R.-L.; Hsu, D.-T. The effect of pressure on cation–cellulose interactions in cellulose/ionic liquid mixtures. *Phys. Chem. Chem. Phys.* **2015**, *17*, 27573–27578. [CrossRef]
53. Liu, H.; Sale, K.L.; Holmes, B.M.; Simmons, B.A.; Singh, S. Understanding the interactions of cellulose with ionic liquids: A molecular dynamics study. *J. Phys. Chem. B* **2010**, *114*, 4293–4301. [CrossRef]
54. Pang, Z.; Dong, C.; Pan, X. Enhanced deconstruction and dissolution of lignocellulosic biomass in ionic liquid at high water content by lithium chloride. *Cellulose* **2015**, *23*, 323–338. [CrossRef]

processes

MDPI

Article

Encapsulation of Lactoferrin for Sustained Release Using Particles from Gas-Saturated Solutions

Kento Ono [1], Hiroki Sakai [1], Shinichi Tokunaga [1], Tanjina Sharmin [1,2], Taku Michael Aida [1,2] and Kenji Mishima [1,2,*]

[1] Department of Chemical Engineering, Faculty of Engineering, Fukuoka University, 8-19-1 Nanakuma Jonan-ku, Fukuoka 814-0180, Japan; td206501@cis.fukuoka-u.ac.jp (K.O.); td193008@cis.fukuoka-u.ac.jp (H.S.); td196501@cis.fukuoka-u.ac.jp (S.T.); sharmin@fukuoka-u.ac.jp (T.S.); tmaida@fukuoka-u.ac.jp (T.M.A.)
[2] Research Institute of Composite Materials, Fukuoka University, 8-19-1 Nanakuma Jonan-ku, Fukuoka 814-0180, Japan
* Correspondence: mishima@fukuoka-u.ac.jp; Tel.: +81-92-871-6631 (ext. 6428); Fax: +81-92-865-6031

Abstract: The particles from gas saturated solutions (PGSS) process were performed to encapsulate lactofer-rin, an iron-binding milk glycoprotein, using supercritical carbon dioxide ($scCO_2$). A natural en-teric polymer, shellac, was used as a coating material of lactoferrin carried out by the PGSS pro-cess. Conditions were optimized by applying different temperatures (20–50 °C) and pressures (8–10 MPa) and the particles were evaluated for particle shape and size, lactoferrin encapsulation ef-ficiency, Fourier transform infrared (FTIR) spectroscopy to confirm lactoferrin entrapment and in vitro dissolution studies at different pH values. Particles with an average diameter of 75.5 ± 7 μm were produced with encapsulation efficiency up to 71 ± 2%. Furthermore, particles that showed high stability in low pH (pH 1.2) and a sustained release over time (t_{2h} = 75%) in higher pH (pH 7.4) suggested an effective encapsulation process for the protection of lactoferrin from gastric di-gestion.

Keywords: enteric polymer; gastric digestion; PGSS; lactoferrin; shellac

Citation: Ono, K.; Sakai, H.; Tokunaga, S.; Sharmin, T.; Aida, T.M.; Mishima, K. Encapsulation of Lactoferrin for Sustained Release Using Particles from Gas-Saturated Solutions. *Processes* **2021**, 9, 73. https://doi.org/10.3390/pr9010073

Received: 8 December 2020
Accepted: 28 December 2020
Published: 31 December 2020

1. Introduction

Lactoferrin (Lf) is an iron binding single-chain glycoprotein (MW~80 kDa, Figure 1a) which is naturally found in the milk of many mammals including humans and cows, as well as in the saliva, tears, and other secretions and in the secondary granules of neutrophils [1–5]. Lf is abundantly present in the colostrum and milk and supports new born immune defence mechanisms [6,7]. Lf also represents a powerful tool in adult host defence mechanism to stimulate the immune system and enhance the body's protection against virus and bacteria [8–10]. Most recently, a lot of attention is being paid to Lf for its in vitro antiviral activity against SARS-CoV, which is likely similar against SARS-CoV-2 as both viruses depend on the same angiotensin-converting enzyme 2 (ACE2) receptor for cell entry [8]. Other physiological functions of Lf involve preventing tissue damage related to aging, promoting healthy intestinal bacteria [11], maintaining vaginal acidity by promoting growth of selected strains of probiotics [12], preventing cancer, etc. [13]. However, low availability during development and aging, Lf is often taken as an oral supplement. In infants, Lf survives gastric digestion because of the immature state of the neonatal gastrointestinal system, however, Lf is rapidly digested in the adult stomach due to enzymatic hydrolysis and fails to reach the Lf receptors (hLfRs) present in the small intestine. Therefore, protecting Lf from the gastric acidic environment and ensuring its delivery to the targeted site is a crucial issue.

Microencapsulation is one of the promising techniques in protecting functional ingredients in food, cosmetic and pharmaceutical applications. Several encapsulation technologies have been developed for the encapsulation of Lf such as calcium-alginate nanospheres via emulsification [14], or multilayer microcapsules via Layer-by-Layer assembly of bovine serum albumin and tanic acid [15]. However, alginate is sensitive to gastric acid which causes alginate to precipitate leading to pre-mature release of Lf which is supposed to be released in the intestine. On the other hand, the amount of Lf absorbed by $CaCO_3$ particles is significantly low in multilayer capsules composed of Lf in bovine serum albumin and tannic acid, performed through absorption of Lf by porous $CaCO_3$ microparticles. Therefore, a more effective encapsulation technique needs to be designed for optimal delivery across the gastro-intestinal route that not only protects Lf from enzymatic degradation but also enables sustained release to the targeted area.

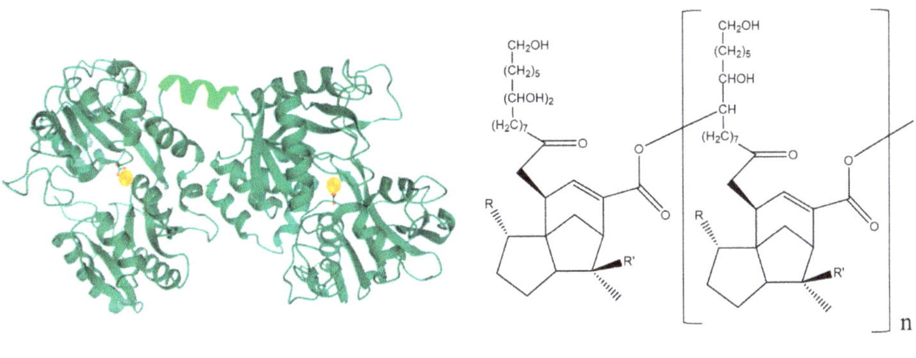

(**a**) (**b**)

Figure 1. (**a**) Crystal structure of diferric bovine lactoferrin, Lf, MW ~80 kDa (PDB entry 1BLF). The N-lobe and C-lobe connecting α-helix is highlighted in light green (aa. 334–344), and the ferric irons are depicted as orange spheres; and (**b**) Chemical structure of shellac, MW ~586.7 g/mol.

In this work, Lf encapsulation initiated with an alternative encapsulation technology, particles from gas saturated solutions (PGSS) process, based on the use of supercritical carbon dioxide ($scCO_2$). By this technique, a gas saturated solution of the coating material is expanded through a nozzle. This sudden depressurization promotes the rapid vaporization of the gas dissolved, that together with the intense cooling due to the Joule-Thomson effect that promotes CO_2 expansion [16], enhances encapsulated particle formation. Shellac, a kind of resin secreted by the female lac bug on trees in the forests of southeastern Asia, is used in this work as a coating material. The chemical structure of shellac is shown in Figure 1b. Shellac is widely used in colon drug delivery system (DDS) as a natural enteric polymer because it cannot be dissolved with gastric acid [17].

Therefore, the aim of this work was to study the encapsulation of Lf with shellac using particles from gas-saturated solutions (PGSS) process: first, conditions were optimized by applying different temperatures (20–50 °C) and pressures (8–10 MPa) and the particles shape, size and encapsulation efficiency were evaluated; and second, the in vitro studies of dissolution of Lf encapsulated in shellac microcapsules at different pH values were studied with the aim to evaluate the sustained release.

2. Methods

2.1. Materials

Lactoferrin, from bovine milk (>95.0 wt%, Iron (Fe) 0.005–0.035 wt%, FUJIFILM Wako Pure Chemical Co, Ltd., Osaka, Japan), liquid Shellac (Shellac 10%, Ethanol 90%), CO_2 (>99.9 vol.%, Fukuoka Sanso Co., Ltd., Fukuoka, Japan), were purchased and used as received without further purification.

2.2. *Preparation of Shellac Microcapsules of Lactoferrin*

The process flow diagram of the PGSS apparatus equipped with a high-pressure cell (inner volume of 500 cm^3, SCV500A, Akico Co., Tokyo, Japan) used to produce microcapsules of Lf is shown in Figure 2. At first, 0.2 g of Lf and 10 mL of shellac was loaded into the pre-heated high-pressure stirred cell (max. 940 rpm) and sealed. The high-pressure autoclave was kept under desired temperature (from 20 to 50 °C). CO_2 was loaded into the high-pressure autoclave to the desired pressure and mixed extensively for 5 min with the pre-loaded materials to form CO_2 saturated solution. CO_2 saturated solution was then charged through outer tube by opening valve V4 and expanded into a chamber with water (100 mL) at atmospheric pressure leading to the formation of microcapsules by precipitation. The recovered microcapsules were separated from water using centrifugation (10 min, 4000 rpm). The concentration of uncoated Lf dissolved in the water was measured by a UV-vis spectrophotometer at 266 nm. Encapsulation efficiency (EE%) was expressed by the following Equation (1).

$$\text{Encapsulation efficiency (\%)} = (W_1 - W_2)/W_1 \times 100 \quad (1)$$

The difference between the amount of loaded Lf in initial loading solution before encapsulation (W1) and the amount of uncoated Lf remained in supernatant after centrifugation (W2) divided by initial protein concentration is equivalent to the encapsulation efficiency (Equation (1)).

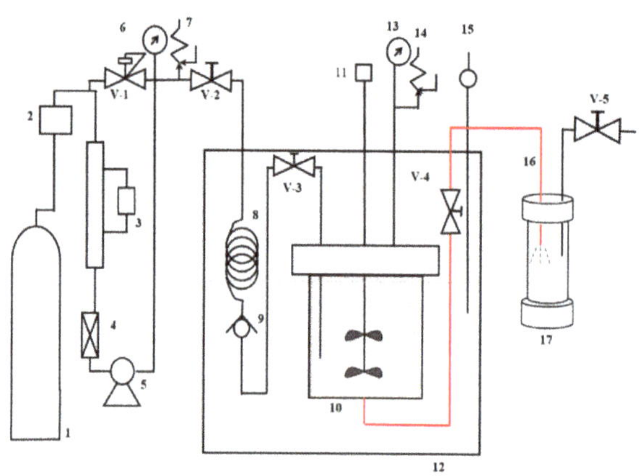

Figure 2. Schematic diagram of apparatus used in PGSS process. 1: gas cylinder, 2: dryer, 3: cooling unit, 4: filter, 5: pump, 6: pressure gauge, 7: safety valve, 8: preheater, 9: check valve, 10: high-pressure cell, 11: agitator, 12: water bath, 13: pressure gauge, 14: safety valve, 15: thermometer, 16: depressurization tunnel, and 17: atmospheric collector vessel with water. V-1 indicates a back-pressure regulator, and V-2 to 5: stop valves.

2.3. *Study of Microcapsule Morphology and Size Distribution*

The structure and morphology of the products were analyzed before and after the PGSS process using a scanning electron microscope (SEM, JEOL JSM6060) and the particle size were investigated by using laser diffraction particle size analyzer (SALD-2000). The release studies of Lf from shellac microcapsules were measured via a UV-vis spectrophotometer (JASCO, V-550, Tokyo, Japan). Fourier transform infrared (FTIR) spectra of pure components, and the shellac microcapsules of Lf were recorded via attenuated total reflectance FTIR spectrometer (FT/IR-4600, JASCO, Tokyo, Japan), and infrared measurements were performed in transmission in the scanning range of 500–4000 cm^{-1} at room temperature.

2.4. Lf microcapsule In Vitro Release

The sustained release rate of Lf from shellac microcapsules was investigated via in vitro release study using vertical Franz-type diffusion cells (VIDTEK, Iwaki, Fukuoka, Japan) for 2 h in simulated gastric fluid (SGF) of pH 1.2 (representing stomach pH) followed by simulated intestinal fluid (SIF) of pH 7.4 (representing small intestine pH) for 6 h. A hydrophilic PTFE membrane (0.45 um pore size, thickness 65 µm, 47 mm diameter; Merck Millipore, Tokyo, Japan) was placed between the upper donor chamber and the receptor chamber of the diffusion cell. SGF of pH 1.2 was prepared by adding 2 g of NaCl and 7 mL of HCl in sufficient MilliQ water to make 1000 mL. SIF of phosphate buffer of pH 7.4 was prepared by dissolving 8 g of NaCl, 0.2 g of KCl, 1.44 g of Na_2HPO_4 and 0.24 g of KH_2PO_4 in 800 mL of MilliQ water followed by adjusting the pH to 7.4 with HCl and adding sufficient MilliQ water to make 1000 mL. Exactly 10 mg of shellac microcapsules was placed on the membrane and filled up the receptor chamber with SGF followed by SIG as required wetting the membrane and the microparticles. The top plate was tightly sealed to avoid evaporation. At fixed intervals (0–480 min.), samples of release liquid were drawn and, the UV absorbance of the microcapsule release medium was measured at 266 nm with a UV-vis detector (JASCO, V-550, Tokyo, Japan). The release medium was returned back to the receptor chamber after measurement.

2.5. Statistical Analysis

All analyses were performed in triplicate. The data were analyzed by one-way analysis of variance (ANOVA) accompanied by Turkey's post hoc. The level of significance was set at $p < 0.05$.

3. Results and Discussion

3.1. Encapsulation of Lf in Shellac by PGSS

Lf encapsulation in shellac by the PGSS process was studied using varying pre-expansion pressures (8–10 MPa) and pre-expansion temperatures (20–50 °C). Figure 3a,b provides the SEM imaged of crude Lf and Lf encapsulated shellac microcapsules produced by PGSS process of $scCO_2$ solutions at 40 °C and 10 MPa, and Figure 3c provides the particle size distribution (PSD) of the Lf encapsulated shellac microcapsules at 10 MPa and 40 °C. As shown in Figure 4, produced particles displayed very similar morphology with irregular shapes which is a very common feature in PGSS.

Table 1 presents a summary of the experimental conditions tested together with PSD and encapsulation efficiency (EE%). As shown in Table 1, Lf encapsulated in shellac microcapsules successfully produced at different pre-expansion pressures (8–10 MPa) and 40 °C pre-expansion temperature. At this condition, the particle size was significantly reduced as the pressure increased from 8 to 10 MPa from 143.0 ± 4 to 75.4 ± 7, respectively, with relatively narrow particle size distributions ($d_{0.1}$ = 61.8 to 36.8 µm and $d_{0.9}$ = 338.5 to 161.3 µm, respectively). This trend can be related to the increase of the solubility of $scCO_2$ in the polymer, which increases as pre-expansion pressure is increased [18]. Polymer plasticity increases with a higher amount of CO_2 dissolved into the polymer at a higher pre-expansion pressure, the Joule-Thomsone effect produced by the release of CO_2 from the polymer during the expansion is stronger, which produces significantly reduced particle size. It can also be seen that encapsulation efficiency increased from 51 ± 5% to 71 ± 2% when expansion pressure increased from 8 to 10 MPa, respectively at 40 °C. However, the solubility of CO_2 into the polymer decreased at lower temperatures (20 and 30 °C). This may occur because of the lower saturation power of CO_2 in liquid (at 20 °C) or sub-critical (30 °C) conditions and caused polymer particles to precipitate and thereby, aggregated inside the high-pressure cell upon depressurization (Figure S1a,b). On the other hand, particle agglomeration was also noted at higher temperatures (i.e., 50 °C, Figure S1c) which may be attributed to the susceptibility of Lf molecules to higher temperature [19] as fine particles were successfully produced using only shellac (without Lf) at higher temperatures (50 °C, Figure S1d).

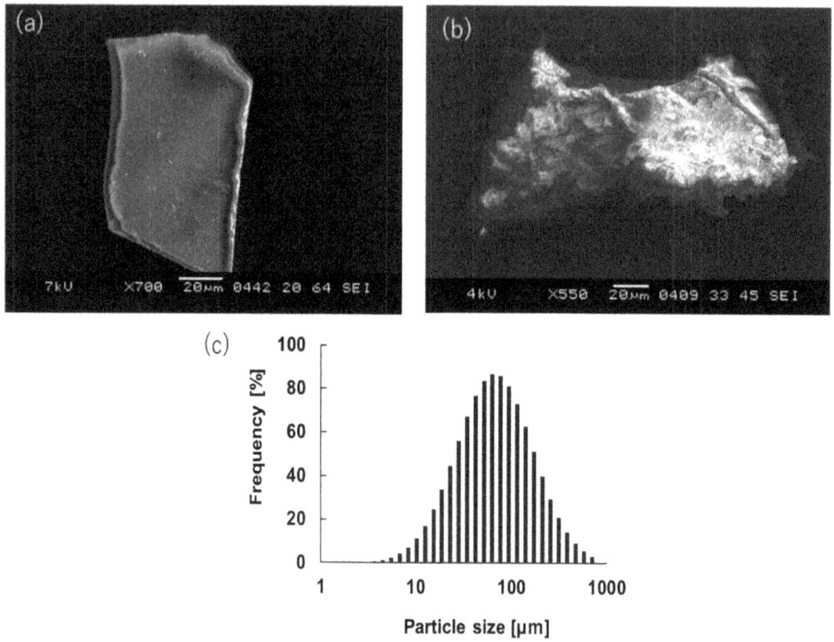

Figure 3. SEM images of (**a**) crude Lf; (**b**) Lf encapsulated shellac microcapsules produced by PGSS process of $scCO_2$ solutions at 40 °C and 10 MPa. (**c**) Particle size distribution of the Lf encapsulated shellac microcapsules produced by PGSS process of $scCO_2$ solutions at 40 °C and 10 MPa.

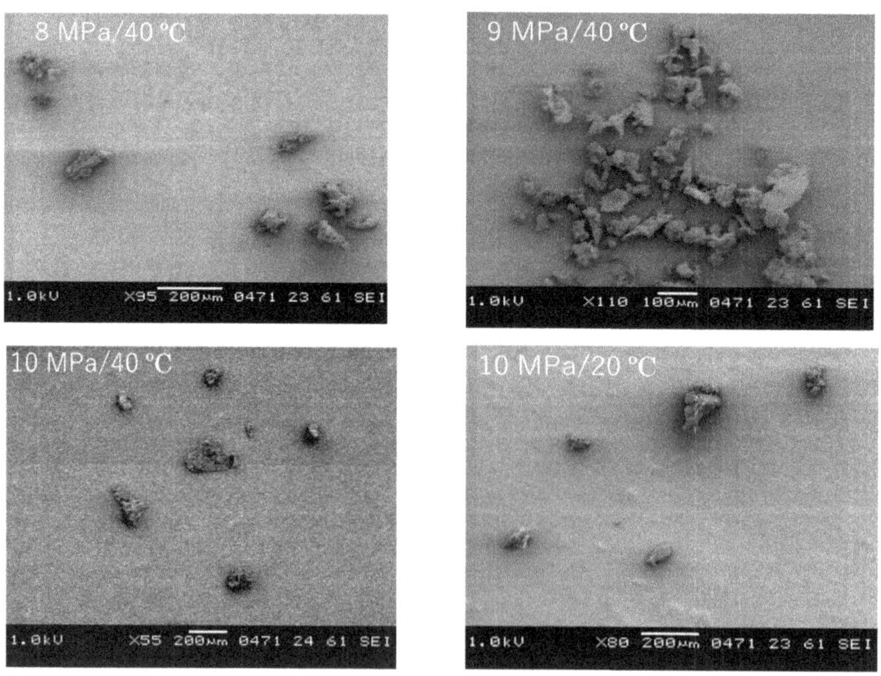

Figure 4. *Cont.*

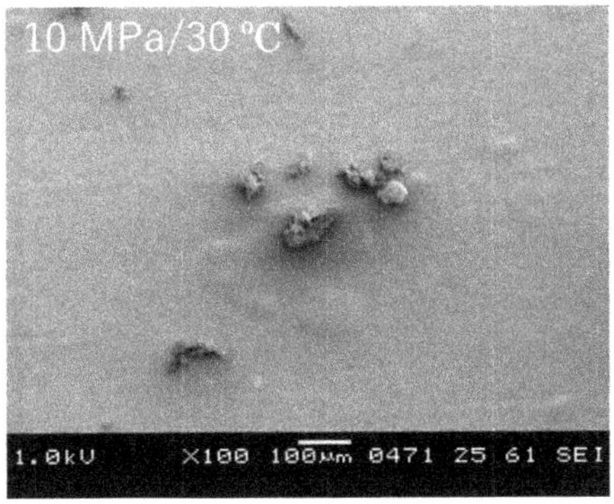

Figure 4. Morphology of Lf loaded shellac microparticles produced at different P and T conditions of supercritical CO_2 in the PGSS process.

Table 1. Influence of process parameters on particle size distribution and encapsulation efficiency of Lf microcapsules.

Experimental Conditions				Avg. Particle Size [μm]	Particle Size Distribution (PSD)				EE [%]
P [MPa]	*T* [°C]	Lactoferrin [g]	Shellac EtOH Ratio		$d_{0.1}$	$d_{0.5}$	$d_{0.9}$	Span	
8	40	0.2	1:10	143.0 ± 4^{b}	61.8	144.3	338.5	1.91	51 ± 5^{a}
9	40	0.2	1:10	132.6 ± 10^{b}	39.8	153.0	592.0	3.61	
10	40	0.2	1:10	75.4 ± 7^{a}	36.8	77.2	161.3	1.61	71 ± 2^{b}
10	50	0.2	1:10	Agglomerated	-	-	-	-	-
10	30	0.2	1:10	59.8 ± 2^{a}	5.09	59.9	559.8	9.92	-
10	20	0.2	1:10	71.4 ± 3^{a}	12.17	77.56	423.3	5.30	-

Here, $d_{0.5}$ refers to the median particle size; whereas $d_{0.1}$ and $d_{0.9}$ refer to the maximum particle diameter below which 10% and 90% of the sample volume exists, respectively; Span = $d_{0.9} - d_{0.1}/d_{0.5}$. Mean values in the same column with different letters are significantly different at $p < 0.05$.

3.2. Fourier Transform Infra-Red Spectroscopy (FT-IR) Analysis

FT-IR analysis was conducted to confirm the entrapment of Lf in the shellac microparticles produced by PGSS process of $scCO_2$ solutions at 40 °C and 10 MPa. Figure 5 presents the FTIR spectra of Lf, shellac, and Lf encapsulated shellac microcapsules. The FTIR spectra of Lf (Figure 5a) showed the stretching and bending vibrations of amide I (1637 cm^{-1}), amide II (1528 cm^{-1}), and C-O-C stretch (1073 cm^{-1}) which are the major characteristics of Lf and sensitive to secondary structure.

The amide I is more commonly used for characterizing the secondary structure and is due to C=O stretching vibrations of the peptide bonds, which are modulated by the secondary structure (α-helix, β-sheet, etc.). The same pattern of stretching frequencies for amide I (1637 cm^{-1}), and amide II (1528 cm^{-1}) were observed in the same position in the shellac microparticles (Figure 5c) suggesting that there was no shift in stretching frequency between Lf and shellac microcapsules. FTIR spectra of shellac (Figure 5b) showed strong symmetric and asymmetric stretching vibrations of CH_2 at 2855 cm^{-1} and 2927 cm^{-1}, respectively. The other strong vibrations peaked at 1710 cm^{-1} are attributed to the C=O stretching vibration of esters. The large band in the range 3100–3600 cm^{-1} with a maximum at about 3428 cm^{-1} is attributed to the O–H stretching vibration, while the O–H bending vibration is identified at 1250 cm^{-1}. The absorption bands at 1167 and 1042 cm^{-1} are due to stretching vibrations of C–O and C–C bonds. The characteristic peaks of shellac in microparticles can be observed at 2927 cm^{-1}, 2855 cm^{-1}, 1710 cm^{-1}, and 1167 cm^{-1} (Figure 5c). Therefore, as the shellac microcapsules exhibited characteristic peaks of both Lf (amide I and II at 1637 cm^{-1} and 1528 cm^{-1}, respectively) and shellac (C=O at 1710 cm^{-1} and CH_2 at 2855 cm^{-1} and 2927 cm^{-1}), suggesting successful encapsulation of Lf without any interaction within the polymer.

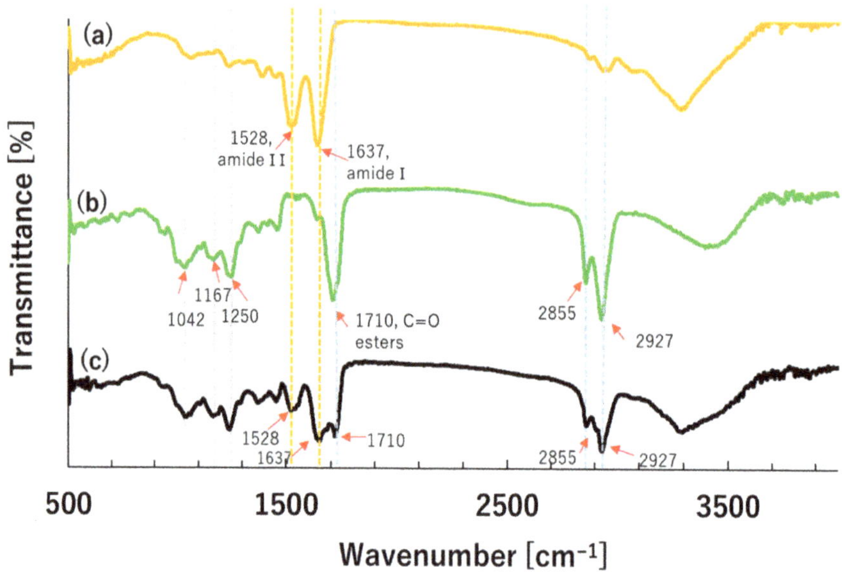

Figure 5. FTIR spectra of (**a**) crude Lf; (**b**) shellac and (**c**) Lf encapsulated shellac microcapsules produced by PGSS process of $scCO_2$ solutions at 40 °C and 10 MPa.

3.3. In Vitro Release

The sustained release profiles of Lf from shellac microcapsules in SGF (pH 1.2) followed by SIF (pH 7.4) are shown in Figure 6. The microcapsules provided a persistent behavior in SGF environment with a very small percentage of Lf release, not more than 11% over time (120 min) establishing that the shellac microcapsules have acidic environment-resistant characteristics. On the other hand, a biphasic Lf release was recorded in the SIF environment: an initial rapid Lf release phase (73% burst in 50 min) was followed by the slow and prolonged phase. The initial burst effect which is frequently observed in protein loaded microparticles because of their high solubility [20] was successfully delayed by the encapsulation of Lf in shellac microcapsules produced by PGSS. The highly hydrophobic nature of shellac in aqueous solution can be attributed to the delayed release observed in the shellac microcapsules. The third-slower release phase was thought to involve the diffusion of Lf entrapped within the inner part of the shellac matrix by means of ester channels of a network of pores. Furthermore, the burst effect may be favorable because a high initial release produces an instant effect which can be subsequently maintained for a prolonged period by a sustained release.

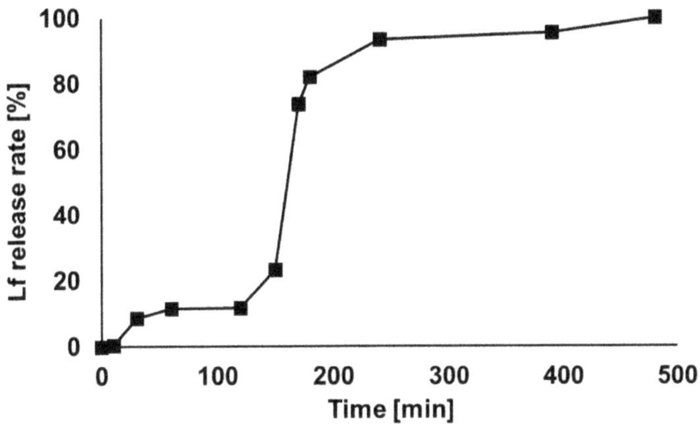

Figure 6. Sustained release profiles of Lf from shellac microcapsules produced by PGSS process of $scCO_2$ solutions at 40 °C and 10 MPa in simulated gastric fluid (SGF) (0–120 min in pH 1.2) followed by simulated intestinal fluid (SIF) (120–480 min in pH 7.4).

4. Conclusions

Sustained release Lf microcapsules were successfully prepared by PGSS process using natural enteric polymer shellac which can remain in the stomach for a long time. Particles produced at 40 °C and 10 MPa showed the highest encapsulation efficiency (71 ± 2%). In vitro release study revealed that shellac coated Lf microcapsules could be resistant against acidic environment, and they would rapidly release Lf under mild alkali conditions with an initial high followed by the slow and prolonged phase release. It could be concluded that shellac coated Lf microcapsules could be a good way to enhance the stability and sustained release of Lf favorable for targeting colon-specific drug delivery system. This study provides the basis for the application of shellac as a potential sustained drug delivery vehicle.

Supplementary Materials: The following are available online at https://www.mdpi.com/2227-9717/9/1/73/s1, Figure S1. Images of Lf loaded shellac microparticle-agglomeration by PGSS process at (a) 20 °C, (b) 30 °C, and (c) 50 °C, at 10 MPa. SEM images of shellac (sole) microparticles produced by PGSS process at 10 MPa and at 50 °C (d).

Author Contributions: Conceptualization, T.S., T.M.A. and K.M.; Data curation, K.O.; Formal analysis, K.O. and H.S.; Funding acquisition, K.M.; Investigation, K.O.; Methodology, K.O.; Project administration, K.M.; Resources, K.O., H.S. and S.T.; Software, K.O. and H.S.; Supervision, K.M.; Validation, K.O., H.S. and S.T.; Visualization, K.O. and H.S.; Writing—original draft, K.O.; Writing—review & editing, T.S., T.M.A. and K.M. All authors have read and agreed to the published version of the manuscript.

Funding: This work was funded by JSPS KAKENHI Grant Number 17K06899.

Institutional Review Board Statement: Not applicable.

Informed Consent Statement: Not applicable.

Data Availability Statement: No new data were created or analyzed in this study. Data sharing is not applicable to this article.

Conflicts of Interest: The authors declare no conflict of interest.

References

1. Gonzalez-Chavez, S.A.; Arevalo-Gallegos, S.; Rascon-Cruz, Q. Lactoferrin: Structure, function and applications. *Int. J. Antimicrob. Agents* **2009**, *33*, 301–308. [CrossRef] [PubMed]
2. Albar, A.H.; Almehdar, H.A.; Uversky, V.N.; Redwan, E.M. Structural heterogeneity and multifunctionality of lactoferrin. *Curr. Protein Pept. Sci* **2014**, *15*, 778–797. [CrossRef] [PubMed]
3. Park, Y.W.; Nam, M.S. Bioactive Peptides in Milk and Dairy Products: A Review. *Korean J. Food Sci. Anim. Resour.* **2015**, *35*, 831–840. [CrossRef] [PubMed]
4. Vogel, H.J. Lactoferrin, a bird's eye view. *Biochem. Cell Biol.* **2012**, *90*, 233–244. [CrossRef] [PubMed]
5. Giansanti, F.; Panella, G.; Leboe, L.; Antonini, G. Lactoferrin from Milk: Nutraceutical and Pharmacological Properties. *Pharmaceuticals* **2016**, *9*, 61. [CrossRef] [PubMed]
6. Sánchez, L.; Calvo, M.; Brock, J.H. Biological role of lactoferrin. *Arch. Dis. Child.* **1992**, *67*, 657–661. [CrossRef] [PubMed]
7. Brock, J. Lactoferrin in human milk: Its role in iron absorption and protection against enteric infection in the newborn infant. *Arch. Dis. Child.* **1980**, *55*, 417–421. [CrossRef] [PubMed]
8. Kell, D.B.; Heyden, E.L.; Pretorius, E. The Biology of Lactoferrin, an Iron-Binding Protein That Can Help Defend Against Viruses and Bacteria. *Front. Immunol.* **2020**, *11*, 1221. [CrossRef] [PubMed]
9. Chang, R.; Ng, T.B.; Sun, W.-Z. Lactoferrin as potential preventative and adjunct treatment for COVID-19. *Int. J. Antimicrob. Agents* **2020**, *56*, 106118. [CrossRef] [PubMed]
10. Valenti, P.; Antonini, G. Lactoferrin: An important host defence against microbial and viral attack. *Cell. Mol. Life Sci.* **2005**, *62*, 2576–2587. [CrossRef] [PubMed]
11. Vega-Bautista, A.; Garza de la, M.; Carrero, J.C.; Campos-Rodríguez, R.; Godínez-Victoria, M.; Drago-Serrano, M.E. The Impact of Lactoferrin on the Growth of Intestinal Inhabitant Bacteria. *Int. J. Mol. Sci.* **2019**, *20*, 4707. [CrossRef] [PubMed]
12. Valenti, P.; Rosa, L.; Capobianco, D.; Lepanto, M.S.; Schiavi, E.; Cutone, A.; Paesano, R.; Mastromarino, P. Role of Lactobacilli and Lactoferrin in the Mucosal Cervicovaginal Defense. *Front. Immunol.* **2018**, *9*, 376. [CrossRef] [PubMed]
13. Yoo, Y.C.; Watanabe, R.; Koike, Y.; Mitobe, M.; Shimazaki, K.; Watanabe, S.; Azuma, I. Apoptosis in human leukemic cells induced by lactoferricin, a bovine milk protein-derived peptide: Involvement of reactive oxygen species. *Biochem. Biophys. Res. Commun.* **1997**, *237*, 624–628. [CrossRef] [PubMed]
14. De Etchepare, M.A.; Barin, J.S.; Cichoski, A.J.; Jacob-Lopes, E.; Wagner, R.; Fries, L.L.M.; Menezes, C.R. Microencapsulation of probiotics using sodium alginate. *Ciência Rural* **2015**, *45*, 1319–1326. [CrossRef]
15. Kilic, E.; Novoselova, M.; Lim, S.; Pyataev, N.; Pinyaev, S.; Kulikov, O.; Sindeeva, O.; Mayorova, O.; Murney, R.; Antipina, M.; et al. Formulation for Oral Delivery of Lactoferrin Based on Bovine Serum Albumin and Tannic Acid Multilayer Microcapsules. *Sci. Rep.* **2017**, *7*, 44159. [CrossRef] [PubMed]
16. Mishima, K. Biodegradable particle formation for drug and gene delivery using supercritical fluid and dense gas. *Adv. Drug Deliv. Rev.* **2008**, *60*, 411–432. [CrossRef] [PubMed]
17. Limmatvapirat, S.; Panchapornpon, D.; Limmatvapirat, C.; Nunthanid, J.; Luangtana-Anan, M.; Puttipipatkhachorn, S. Formation of shellac succinate having improved enteric film properties through dry media reaction. *Eur. J. Pharm. Biopharm.* **2008**, *70*, 335–344. [CrossRef] [PubMed]
18. Labuschagne, P.W.; Naicker, B.; Kalombo, L. Micronization, characterization and in-vitro dissolution of shellac from PGSS supercritical CO_2 technique. *Int. J. Pharm.* **2016**, *499*, 205–216. [CrossRef] [PubMed]
19. Brisson, G.; Britten, M.; Pouliot, Y. Heat-induced aggregation of bovine lactoferrin at neutral pH: Effect of iron saturation. *Int. Dairy J.* **2007**, *17*, 617–624. [CrossRef]
20. White, L.J.; Kirby, G.T.; Cox, H.C.; Qodratnama, R.; Qutachi, O.; Rose, F.R.; Shakesheff, K.M. Accelerating protein release from microparticles for regenerative medicine applications. *Mater. Sci. Eng. C Mater. Biol. Appl.* **2013**, *33*, 2578–2583. [CrossRef] [PubMed]

processes

MDPI

Article

Optimising Brewery-Wastewater-Supported Acid Mine Drainage Treatment vis-à-vis Response Surface Methodology and Artificial Neural Network

Enoch A. Akinpelu [1,*], Seteno K. O. Ntwampe [2,*], Abiola E. Taiwo [3] and Felix Nchu [1,4]

1 Bioresource Engineering Research Group (*BioERG*), Cape Peninsula University of Technology, P.O. Box 652, Cape Town 8000, South Africa; NchuF@cput.ac.za
2 Water Pollution Monitoring and Remediation Initiatives Research Group, School of Chemical and Minerals Engineering, North-West University, P. Bag X60001, Potchefstroom 2520, South Africa
3 Department of Chemical Engineering, College of Engineering, Landmark University, PMB 1001, Omu Aran 240243, Nigeria; taiwoabiolaezekiel@gmail.com
4 Department of Horticultural Sciences, Bellville Campus, Cape Peninsula University of Technology, Symphony Way, PO Box 1906, Bellville 7535, South Africa
* Correspondence: AkinpeluE@cput.ac.za (E.A.A.); karabo.ntwampe@nwu.ac.za (S.K.O.N.); Tel.: +27-18-299-1762 (E.A.A.); Tel.: +27-18-299-1762 (S.K.O.N.)

Received: 23 October 2020; Accepted: 9 November 2020; Published: 18 November 2020

Abstract: This study investigated the use of brewing wastewater (BW) as the primary carbon source in the Postgate medium for the optimisation of sulphate reduction in acid mine drainage (AMD). The results showed that the sulphate-reducing bacteria (SRB) consortium was able to utilise BW for sulphate reduction. The response surface methodology (RSM)/Box–Behnken design optimum conditions found for sulphate reduction were a pH of 6.99, COD/SO_4^{2-} of 2.87, and BW concentration of 200.24 mg/L with predicted sulphate reduction of 91.58%. Furthermore, by using an artificial neural network (ANN), a multilayer full feedforward (MFFF) connection with an incremental backpropagation network and hyperbolic tangent as the transfer function gave the best predictive model for sulphate reduction. The ANN optimum conditions were a pH of 6.99, COD/SO_4^{2-} of 0.50, and BW concentration of 200.31 mg/L with predicted sulphate reduction of 89.56%. The coefficient of determination (R^2) and absolute average deviation (AAD) were estimated as 0.97 and 0.046, respectively, for RSM and 0.99 and 0.011, respectively, for ANN. Consequently, ANN was a better predictor than RSM. This study revealed that the exclusive use of BW without supplementation with refined carbon sources in the Postgate medium is feasible and could ensure the economic sustainability of biological sulphate reduction in the South African environment, or in any semi-arid country with significant brewing activity and AMD challenges.

Keywords: acid mine drainage; artificial neural network; brewing wastewater; response surface methodology; sulphate reduction; optimisation

---

1. Introduction

Of the numerous wastewaters from different food and beverage industries, malting and brewing wastewaters are especially nutrient-rich. In South Africa alone, the brewing industry's capacity is more than 3.1 billion litres of beer per annum [1], which culminates in an industry which is potable-water-intensive with up to 10 litres of water being used to produce 1 litre of beer [2]. The composition and characteristics of brewery wastewater (BW) vary, albeit with high concentrations of crude protein (up to 754 mg/L), BOD_5 (up to 3980 mg/L), COD_{total} (up to 8926 mg/L), and total nitrogen (up to 1305 mg/L) with minute traces of heavy metals such as lead, nickel, iron, manganese, and copper; all of these are suitable for effective biomass growth and to support other beneficial biological processes [3,4]. Some of these nutrients are from extracts of spent biowaste, as for 1 hectolitre (hl) of beer produced, approximately 20 kg of spent bio-waste in the form of grains, excess yeast, and hops ends up in the wastewater. This is equivalent to 85% of all by-products; hence, 15% of the bio-waste is attributed to surplus hops and yeast [5]. Overall, malting and beer racking have been determined to contribute the largest COD_{total} (up to 3000 mg/L) to BW [6]. Therefore, the disposal of such nutrient-rich wastewater into municipal wastewater treatment plants (MWTPs) increases nutrient loading, which can negatively affect the operation of such works. It is advisable to repurpose such wastewater as a bio-waste-type medium for the remediation of other environmental challenges, while others prefer that the water be treated—an undertaking which can result in additional costs to plant operations. Redirecting such wastewater to low-performance and passive processes as a medium could provide a sustainable mitigation strategy for the bio-waste-containing wastewater while providing for an environmentally benign approach to remediate significant environmental pollution [7–9]. A recent example was when BW was used as a fertiliser treatment for crop production, resulting in yields that resembled those under inorganic fertiliser [10]. This means that BW can provide an alternative sustainable nutrient source to passively support biological and environmentally benign processes in other industries while providing remedial action for other unrelated brewery operation challenges. It was further suggested that BW, when at ambient temperature, has high biodegradability; however, this assertion was made for aerobic conditions [6].

South Africa is a semi-arid country with limited water resources, with increasingly reduced rainfall and number of rainy days during autumn, a season associated with high rainfall [11]; current groundwater-sourced acid mine drainage (AMD) water production is high due to previous mining activity, with an example being the 2500 m^3 AMD/day draining into the West Rand Goldfields [12]. Instead of disposing of the BW into MWTPs, it can be redirected to support the passive treatment of AMD for the effective removal of sulphates. Current technologies and research studies in the nexus of BW and AMD are largely focusing on (1) the use of sludge containing sulphate-reducing bacteria treating BW for the anaerobic treatment of AMD in bioreactors [13] and (2) mine tailing treatment in bioreactors supported by anaerobic sludge generated using BW [14], with no studies focusing on the BW being used directly to support a passive system for the treatment of AMD—in particular, sulphate reduction using bacteria supported on BW while remediating AMD for water recovery to be used elsewhere, of which the primary step is sulphate removal. Furthermore, optimising such treatment systems adapted for local conditions, i.e., taking into account the unique characteristics of the AMD in South Africa, has never been attempted. For biological system model development, artificial neural networks (ANNs) and response surface methodology (RSM) have been reported as effective tools for process optimisation [15,16]. Therefore, an artificial neural network and a Box–Behnken design RSM were used to determine the better predictor of the performance of an anaerobic bioreactor designed specifically to treat South African AMD using BW as the primary carbon source in the Postgate medium. Furthermore, the novelty of using BW lies in the fact that it is usually discarded and not used for AMD remediation.

2. Materials and Methods

2.1. Chemical Reagents

Postgate medium B, herein referred to as Postgate medium, constitutes monopotassium phosphate (0.5 g/L), ammonium chloride (1.0 g/L), sodium sulphate (1.0 g/L), calcium chloride dihydrate (0.1 g/L), magnesium sulfate (2.0 g/L), yeast extract (1.0 g/L), ascorbic acid (0.1 g/L), thioglycolic acid (0.1 g/L), ferrous sulphate heptahydrate (0.5 g/L), sodium chloride (26 g/L), and sodium lactate (5 mL). The medium pH was 7–7.5. Sodium bromoethane sulphonate (98%, Merck, Modderfontein, South Africa) and sodium DL-lactate solution (60% *w*/*w*, Sigma-Aldrich, Modderfontein, South Africa) were used as received from the suppliers.

2.2. Bacterial Inoculum

The AMD sample was collected using a previously described procedure from a mining facility in South Africa [17]. The sulphate-reducing consortium was activated using selective modified Postgate isolation medium for the propagation of sulphate-reducing bacteria (SRB) at 35 °C and pH of 7 ± 0.2 in an anaerobic reactor. The constituents of the modified Postgate medium were as described earlier [17]. The black-grey colour of the medium signified positive growth of the sulphate-reducing consortium. The experiments were conducted in triplicate.

2.3. Carbon Source Limiting Growth Test

A sterile bioreactor containing 100 mL AMD (8080 mg SO_4^{2-}/L) was inoculated with 20% (*v/v*) SRB consortium in the Postgate medium. The characteristics of the AMD were as described in Akinpelu et al. [17]; i.e., the AMD had an E^0 value of 229.5 mV, which is high, a fairly acidic pH of 2.98, and electrical conductivity of 7.84 mS/cm. SRB growth was observed on both the lactate (L) and BW, used as carbon sources at a feed rate of 0.05 g/L h in the Postgate medium. To minimise methanogenic activities, sodium bromoethane sulphonate (3.2 g/L) was added to the bioreactors. The bioreactors were incubated anaerobically at 35 °C, pH of 7 ± 0.2, in a rotary ZHICHENG shaker (model ZHYWY-1102) at 160 rpm (ZHICHENG Analytical Instruments Manufacturing Co. Ltd, Shanghai, China). The bioreactors were purged with nitrogen gas to displace the dissolved oxygen and then sealed with Parafilm tape to sustain the anaerobic conditions. Bioreactors without inoculum served as a control. The BW, collected from a beer plant in South Africa, contained protein (3.3 ± 0.01 mg/mL), glucose (5 ± 0.02 mg/mL), fat (2.6 ± 0.03 mg/mL), lactose (5 ± 0.02 mg/mL), maltose (5 ± 0.03 mg/mL), sucrose (5 ± 0.02 mg/mL), fructose (5 ± 0.03 mg/mL), and dry matter (4.2 ± 0.02 mg/mL), as measured using a high-performance liquid chromatograph (HPLC, Agilent 1290 Infinity) (Chemetrix Export (Pty) Ltd, Johannesburg, South Africa); equipped with a 300 m × 7.8 mm Aminex HPX-97H column. The HPLC columns used isocratic conditions with water as the mobile phase at 0.6 mL/min and temperature as the main variable for control of the resolution (60–70 °C), according to the manufacturer's manual for the separation of carbohydrates. Samples were taken at predetermined intervals for turbidity measurement in a GENESYSTM 10S UV/Visible spectrophotometer (Thermo Fisher Scientific, Waltham, MA, USA) at a wavelength of 600 nm. All procedures were performed in triplicate. Experimental errors were estimated as the standard deviation of a set of data ($n = 3$).

2.4. Experimental Set-Up

The experiments were done in 250 mL anaerobic bioreactors (35 °C, 160 rpm, N_2 purged) and were initiated with 20% (*v/v*) inoculum in 100 mL Postgate medium for 48 h. Sodium bromoethane sulphonate (3.2 g/L) was added to the bioreactor to minimise methanogenic activity. Subsequently, 50 mL of AMD (8080 mg SO_4^{2-}/L) was added, and the bioreactor was operated for 72 h at various pH, COD/SO_4^{2-}, and BW levels specified by the Box–Behnken design in Table 1. All bioreactors, with or without inoculum, were conditioned to the various specified pre-set testing variables, and the pH of the cultures was adjusted using 0.1 M HCl (Lasec SA, Cape Town, South Africa); or 0.1 M NaOH (Lasec SA, Cape Town, South Africa); accordingly. The sulphate removal efficiency was estimated using Equation (1):

$$\text{Removal}\ (\%) = [(C_i - C_f)/C_i)] \times 100, \tag{1}$$

where C_i and C_f are concentrations of sulphate (mg/L) in the initial and treated AMD, respectively.

Table 1. Experimental design variables.

Variable	Code	Unit	Coded Factor Level		
			1	0	−1
pH	A	-	7	5	3
COD/SO_4^{2-}	B	-	3	1.75	0.5
Brewing wastewater	C	mg/L	500	350	200

2.5. Design of Experiment—Box–Behnken

To optimise the process and investigate the dynamics among the key process parameters for biological sulphate reduction, RSM was used. It has been previously used in optimising numerous environmental processes, having been determined to be superior in the extended optimisation of such processes [18]. RSM is a collection of statistical methods for the design of an experiment involving several variables, leading to peak system performance at minimal cost to establish the optimum response [19,20]. Amongst the numerous RSM designs, there is the Box–Behnken design (BBD), which is an efficient quadratic design where variable combinations are at the lower, centre and higher levels, usually coded as −1, 0, and +1, respectively, to give a minimum number of experimental runs. The centre points are used to calculate the experimental deviations [21,22]. Three operational parameters (pH, COD/SO_4^{2-}, and BW concentration) were designated as input variables, and the sulphate removal efficiency after 120 h was the dependent parameter. The ranges of the variables (pH and COD/SO_4^{2-}) were determined based on the optimum values reported for most biological sulphate-reducing systems [19,23]. The concentration of brewing wastewater was chosen based on the exponential phase in substrate limitation of the SRB consortium. Design-Expert® software version 12 (Stat-Ease Inc., Minneapolis, MN, USA, 2019) was used for the design of experiments (DoE) and data analysis. Seventeen points, comprising 5 centre points and 12 factorial points, were examined, and bioreactors without inoculum served as controls at various specified conditions (see Table 2). A COD meter and Multiparameter Bench Photometer HI 83,099 (Hanna Instruments Inc., Woonsocket, RI, USA) were used to measure both the COD and sulphate concentration in the AMD samples. This was done via a photometric chemical procedure that is based on the absorption of a compound using special subminiature tungsten lamps and a narrow bandwidth from a specific chemical reaction between the sample and reagents, so that high performance and reliable results can be guaranteed. Based on the Lambert–Beer Law, the molar concentration of the sample can be estimated. All reagents were of analytical grade. The COD and sulphate were measured using Hanna's reagents HI 93754C-25 and HI 93751-01, respectively. All experiments were performed in triplicate, and the mean values of experimental data were fitted into the following polynomial quadratic model (Y):

$$Y = \beta_o + \sum \beta_i x_i + \sum \beta_{ii} x_{ii}^2 + \sum \beta_{ij} x_i x_j + \varepsilon, \quad (2)$$

where x_i, x_j, and x_{ii} are independent coded factors; β_o is the offset term; β_i, β_{ii}, and β_{ij} are linear, square, and interaction effects, respectively; and ε is the error.

Table 2. Design matrix with actual and predicted response values via response surface methodology (RSM) and artificial neural network (ANN).

Run	Variables			Sulphate Removal (%)		
	A	B	C	Actual	RSM Predicted	ANN Predicted
1	**1**	**1**	**0**	**77.80**	**80.56**	**79.27**
2	−1	−1	0	30.70	27.94	29.40
3	0	1	−1	68.80	68.43	68.81
4	0	1	1	61.60	64.25	60.02
5	1	0	1	84.60	79.19	84.94
6	1	0	−1	87.90	85.51	87.58
7	**−1**	**0**	**−1**	**25.70**	**31.11**	**25.63**
8	**0**	**0**	**0**	**51.62**	**51.62**	**51.95**
9	0	0	0	51.62	51.62	51.93
10	−1	1	0	42.90	37.86	42.95
11	0	−1	1	63.50	63.88	63.21
12	0	0	0	51.62	51.62	51.93
13	0	0	0	51.62	51.62	51.93
14	−1	0	1	35.60	37.99	37.19

Table 2. *Cont.*

Run	Variables			Sulphate Removal (%)		
	A	B	C	Actual	RSM Predicted	ANN Predicted
15	1	−1	0	75.80	80.84	75.82
16	0	0	0	51.62	51.62	51.93
17	0	−1	−1	61.80	59.15	62.08

ANN training set: normal numbers, ANN testing set: bold numbers.

2.6. Artificial Neural Network (ANN) Analysis

Neural Power (CPC-X Software, version 2.5, Carnegie, PA, USA, 2019) was used for the neural network analysis in this study. This software is a Windows-based package that supports several types of training algorithms. The optimum artificial neural network conditions reported by Betiku and Taiwo [16] were used to predict the percentage of sulphate reduction in the treatment of South African AMD. The ANN training algorithm chosen in the present study was a multilayer full feedforward (MFFF) connection with an incremental backpropagation network and hyperbolic tangent (tanh) as the transfer function. The ANN framework consisted of three input layers, three hidden neurons, and one output layer, forming a 3-3-1 topology (Figure 1). After several trials, an optimal network topology, with only one hidden layer, three hidden neurons with tanh as the transfer function, and an output layer with Linear as the transfer function, was iteratively used in training the network. The neural network was trained until the root-mean-square error (RMSE) was less than 0.0001, with the average correlation coefficient (R^2) and the average determination coefficient (DC) tending towards unity (1). Other ANN parameters were chosen as the standard settings of the software. In this study, data generated from the BBD (Table 1) were deployed in ANN model development. Experimental data sets from the Box–Behnken design (BBD) were split into training and testing sets. The data were divided in a ratio of 1 to 5 with training sets comprising 14 data sets (70%) and testing sets (30%) comprising 3 data sets, making a total of 17 experimental runs. The details of all other conditions and techniques used in the ANN analysis for prediction and optimisation in bioprocessing have been reported in detail elsewhere [16].

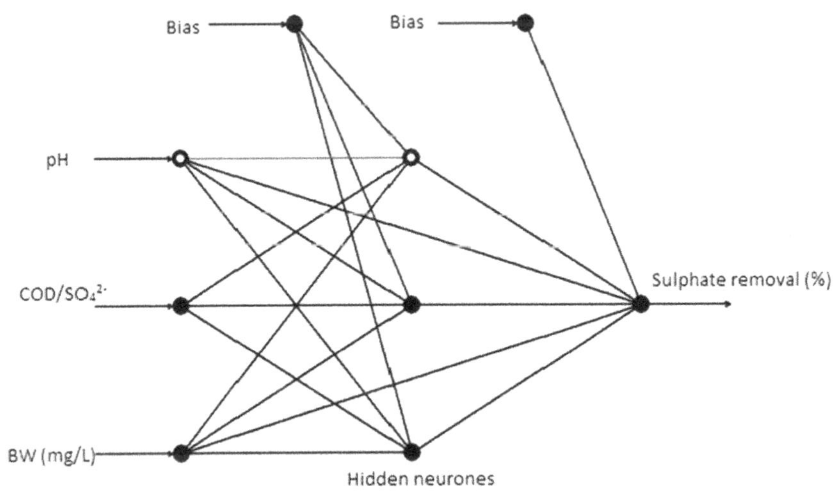

Figure 1. Artificial neural network (ANN) architecture for the estimation of sulphate reduction in the treatment of acid mine drainage (AMD) using brewing wastewater (BW) as carbon source.

2.7. Appraisal of Artificial Neural Network Predictability

To measure the model fit for predictability in the ANN, it was necessary to do a controlled training for estimation of the ANN output error between the actual and predicted outputs. Equations (3)–(6) show the predictive measures used for the model evaluation: the mean squared error (MSE), root mean square error (RMSE), R^2, and absolute average deviation (AAD) [24]. The predicted and actual responses are denoted y_i and y_{di}, respectively. The number of the experimental runs is n, and y_m is the average of the experimental values. The network having minimum errors (MSE, RMSE, AAD) and R^2 closest to unity was taken as the best neural network model [25–27].

$$\mathrm{MSE} = \frac{1}{\mathrm{n}}\sum_{\mathrm{i}=1}^{\mathrm{n}}\left(\mathrm{y_i} - \mathrm{y_{di}}\right)^2 \tag{3}$$

$$\mathrm{RMSE} = \mathrm{MSE}^{1/2} \tag{4}$$

$$AAD = \left\{\left[\sum_{i=1}^{n}\left(|y_i - y_{di}|/y_{di}\right)\right]/n\right\} \tag{5}$$

$$R^2 = 1 - \sum_{i=1}^{n}\left(\frac{(y_i - y_{di})^2}{(y_{di} - y_m)^2}\right) \tag{6}$$

3. Results and Discussion

3.1. Effect of Carbon Substrate Limitation on the SRB Consortium

The SRB consortium's growth increased with increasing substrate concentration up to a saturation point of 0.3 g/L for both lactate and BW used as a carbon source. Nevertheless, the maximum biomass growth in BW was greater than the growth in lactate (Figure 2). Although previous reports have shown that lactate shows superior biomass growth when compared to other carbon sources, the scenario in this study can be attributed to the several readily available reducible sugars in the BW for sustained microbial growth [19,28]. Hence, BW was assessed as being important for its interaction and ability in combination with other factors and environmental conditions in the optimisation of sulphate removal in AMD.

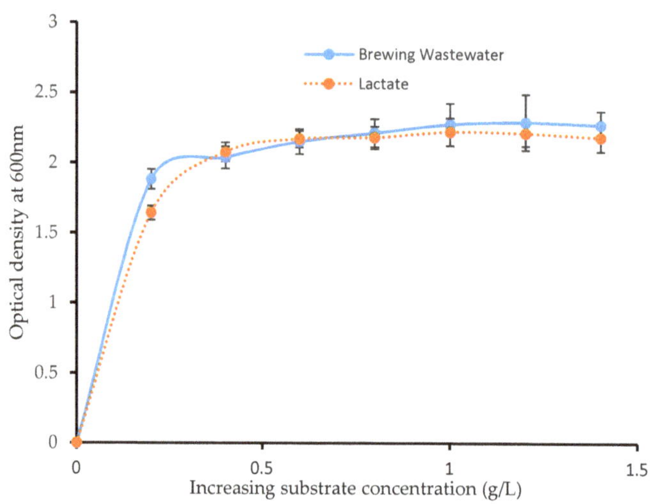

Figure 2. Effect of carbon substrate limitation on the sulphate-reducing bacteria (SRB) consortium's growth.

3.2. RSM Modelling: Box–Behnken Design

The result of individual factors and their interactive effects on the biological sulphate removal is given in Table 2. The responses demonstrated random variations in the results measured that indicated the effect of operational parameters on the SRB consortium's metabolic activity and, perhaps, process instability. The highest sulphate removal was observed at a factorial point (Run 6), at pH of 7, COD/SO_4^{2-} of 1.75, and BW of 200 mg/L, where sulphate removal of 87.9% from an initial sulphate concentration of 8080 mg SO_4^{2-}/L was observed 72 h after introducing the AMD. The residual sulphate concentration of 978 mg SO_4^{2-}/L was well below the discharge standard (1500 mg SO_4^{2-}/L) for South Africa. This is also within the range reported for sulphate reduction by diverse carbon sources in Postgate medium, including cheese whey supplemented with calcite tailings, which resulted in 80% sulphate removal [23,29]. Most of the peak performance (removal percent > 75%) corresponds to a neutral pH, and the BW concentration was within the exponential phase of the consortium's growth, as shown in Figure 2. Studies have shown that optimum sulphate reduction can only be achieved when the pH is >5 in an anaerobic environment [23,29]. Since the pH criterion was met, it created amiable conditions for SRB activity, allowing consumption of the carbon source, i.e., BW and sulphate reduction. However, some runs (2, 7, 10, and 14) displayed extremely low sulphate reduction, a confirmation of slow microbial activity as a result of low-pH conditions.

Statistical analysis of the data was performed using analysis of variance (ANOVA) to assess the significance of each variable in the model developed (Table 3). The quadratic polynomial model of sulphate removal in terms of coded values was as follows:

$$Y = 51.62 + 23.9A + 2.41B + 0.14C - 0.15A^2 + 5.33B^2 + 6.89C^2 - 2.55AB - 3.3AC - 2.23BC \quad (7)$$

where A, B, and C are coded values for pH, COD/SO_4^{2-}, and brewing wastewater concentration, respectively.

Table 3. ANOVA of the sulphate removal response quadratic model.

Source	Sum of Squares	df	Mean Square	*F*-Value	*p*-Value	
Model	5048.90	9	560.99	26.12	0.0001	significant
A—pH	4569.68	1	4569.68	212.78	<0.0001	significant
B—COD/SO_4	46.56	1	46.56	2.17	0.1844	
C—BW	0.1512	1	0.1512	0.0070	0.9355	
AB	26.01	1	26.01	1.21	0.3075	
AC	43.56	1	43.56	2.03	0.1974	
BC	19.80	1	19.80	0.9221	0.3689	
A^2	0.0916	1	0.0916	0.0043	0.9498	
B^2	119.50	1	119.50	5.56	0.0504	
C^2	204.99	1	204.99	9.55	0.0176	significant
Residual	150.33	7	21.48			
Lack of Fit	150.33	3	50.11			
Pure Error	0.0000	4	0.0000			
Cor Total	5199.23	16				

The R^2 (0.9711), adjusted R^2 (0.9339), predicted R^2 (0.5374), and adequate precision (16.199), including the low standard deviation (4.63). "Prob > F" of less than 0.05 indicated model terms that were significant, with any values greater than 0.1 indicating model terms that were not significant.

Similarly, the coefficient of interaction was estimated from the mean of the two confidence levels. Since the predicted R^2 was not close to the adjusted R^2, a model reduction was considered, i.e., removal of terms with p-values of >0.1, to improve the model to the following:

$$Y = 51.62 + 23.9A + 5.33B^2 + 6.89C^2. \quad (8)$$

The p-value and the F-value of 26.12 implied that the model was significantly relevant, and there was only a 0.01% probability that an F-value this large could occur due to noise. The R^2 further showed the precision of the fitted model to the data, as a model with $R^2 > 0.8$ is considered to be well fitted to the experimental data [30]. In this study, the R^2 of 0.9711 implies that 97% of variations in the predicted and actual values are clarified by the model. This indicated the applicability and accuracy of the model for forecasting biological sulphate removal using BW. The adequate precision ratio of 16.199 is a sufficient signal that can be used to explore the design space in this model. The non-significance of the F-value of lack of fit further supported the suitability of the model. The diagnostic investigation of the model showed a normal distribution of errors, shown in Figure 3a, with an average value of zero. Similarly, a study of the residual showed evenly distributed scatters above and below the horizontal axis, validating the sufficiency of the model (Figure 3b). Also, the plot of predicted values against the actual experimental value showed clustering of points around the diagonal line, an indication of the high correlation and robustness of the model (Figure 3c). To validate the model, a plot of the standard error in sulphate removal as a function of the pH and COD/SO_4^{2-} is shown in Figure 3d. The shape of the standard error plot was well fitted on the design points, as well as presenting symmetrical shapes and circular contours around the centre point, signifying ideal conditions. The standard error value around the centre point was 0.5 and increased away from the optimisation point.

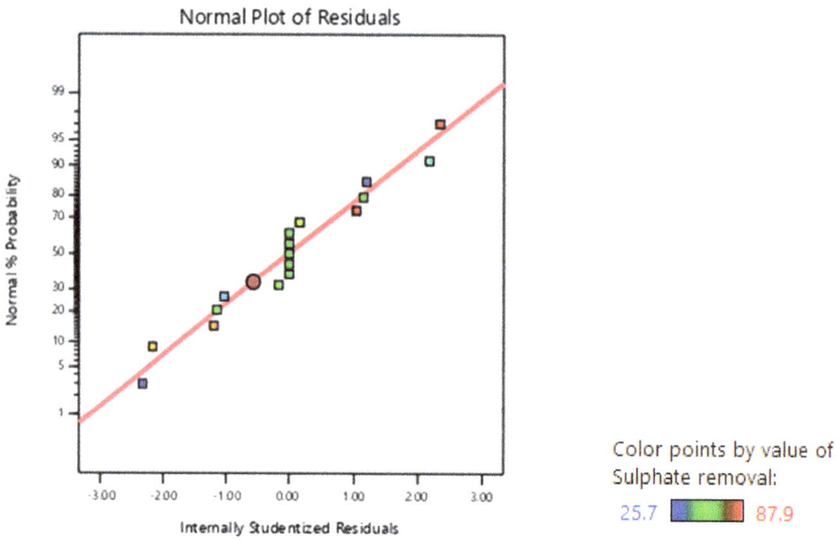

(a)

Figure 3. *Cont.*

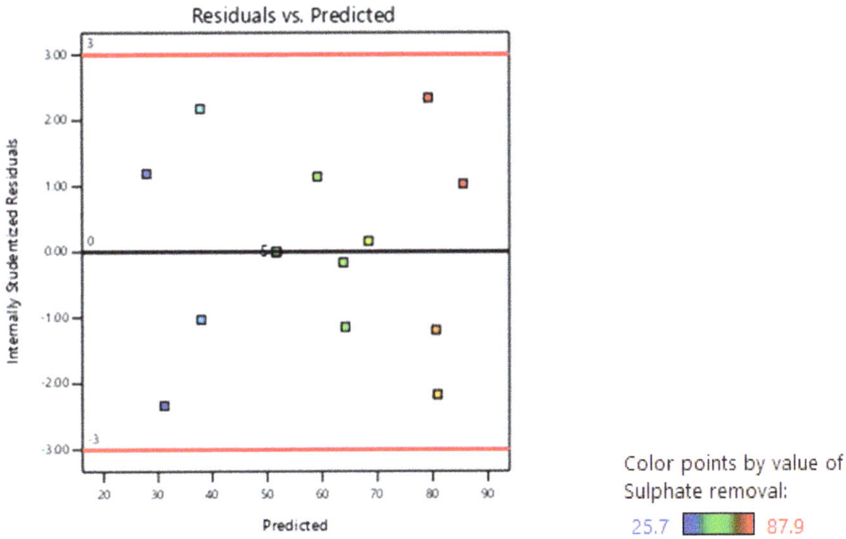

(**b**)

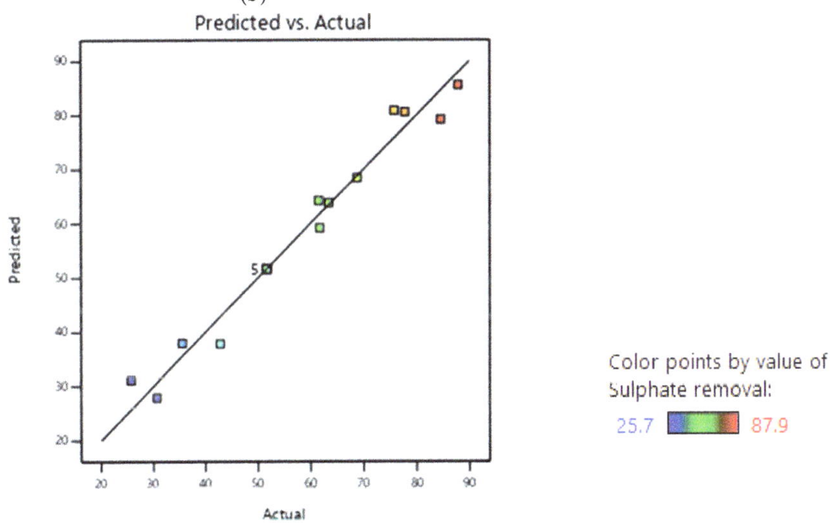

(**c**)

Figure 3. *Cont.*

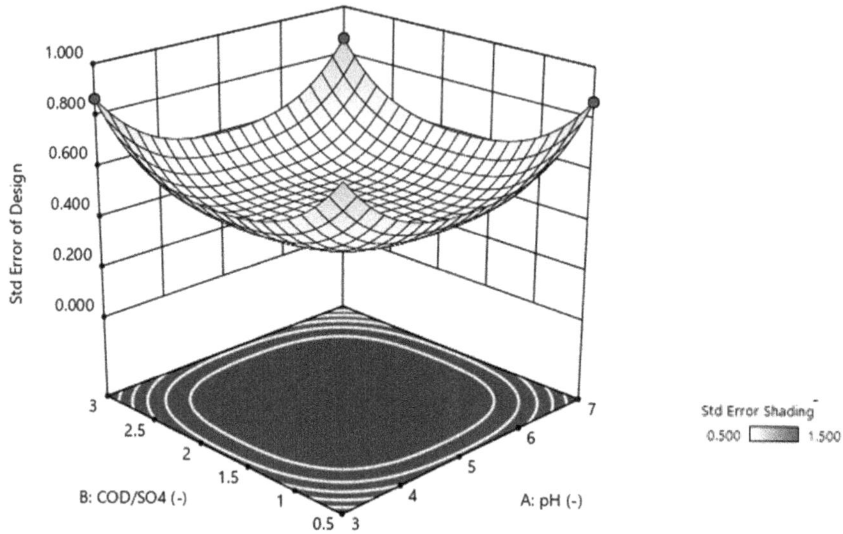

(d)

Figure 3. (**a**) Normal probability plot of residuals, (**b**) residuals vs. predicted, (**c**) predicted vs. actual output of the model, and (**d**) 3-D standard error plot for sulphate removal.

3.3. Graphical Representation of the Model

The effects of changes in all the three variables on sulphate removal are shown in the perturbation diagram (Figure 4a). The effect of Parameter A on sulphate removal was shown to be close to linear and deviate farther from the centre point, though predictably. The influence of Parameters B and C is non-linear, and their values are similar and of the same order of magnitude; however, the effect of Parameter C is more notable in lower values. Besides this, for ease of clarification of these results and the forecasting of optimal conditions, 3-D plots of the system response were analysed by comparing any two factors while keeping the third factor constant, which allowed for the analysis of any interactive effects the three independent factors have on the system's response. For the pairs involving pH, the sulphate removal increased with increasing pH value, while for the two other factors, sulphate removal increased away from the centre point (Figure 4b–d). The pair pH and BW at constant COD/SO_4^{2-} resulted in the highest sulphate removal.

In certain instances, RSM was determined to have a lesser prediction ability of biologics, with ANN models presenting higher predictive coefficient of determination and minor RMSE and mean absolute deviation (MAD), thus having a higher model resolution than RSM [31]. Therefore, ANN analysis of sulphate reduction is warranted, particularly if BW in AMD remediation is to be used on an industrial scale.

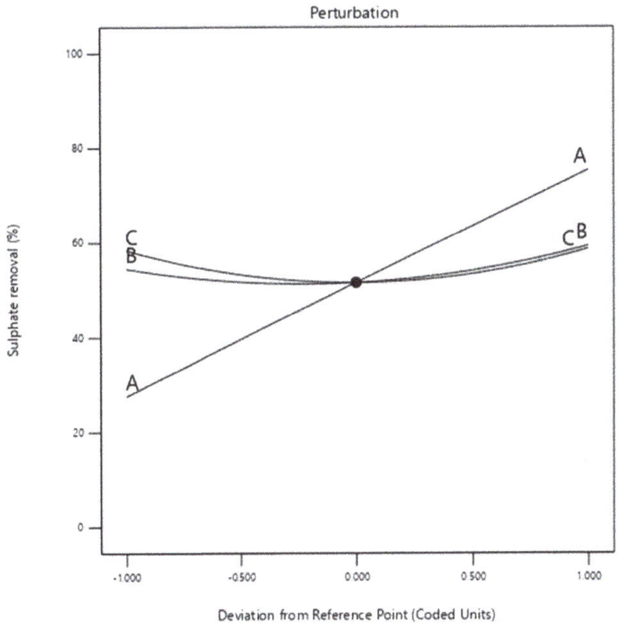

(**a**)

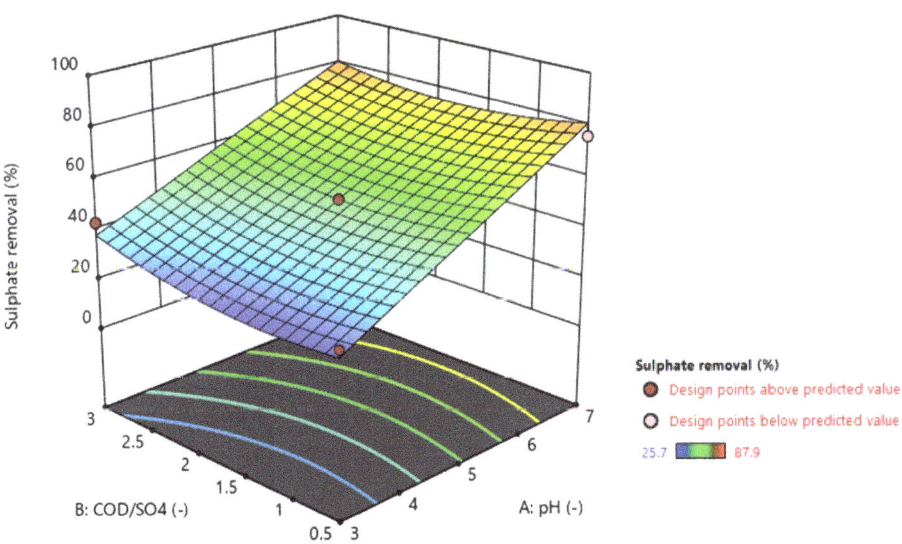

(**b**)

Figure 4. *Cont.*

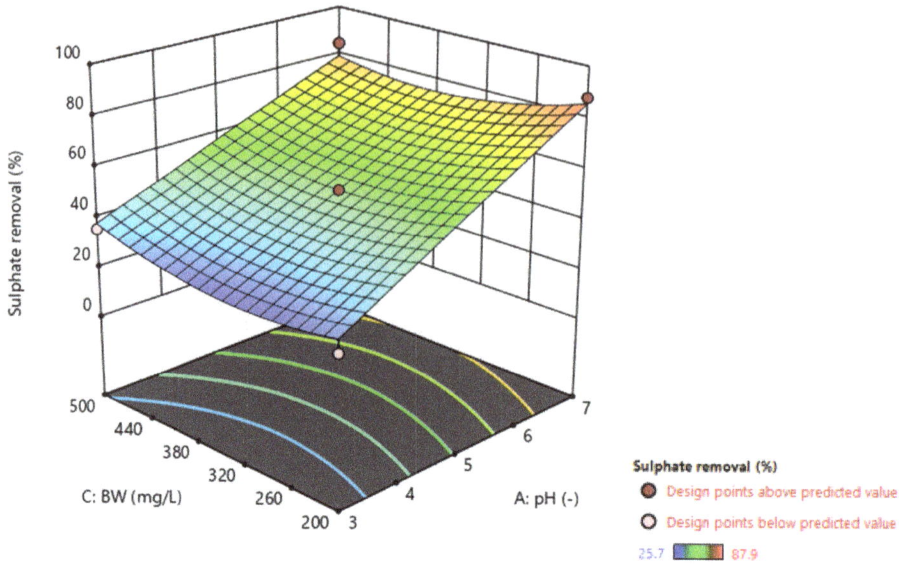

(c)

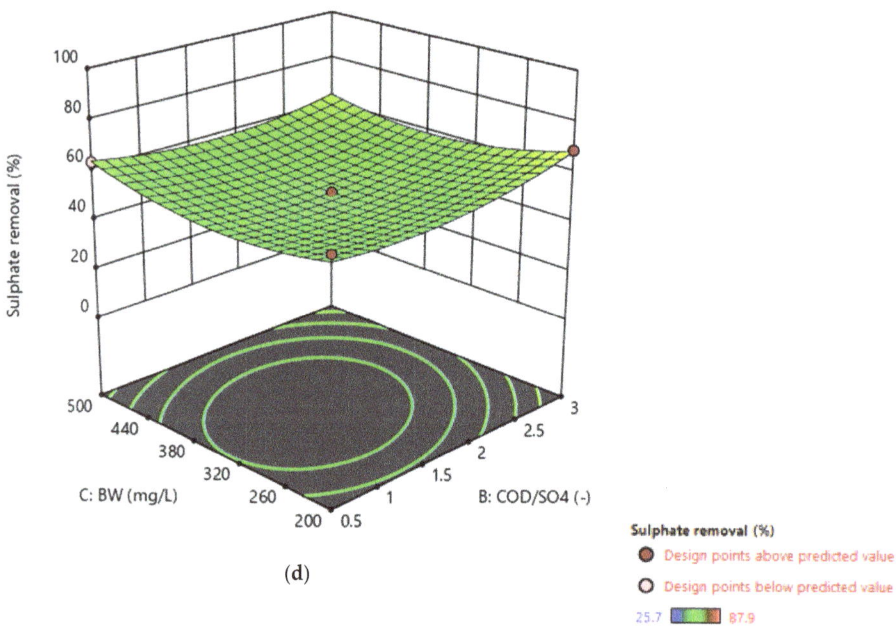

(d)

Figure 4. Perturbation of factors as a function of coded values (**a**) and 3-D plots (**b–d**) showing the effect of independent variables on sulphate removal.

3.4. Artificial Neural Network Analysis of Sulphate Reduction Using Brewery Wastewater

The multilayer full feedforward (MFFF) interaction with an incremental backpropagation network and hyperbolic tangent (tanh) as the transfer function used for the ANN modelling proved to be effective, with the coefficient of determination being close to unity. ANN analysis showed that the percentages of controlled variable contributions for sulphate removal in the treatment of the South African AMD were 44.3, 32.09, and 23.61%, for BW, pH, and COD/SO_4^{2-}, respectively. This shows that BW was the most impactful variable in the experimental studies. Table 4 shows that all the error measurements (MSE, RMSE, and AAD) for both the training and testing datasets were less than 1, while R^2 for the model fitness neared unity for both datasets. A comparison of the actual and predicted values showed the competency of the ANN in the prediction of unknown data [16] (Figure 5). This suggests that the model generated by the ANN can sufficiently explain the relationship between the operating parameters and sulphate reduction.

Table 4. ANN model predictive measurements for model fit in the training and testing sets.

ANN Model Predictive Tool	Training Set	Testing Set
MSE	0.53	0.76
RMSE	0.73	0.87
AAD	0.011	0.0093
R^2	0.99	0.99

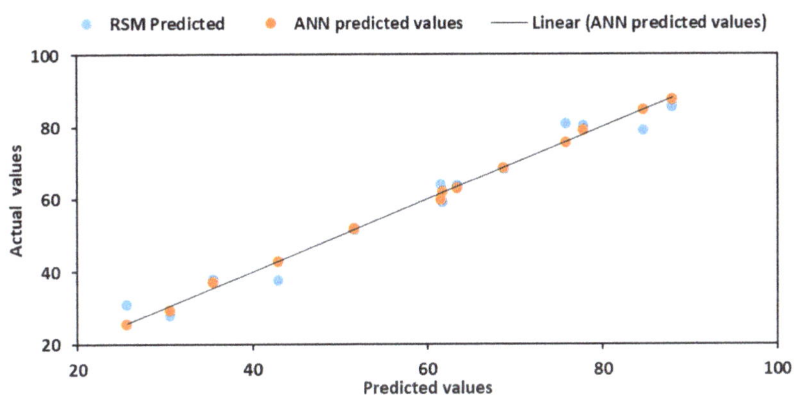

Figure 5. Parity plots of RSM and ANN predictions for sulphate removal in the treatment of South African acid mine drainage.

3.5. Optimum Comparison of RSM and ANN

For RSM, numerical analysis in Design-Expert® version 12 software was deployed for the optimisation of the sulphate removal. Independent variables COD/SO_4^{2-} and BW were set within range, with pH being set between 5 and 7 for a maximum outcome, i.e., sulphate removal. Design-Expert® generated a set of 89 solutions that matched the criteria from most to least desirable. The optimum point with maximum sulphate reduction of 91.59% was found at a pH of 6.99 (which is within the South African discharge standards), COD/SO_4^{2-} of 2.874, and BW concentration of 200.24 mg/L. The experiment conducted at this optimum culminated in the reduction of the sulphate concentration by 90%.

Table 5 shows the statistical tool used in model evaluation between the actual and predictive data sets for both RSM and ANN. Although both models performed well with respect to the R^2 and AAD values, with stable responses, the ANN-based approach was superior in terms of data estimation and fitting when MSE and RMSE were considered. This was also confirmed by the parity plots between the predicted and actual values (Figure 5). Genetic optimisation (GA) was used as the optimal algorithm in the ANN analysis, while RSM optimisation was carried out as embedded in the BBD design. The optimal results, presented also in Table 5, show about a 2% difference in the percentage of sulphate reduction in RSM when compared to the ANN. However, it is worth noting that there was a 2% decrease in the optimal value of COD/SO_4^{2-} in the ANN optimisation variables.

Table 5. Statistical measurement of the RSM and ANN data sets for model fitting.

Statistical Tool	RSM Whole Data Set	ANN Whole Data Set	Optimisation Variable	RSM	ANN
MSE	8.84	0.57	Sulphate reduction (%)	91.59	89.56
RMSE	2.97	0.75	pH	6.99	6.99
AAD	0.046	0.011	COD/SO_4^{2-}	2.87	0.50
R^2	0.97	0.99	BW (mg/L)	200.24	200.31

3.6. Overall Effect of Individual Parameters

The effect of pH can be seen in both Table 3 and Figure 4. An increase in pH increased the sulphate removal efficiency, and the maximum removal rate was observed at pH around 7. This aligns with the available data in the literature showing that most SRB grow optimally at pH values between 6 and 8 [29,32], as a reduced level of removal was observed in low pH. This may be attributed to acidotolerant and acidophilic SRB that have been reported [33–35], which made it possible to treat AMD without prior neutralisation. Similarly, at low pH, there are more protons than at neutral pH, which causes diffusion pressure on the cell membrane due to higher Gibbs free energy at low pH during sulphate reduction; hence, growth can be achieved at low pH [36]. However, a higher retention time is required to achieve substantial sulphate removal at low pH.

The COD/SO_4^{2-} plays a major role in the oxidation–reduction reactions during sulphate removal. The SRB oxidise the BW, and the released electron is used to reduce the sulphate. Hence, the number of electrons transferred between the two reactions determines the sulphate removal efficiency [37]. In this study, the sulphate removal increased away from the centre point for both COD/SO_4^{2-} and BW. Although there are reports that have indicated that COD/SO_4^{2-} should not be greater than 2.72 to prevent methanogenic activity [38,39], sodium bromoethane sulphonate was added to the reactor, and the COD/SO_4^{2-} corresponding to optimum sulphate reduction in this study was 1.75. Since the carbon source determines bacterial proliferation, irrespective of the carbon source selected, the minimum amount of organic compound needed for reducing sulphate should be a little more than the theoretical stoichiometric ratio since a portion of the energy dissipates into microbial maintenance and growth [40]. The BW amount (carbon source) for optimum sulphate reduction in this study was 200 mg/L.

4. Conclusions

The SRB consortium growth on BW containing Postgate medium showed that BW is a suitable carbon source that can be deployed in a system designed for environmental remediation of AMD and the recovery of such water for other purposes. The analysis of response from Box–Behnken RSM showed that pH is the most significant factor, while ANN analysis indicated carbon source (BW) as the most significant factor. The optimised conditions for BBD in RSM were pH of 6.99, COD/SO_4^{2-} of 2.87, and BW concentration of 200.24 mg/L with sulphate reduction of 91.58%, while those for ANN were pH of 6.99, COD/SO_4^{2-} of 0.50, and BW concentration of 200.31 with sulphate reduction of 89.56%. The performance of both models was very good with respect to the R^2 and AAD values, with stable responses; nonetheless, the ANN-based approach was superior in terms of data estimation and fitting. The residual sulphate concentration (680 mg SO_4^{2-}/L) of the optimised conditions is within the discharge limit in South Africa. This study recognises that BW could function as a suitable feedstock in biological sulphate reduction in South Africa and in semi-arid regions with high BW production and AMD challenges.

Author Contributions: E.A.A. conceptualised the research, performed the experiments, and wrote the first draft; S.K.O.N. conceptualised the research, reviewed the manuscript, and provided supervision; A.E.T. performed ANN experimentation and analysis of the result; F.N. reviewed the manuscript and is a postdoctoral host (supervisor) of E.A.A. All authors have read and agreed to the published version of the manuscript.

Funding: This work is based on research supported wholly by the National Research Foundation of South Africa (Grant Number: 111993). Any opinion, findings, and conclusions or recommendations expressed are those of the authors, and NRF accepts no liability whatsoever in this regard.

Conflicts of Interest: The authors declare no conflict of interest. Furthermore, the funders had no role in the design of the study; in the collection, analyses, or interpretation of data; in the writing of the manuscript; or in the decision to publish the results.

References

1. SAB. South African Breweries Heritage 2020. Available online: http://www.sab.co.za/heritage (accessed on 30 July 2020).
2. Seluy, L.G.; Isla, M.A. A process to treat high-strength brewery wastewater via ethanol recovery and vinasse fermentation. *Ind. Eng. Chem. Res.* **2014**, *53*, 17043–17050. [CrossRef]
3. Ma, L.; Liang, J.; Liu, Y.; Zhang, Y.; Ma, P.; Pan, Z.; Jiang, W. Production of a bioflocculant from *Enterobacter* sp. P3 using brewery wastewater as substrate and its application in fracturing flowback water treatment. *Environ. Sci. Pollut. Res.* **2020**, *27*, 18242–18253. [CrossRef] [PubMed]
4. Hultberg, M.; Bodin, H. Fungi-based treatment of brewery wastewater—Biomass production and nutrient reduction. *Appl. Microbiol. Biotechnol.* **2017**, *101*, 4791–4798. [CrossRef] [PubMed]
5. Nassary, E.K.; Nasolwa, E.R. Unravelling disposal benefits derived from underutilized brewing spent products in Tanzania. *J. Environ. Manag.* **2019**, *242*, 430–439. [CrossRef]
6. Karlović, A.; Jurić, A.; Ćorić, N.; Habschied, K.; Krstanović, V.; Mastanjević, K. By-Products in the malting and brewing industries—Re-usage possibilities. *Fermentation* **2020**, *6*, 82. [CrossRef]
7. Pham, T.-L.; Bui, M.H. Removal of nutrients from fertilizer plant wastewater using *Scenedesmus* sp.: Formation of bioflocculation and enhancement of removal efficiency. *J. Chem.* **2020**, *2020*, 8094272.C. [CrossRef]
8. Hawrot-Paw, M.; Koniuszy, A.; Gałczyńska, M.; Zając, G.; Szyszlak-Bargłowicz, J. Production of microalgal biomass using aquaculture wastewater as growth medium. *Water* **2020**, *12*, 106. [CrossRef]
9. Simate, G.S.; Cluett, J.; Iyuke, S.E.; Musapatika, E.T.; Ndlovu, S.; Walubita, L.F.; Alvarez, A.E. The treatment of brewery wastewater for reuse: State of the art. *Desalination* **2011**, *273*, 235–247. [CrossRef]
10. Riera-Vila, I.; Anderson, N.O.; Hodge, C.F.; Rogers, M. Anaerobically-digested brewery wastewater as a nutrient solution for substrate-based food production. *Horticulture* **2019**, *5*, 43. [CrossRef]
11. Nicholson, S.E.; Funk, C.; Fink, A.H. Rainfall over the African continent from the 19th through the 21st century. *Glob. Planet Chang.* **2018**, *165*, 114–127. [CrossRef]
12. Auditor-General SA. *Report of the Auditor-General to Parliament on a Performance Audit of the Rehabilitation of Abandoned Mines at the Department of Minerals and Energy*; Auditor-General: Pretoria, South Africa, 2009.
13. La, H.-J.; Kim, K.-H.; Quan, Z.-X.; Cho, Y.-G.; Lee, S.-T. Enhancement of sulfate reduction activity using granular sludge in anaerobic treatment of acid mine drainage. *Biotechnol. Lett.* **2003**, *25*, 503–508. [CrossRef] [PubMed]
14. Valenzuela, E.I.; García-Figueroa, A.C.; Amábilis-Sosa, L.E.; Molina-Freaner, F.E.; Pat-Espadas, A.M. Stabilization of potentially toxic elements contained in mine waste: A microbiological approach for the environmental management of mine tailings. *J. Environ. Manag.* **2020**, *270*, 110873. [CrossRef] [PubMed]
15. Sivamani, S.; Selvakumar, S.; Rajendran, K.; Muthusamy, S. Artificial neural network-genetic algorithm-based optimization of biodiesel production from *Simarouba glauca*. *Biofuels* **2019**, *10*, 393–401. [CrossRef]
16. Betiku, E.; Taiwo, A.E. Modeling and optimization of bioethanol production from breadfruit starch hydrolyzate vis-à-vis response surface methodology and artificial neural network. *Renew. Energy* **2015**, *74*, 87–94. [CrossRef]
17. Akinpelu, E.A.; Fosso-Kankeu, E.; Waanders, F.; Angadam, J.O.; Ntwampe, S.K.O. Diversity and Performance of sulphate-reducing bacteria in acid mine drainage remediation systems. In *Frontiers in Water-Energy-Nexus—Nature-Based Solutions, Advanced Technologies and Best Practices for Environmental Sustainability*; Naddeo, V., Balakrishnan, M., Choo, K.-H., Eds.; Springer: Cham, Switzerland, 2020; pp. 121–123.

18. Kozik, V.; Barbusinski, K.; Thomas, M.; Sroda, A.; Jampilek, J.; Sochanik, A.; Smolinski, A.; Bak, A. Taguchi method and Response Surface Methodology in the treatment of highly contaminated tannery wastewater using commercial potassium ferrate. *Materials* **2019**, *12*, 3784. [CrossRef] [PubMed]
19. Najib, T.; Solgi, M.; Farazmand, A.; Heydarian, S.M.; Nasernejad, B. Optimization of sulfate removal by sulfate reducing bacteria using response surface methodology and heavy metal removal in a sulfidogenic UASB reactor. *J. Environ. Chem. Eng.* **2017**, *5*, 3256–3265. [CrossRef]
20. Akinpelu, E.A.; Ntwampe, S.K.O.; Mekuto, L.; Tombo, E.F.I. Optimizing the bioremediation of free cyanide containing wastewater by *Fusarium oxysporum* grown on beetroot waste using response surface methodology. In *Lecture Notes in Engineering and Computer Science, Proceedings of the World Congress on Engineering and Computer Science, San Francisco, CA, USA, 22–24 October 2019*; Ao, S.I., Douglas, C., Grundfest, W.S., Eds.; Newswood Limited: San Francisco, CA, USA, 2016; pp. 664–670.
21. Mona, S.; Kaushik, A.; Kaushik, C.P. Biosorption of chromium(VI) by spent cyanobacterial biomass from a hydrogen fermentor using Box-Behnken model. *Int. Biodeterior. Biodegrad.* **2011**, *65*, 656–663. [CrossRef]
22. Montgomery, D.C. *Design and Analysis of Experiments*; John Wiley & Sons: Hoboken, NJ, USA, 2008.
23. Martins, M.; Faleiro, M.L.; Barros, R.J.; Veríssimo, A.R.; Costa, M.C. Biological sulphate reduction using food industry wastes as carbon sources. *Biodegradation* **2009**, *20*, 559–567. [CrossRef]
24. Ghaffari, A.; Abdollahi, H.; Khoshayand, M.R.; Bozchalooi, I.S.; Dadgar, A.; Rafiee-Tehrani, M. Performance comparison of neural network training algorithms in modeling of bimodal drug delivery. *Int. J. Pharm.* **2006**, *327*, 126–138. [CrossRef]
25. Taiwo, A.E.; Madzimbamuto, T.N.; Ojumu, T.V. Optimization of corn steep liquor dosage and other fermentation parameters for ethanol production by *Saccharomyces cerevisiae* type 1 and anchor instant yeast. *Energies* **2018**, *11*, 1740. [CrossRef]
26. Sin, H.N.; Yusof, S.; Hamid, N.S.A.; Rahman, R.A. Optimization of enzymatic clarification of sapodilla juice using response surface methodology. *J. Food Eng.* **2006**, *73*, 313–319. [CrossRef]
27. Nath, A.; Chattopadhyay, P.K. Optimization of oven toasting for improving crispness and other quality attributes of ready to eat potato-soy snack using response surface methodology. *J. Food. Eng.* **2007**, *80*, 1282–1292. [CrossRef]
28. White, C.; Gadd, G.M. A comparison of carbon/energy and complex nitrogen sources for bacterial sulphate-reduction: Potential applications to bioprecipitation of toxic metals as sulphides. *J. Ind. Microbiol.* **1996**, *17*, 116–123. [CrossRef] [PubMed]
29. Tan, L.C.; Papirio, S.; Luongo, V.; Nancharaiah, Y.V.; Cennamo, P.; Esposito, G.; van Hullebusch, E.D.; Lens, P.N.L. Comparative performance of anaerobic attached biofilm and granular sludge reactors for the treatment of model mine drainage wastewater containing selenate, sulfate and nickel. *Chem. Eng. J.* **2018**, *345*, 545–555. [CrossRef]
30. Weijma, J.; Bots, E.A.A.; Tandlinger, G.; Stams, A.J.M.; Pol, L.W.H.; Lettinga, G. Optimisation of sulphate reduction in a methanol-fed thermophilic bioreactor. *Water Res.* **2002**, *36*, 1825–1833. [CrossRef]
31. Shafi, J.; Sun, Z.; Ji, M.; Gu, Z.; Ahmad, W. ANN and RSM based modelling for optimization of cell dry mass of *Bacillus* sp. strain B67 and its antifungal activity against *Botrytis cinerea*. *Biotechnol. Biotechnol. Equip.* **2018**, *32*, 58–68. [CrossRef]
32. Sánchez-Andrea, I.; Sanz, J.L.; Bijmans, M.F.M.; Stams, A.J.M. Sulfate reduction at low pH to remediate acid mine drainage. *J. Hazard. Mater.* **2014**, *269*, 98–109. [CrossRef]
33. Senko, J.M.; Zhang, G.; McDonough, J.T.; Bruns, M.A.; Burgos, W.D. Metal Reduction at Low pH by a *Desulfosporosinus* species: Implications for the biological treatment of acidic mine drainage. *Geomicrobiol. J.* **2009**, *26*, 71–82. [CrossRef]
34. Sánchez-Andrea, I.; Stams, A.J.M.; Amils, R.; Sanz, J.L. Enrichment and isolation of acidophilic sulfate-reducing bacteria from Tinto River sediments. *Environ. Microbiol. Rep.* **2013**, *5*, 672–678. [CrossRef]
35. Alazard, D.; Joseph, M.; Battaglia-Brunet, F.; Cayol, J.-L.; Ollivier, B. *Desulfosporosinus acidiphilus* sp. nov.: A moderately acidophilic sulfate-reducing bacterium isolated from acid mining drainage sediments. *Extremophiles* **2010**, *14*, 305–312. [CrossRef]
36. Bijmans, M.F.M.; Dopson, M.; Peeters, T.W.T.; Lens, P.N.L.; Buisman, C.J.N. Sulfate reduction at pH 5 in a high-rate membrane bioreactor: Reactor performance and microbial community analyses. *J. Microbiol. Biotechnol.* **2009**, *19*, 698–708. [CrossRef] [PubMed]

37. Mohan, S.V.; Rao, N.C.; Prasad, K.K.; Sarma, P.N. Bioaugmentation of an anaerobic sequencing batch biofilm reactor (AnSBBR) with immobilized sulphate reducing bacteria (SRB) for the treatment of sulphate bearing chemical wastewater. *Process Biochem.* **2005**, *40*, 2849–2857. [CrossRef]
38. Nagpal, S.; Chuichulcherm, S.; Peeva, L.; Livingston, A. Microbial sulfate reduction in a liquid–solid fluidized bed reactor. *Biotechnol. Bioeng.* **2000**, *70*, 370–380. [CrossRef]
39. Choi, E.; Rim, J.M. Competition and inhibition of sulfate reducers and methane producers in anaerobic treatment. *Water Sci. Technol.* **1991**, *23*, 1259–1264. [CrossRef]
40. Kousi, P.; Remoundaki, E.; Hatzikioseyian, A.; Battaglia-Brunet, F.; Joulian, C.; Kousteni, V.; Tsezos, M. Metal precipitation in an ethanol-fed, fixed-bed sulphate-reducing bioreactor. *J. Hazard. Mater.* **2011**, *189*, 677–684. [CrossRef]

Publisher's Note: MDPI stays neutral with regard to jurisdictional claims in published maps and institutional affiliations.

processes MDPI

Article

Denoising of Hydrogen Evolution Acoustic Emission Signal Based on Non-Decimated Stationary Wavelet Transform

Zazilah May [1,2], Md Khorshed Alam [1,*], Noor A'in A. Rahman [1], Muhammad Shazwan Mahmud [3] and Nazrul Anuar Nayan [2,*]

1 Electrical and Electronic Engineering Department, Universiti Teknologi PETRONAS, Seri Iskandar 32610, Perak Darul Ridzuan, Malaysia; zazilah@utp.edu.my (Z.M.); noorainarahman6@gmail.com (N.A.A.R.)
2 Electrical Electronics and Systems Department, Faculty of Engineering and Built Environment, Universiti Kebangsaan Malaysia, Bangi 43600, Selangor, Malaysia
3 Mechanical Engineering Department, Universiti Teknologi PETRONAS, Seri Iskandar 32610, Perak Darul Ridzuan, Malaysia; shazwan_092@yahoo.com
* Correspondence: md.khorshed_g03456@utp.edu.my (M.K.A.); nazrul@ukm.edu.my (N.A.N.)

Received: 21 October 2020; Accepted: 6 November 2020; Published: 15 November 2020

Abstract: Monitoring the evolution of hydrogen gas on carbon steel pipe using acoustic emission (AE) signal can be a part of a reliable technique in the modern structural health-monitoring (SHM) field. However, the extracted AE signal is always mixed up with random extraneous noise depending on the nature of the service structure and experimental environment. The noisy AE signals often mislead the obtaining of the desired features from the signals for SHM and degrade the performance of the monitoring system. Therefore, there is a need for the signal denoising method to improve the quality of the extracted AE signals without degrading the original properties of the signals before using them for any knowledge discovery. This article proposes a non-decimated stationary wavelet transform (ND-SWT) method based on the variable soft threshold function for denoising hydrogen evolution AE signals. The proposed method filters various types of noises from the acquired AE signal and removes them efficiently without degrading the original properties. The hydrogen evolution experiments on carbon steel pipelines are carried out for AE data acquisition. Simulations on experimentally acquired AE signals and randomly generated synthetic signals with different levels of noise are performed by the ND-SWT method for noise removal. Results show that our proposed method can effectively eliminate Gaussian white noise as well as noise from the vibration and frictional activity and provide efficient noise removal solutions for SHM applications with minimum reconstruction error, to extract meaningful AE signals from the large-scale noisy AE signals during monitoring and inspection.

Keywords: acoustic emission; hydrogen evolution; denoising; stationary wavelet transform; SHM

1. Introduction

Acoustic emission (AE) signal is a phenomenon of transient elastic waves caused by a change in external conditions (stress, temperature, etc.) on the part of a structure [1]. The technique of acquiring and analyzing AE signals to determine the level of internal and external damages in a composite structure is called AE sensor detection technology, which is currently applied for monitoring mechanical and aerospace engineering structure [2,3]. The AE signals are the elastic waves released by energy within a composite material, which can assess the physical phenomena essence of hydrogen-related damage generation (stress corrosion cracking (SCC), hydrogen embrittlement (HE) and hydrogen-induced cracking (HIC)), especially to the metallic structure. Severe hydrogen evolution and absorption may lead to the failure of the structure itself [4–7]. Since the AE signals are released by many other sources such as vibration, friction and temperature, the actual collected hydrogen evolution AE signals have overlapping frequency bands and weak features under a complex noise background [8]. To provide online detection of hydrogen-related damages, it is essential to extract the AE signal of a corrosion cracking source under a complicated noise background. Thus, the denoising of AE signals during the hydrogen evolution process on the welded structure is the key to acquiring the AE detection of hydrogen evolution-related corrosion cracking.

Several noise-reduction techniques have been explored and used outside the data acquisition system as preprocessing tools in the literature [2,9–11]. Fast Fourier Transform (FFT) is one of the most common conventional noise-reduction methods which is usually employed in much commercial AE data acquisition equipment. The FFT represents the AE signals into the time-frequency domain to extract the frequency components and discriminate the unnecessary signals. Though FFT is very effective for noise reduction, it has a low signal resolution, and cannot perform the time domain and frequency domain analysis simultaneously [9]. It is also less suitable for non-stationary and transient signals [9]. On the other hand, wavelet transform (WT) is a linear transformation of a signal in which the fundamental functions are shifted and changed based on the scaling function of a "mother wavelet" [12]. Khamedi et al. in [10] introduced a denoising method named wavelet packet transform (WPT) to eliminate the noise of the AE signals that originated from the fracture events on composite materials. The WPT is an extended model of the conventional wavelet transform, which provides a complete level-to-level resolution for the AE signal. A significant amount of noise reduction of non-stationary signals was achieved using multi-resolution characteristics [2]. Another approach for noise reduction of the AE signal is the Empirical Mode Decomposition (EMD), which decomposes the AE signal into multiple components based on the frequency-amplitude of the signal. By selecting the important components, an amount of noise reduction was performed. However, it does not interpret stationary signals, which often lead to false signals and redundant signals [11]. The work in [2] has been combined with EMD and WPT to denoise the AE signal by enhancing the aliasing mode occurrence. The AE signal for each application is unique and subjective to their nature, including random noise involved. The AE events are a more acoustical active mechanism of hydrogen evolution monitoring because of bubble friction noise along the structure [13]. Various types of noise, including mechanical, external environmental, friction, and Gaussian white noise are usually induced to the AE signals during acquisition. These may hinder discovery of the information from the signals. Thus, the noises should be removed before performing the information-discovery technique.

Several studies have attempted to use the clustering technique as a tool to remove the noise of the hydrogen evolution AE signals [14]. The method managed to classify more than 60% of detected signals as noise with a classification of accuracy around 65%. However, the proposed technique is not considered to be universal for all types of AE data. Another study has also been identified as the useful information of the AE signal from the evolution of gaseous hydrogen events by using statistical procedures [15]. Since an AE signal is a non-linear and non-stationary signal, the conventional noise-reduction methods are based on linear and stationary assumptions which are unable to fully extract the necessary information. The conventional wavelet decomposition can denoise non-stationary signals with multi-resolution characteristics. However, it aims at only a low-frequency signal part which is unable to satisfy both the low- and high-frequency resolutions. The actual hydrogen evolution AE signal under complex background noise has an overlapping high-frequency band and different characteristics of information. Thus, there is a necessity to incorporate a new multi-resolution denoising method to effectively extract the hydrogen evolution AE signals from the complex noise background. In this paper, we propose a non-decimated stationary wavelet transform (ND-SWT) method based on the variable soft threshold function for denoising hydrogen evolution AE signals. The proposed method will concentrate on both low- and high-frequency resolutions and use to suppress the various types of noises from the acquired AE signal efficiently without degrading the original properties.

2. Background Concepts of the Denoising Methods

The fundamentals of the denoising methods that have been adopted in this research are discussed as follows:

2.1. Empirical Mode Decomposition (EMD)

The Empirical Mode Decomposition (EMD) was introduced by Huang et al. [16]. EMD is an adaptive time-frequency decomposition technique using the Hilbert–Huang Transform (HHT) for non-linear and non-stationary time-series data. The main principle of using EMD is to decompose a given time-series signal $x(t)$ into a sum of oscillatory functions, called intrinsic mode function (IMF). The IMF can be obtained by using the sifting process. In the EMD, an IMF should satisfy the following two conditions: (i) the difference between the extrema which is the sum of maxima and minima, and the number of zero crossings must be 1; and (ii) the local average or mean from the maxima and minima should be 0. According to the time scale characteristics, the signal is decomposed into several IMFs. The given original signal $x(t)$ is decomposed by EMD [2] as in Equation (1):

$$x\,(t) = \sum_{i=1}^{n} IMF_i\,(t) + rn\,(t)\,, \tag{1}$$

where $IMF_i(t)$ represents the series of IMF components and $rn(t)$ refers to the residual component. The first IMF retains the high-frequency impact, and it decreases accordingly for the rest of the IMFs until the signal is not smooth. Selecting the good IMFs with high-frequency and residual components, the signal can be reconstructed, which refers to a denoised signal.

2.2. Wavelet Transform (WT)

The wavelet transform (WT) plays a significant role for various SHM applications in the preprocessing of stationary and non-stationary signals. The preprocessing of the signals includes the removal of noises from the signals, detection of abrupt discontinuities, and compression of large amounts of data. This research emphasizes the wavelet denoising method which provides a significant outcome for noise removal for a broad class of signals of varying degrees of smoothness [17] encountered in a diversity of applications. Wavelet denoising method is of a disarming simplicity, yet it achieves many objectives simultaneously, such as removing the noise without significant signal degradation. Wavelet denoising method offers all that we might desire of a technique, from optimality to generality [17].

2.2.1. Discrete Wavelet Transform (DWT)

The DWT is a conventional method of the WT family which discretizes a signal based on downsampling and represents it in a different level of tree-structure manner. The DWT provides an important advantage over the traditional Fourier Transform (FT) method [18]. The DWT transforms a signal into various scales representing different frequency components. At each scale, the position of the WT can be determined as the vital time characteristic that provides the facility to identify the noises and remove them effectively. The general procedures of the DWT-based signal denoising method used in the literature are divided into three steps. Initially, the signal is transformed up to a predefined decomposition level of j to obtain the detail coefficients. Afterwards, the threshold function is performed on the j signal details using different threshold selection rules, either soft or hard threshold functions, by considering a basic noise model. Finally, the signal is reconstructed using the original approximation coefficients of the jth level and the modified detail coefficients of all levels.

2.2.2. Wavelet Packet Transform (WPT)

The WPT is a generalization of wavelet decomposition, which offers better signal preprocessing and analysis functions than the conventional WT methods [2,19]. The WPT introduces the effects of transformation parameters for denoising effect of a signal, such as the effects of mother wavelet, wavelet packet bases, decomposition level, and threshold function selection. The WPT divides the frequency band into multiple decomposition levels and further transforms the high-frequency part that is not subdivided by wavelet analysis. The WPT selects adaptively the corresponding frequency band to match the spectrum characteristic of the signal, which provides better time-frequency resolution. The wavelet packet transforms a signal into the corresponding frequency band components according to the random time-frequency resolution. In wavelet packet analysis, the noise-reduction algorithm of the signal was basically the same as that of the wavelet analysis. The difference is that wavelet packet analysis decomposed the low and high-frequency part at the same time, which had more accurate local analysis ability.

2.2.3. Stationary Wavelet Transform (SWT)

Stationary Wavelet Transform is more flexible in denoising signals compared to the traditional wavelet transform, which is also known as a non-decimated Wavelet Transform. The SWT is slightly different from the standard DWT because of its up-sampling characteristic of filters at various levels [20]. The SWT preserves the invariance property of a translation of the original signal, whereas the translation invariance in DWT is not preserved because of the sub-sampling operations in the pyramidal method. The SWT uses recursively dilated filters instead of sub-sampling operations to divide the bandwidth from one level to another level. There is no requirement to imply a translation of the corresponding wavelet coefficients. The SWT can maintain the same number of coefficients to all levels, while the traditional wavelet transforms lose the coefficients at each level. Thus, the length of approximations and coefficients is the same at each level, which also corresponds with the original signal.

3. Materials and Methods

This section explains the various features and characteristics of the given specimens. The details about the experimental tests for mechanical and AE data acquisition are discussed.

3.1. Hydrogen Evolution System

A portion of a long carbon steel pipe was used in this test. An exposed area approximately 800 mm^2 was made on the segmented pipe (artificial defect). The electrolyte used in this work is 3.5 wt.% of sodium chloride. The cathodic hydrogen charging was done by inducing 3.15 A current using power supply at 25 V to create the hydrogen evolution mechanism on the exposed area through the test. The counter electrode (CE) is a stainless-steel rod, which is used to complete the electrochemical system. The schematic diagram in Figure 1 shows the cathodic hydrogen charging setup, attached with the sensors and location of the active area.

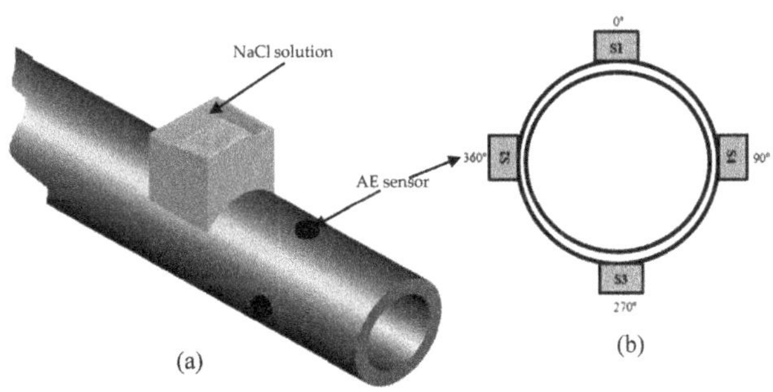

Figure 1. The schematic diagram for (**a**) cathodic hydrogen charging setup and (**b**) cross section pipe with location of sensors.

3.2. Acoustic Emission Signal Acquisition

AE signals were continuously acquired during the test. Signals were monitored via four channels of data acquisition environmental noise test. The sensors are attached to the specimen using the magnetic clamp. A coupling agent was applied between sensors and the specimen to greatly increase the acoustic energy transmission from specimen to the sensor. The whole system, including sensors, are supplied by Physical Acoustics Corporation (USA). Pencil lead-break procedure was used prior to the data acquisition for calibration and to ensure all the sensors were receiving the optimum amplitude from the lead break. Values for peak definition time (PDT), hit definition time (HDT) and hit lockout time (HLT), threshold value, and sample rate were employed in the acquisition setting, as shown in Table 1. The sensors were positioned at 12, 3, 6, and 9 o'clock relative to the pipe specimen as shown in Figure 1b by following the recommended configuration for the cylinder-type structure.

Table 1. AE parameters.

Parameter	Value
Hit definition time (HDT)	2000 μs
Peak definition time (PDT)	1000 μs
Hit lockout value (HLT)	500 μs
Threshold value	40 dB
Sample rate	1 μs per sample

In this work, R6I-AST sensors are used to acquire the AE signal and provide high-sensitivity data acquisition and data recording components. The sensor specification is summarized in Table 2. The normal (ground-truth) and abnormal (cathodic charging) AE signal data were taken before and during the experiment, respectively. The total duration of charging and data acquisition is approximately 146 s, recorded at a sampling rate of 1 μs per sample. Thus, every AE hit was producing a 1024 line of data points. The data were acquired and recorded by AEwin software, including all the waveform features. The AE features were analyzed and processed using the wavelet packet transform, empirical mode decommission, and other methods.

Table 2. Specifications of R6I-AST sensor.

Parameter	Value
Peak sensitivity, ref (V/(m/s))	117 dB
Operating frequency Range	40–100 kHz
Resonant Frequency, ref (V/(m/s))	55 kHz

3.3. Denoising of a Signal Based on ND-SWT

Non-decimated stationary wavelet transform is adopted for denoising signals which are the foundation of the orthogonal wavelet transform. The basic decomposition tree of ND-SWT is presented in Figure 2, where $x[n]$ is an original signal with noise, H, and L are the high-pass and low-pass filters as well as c is a constant, D_i and A_i are the details of coefficients and approximations respectively. This method avoids using decimators because it decreases the length of the approximation and coefficient sequences with the increment of the iteration index in traditional DWT. The ND-SWT uses a series of low-pass and high-pass filters at each iteration level, and the filters are upsampled from the corresponding filters of the previous level. Therefore, the main difference between the SWT and DWT is the filters are upsampled at each decomposition level in SWT instead of sub-sampled. The ND-SWT is an inherently redundant method as each sequence of coefficients contains an equal number of samples as the number of samples in the original signal. For example, the length of the approximations and coefficients is $2lj$, where $l = 1, 2, \cdots, n$ is the length of the approximations or coefficients, and j is the number of levels.

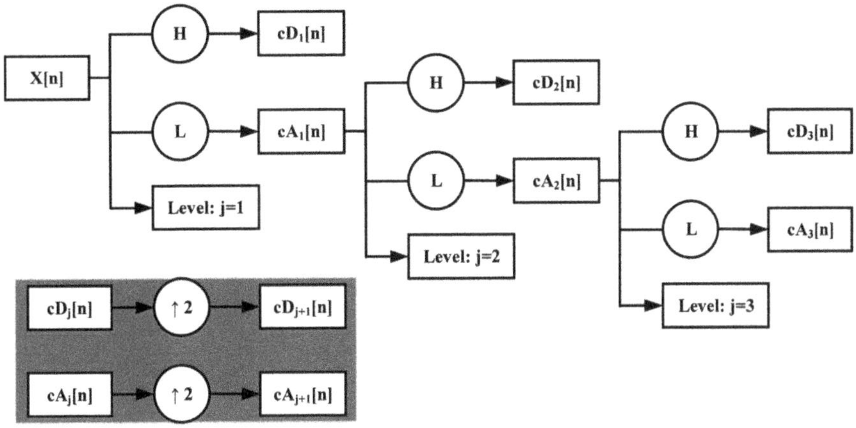

Figure 2. The ND-SWT decomposition tree.

Basic Steps of a Signal Denoising by ND-SWT

A time-series, non-stationary signal with noise is assumed, and one-dimension signal model is described as in Equation (2).

$$x_l = f(t_l) + \omega z_l \tag{2}$$

where f is an unknown function, $t_l = l/n$, $z_l \overset{iid}{\cong} N(0,1)$ is a Gaussian white noise and ω is a noise level. The ND-SWT-based denoising operations can be performed according to the following steps:

Step-1: Wavelet decomposition process is performed on an original one-dimensional signal with noise. This process is also known as multi-resolution analysis of a signal with noise. The original signal is decomposed into jth levels varying the wavelet scale and each level carries the different frequency components of the signal. At each level of decomposition, the signal is divided into two parts—details of coefficients with high-frequency information of the signal and approximations with low-frequency information of the signal. Hence, more decomposition steps provide the depth frequency information of the signal as in Figure 2. Suppose, $A_1(l) =< \psi, \phi_{-1,l} >$ and $D_1(l) =< \phi, \phi_{-1,l} >$, where ϕ is the scale function and ψ is the wavelet function. The non-decimated SWT-based wavelet decomposition can be expressed as in Equations (3) and (4).

$$D_{j,k} = \sum_l H_j^{\uparrow 2}(l-2k)\, D_{j-1,l} \tag{3}$$

$$A_{j,k} = \sum_l L_j^{\uparrow 2}(l-2k)\, A_{j-1,l} \tag{4}$$

where $A_{j,k}$ denotes the scale coefficients of the approximation part, $D_{j,k}$ denotes the wavelet coefficients of the details part. $H_j^{\uparrow 2}$ and $L_j^{\uparrow 2}$ refer to the interpolation of A_{j-1} and D_{j-1} respectively.

Step-2: Threshold functions are performed on the wavelet coefficients at each decomposition level to de-noise the signal without degrading the signal quality. There are two threshold functions, including hard threshold and soft threshold most commonly used in the wavelet transformation-based signal denoising methods [2]. However, one of the main limitations of a hard threshold function is to interrupt the signal continuity and provide a poor smoothness reconstructed signal. On the other hand, the soft threshold function may discard the important characteristics of the decomposed signal resulting may distort the reconstructed signal. Therefore, this paper adopts a dynamic soft threshold function which is computed based on the correlation factor of the wavelet coefficients and it regulates in between 0 to 1, to remove all wavelet coefficients that are greater than or equal to the set threshold value causing the reconstructed signal to be smooth are clean. Different categories of threshold functions are mathematically expressed in Equations (5)–(7).

$$\eta\,(d,\lambda) = \begin{cases} (d) & |d| \geq \lambda \\ 0 & |d| < \lambda \end{cases} \quad \{\text{Hard thresholding}\} \tag{5}$$

$$\eta\,(d,\lambda) = \begin{cases} sign\,(d)\,(|d| - \lambda) & |d| \geq \lambda \\ 0 & |d| < \lambda \end{cases} \quad \{\text{Soft thresholding}\} \tag{6}$$

$$\eta\,(d,\lambda) = \begin{cases} sign\,(d)\,(|d| - \alpha\lambda) & |d| \geq \lambda \\ 0 & |d| < \lambda \end{cases} \quad 0 \leq \alpha \leq 1 \quad \{\text{Dynamic soft thresholding}\} \tag{7}$$

where d is a wavelet coefficient, λ, is a threshold value, and α is the quantified threshold value in between 0 to 1. After applying the ND-SWT, the wavelet coefficients at each decomposition level are the combination of noise-free and noisy characteristics coefficients in the wavelet domain. The noise-free coefficients are coherent and their energy concentration property is a high magnitude of coefficients, whereas the noisy coefficients are incoherent and presented by many coefficients with small magnitudes. Denoising of a signal is complemented by setting the noisy coefficients to near zero or comparatively small while the threshold function is performed.

Step-3: Reconstruction of the decomposed signal. Based on the nth low approximation coefficients and nth high wavelet detail coefficients, the reconstruction process of the decomposed signal is performed. Reconstruction process can be expressed as in Equation (8).

$$A_{j-1,l} = \frac{1}{2}\sum_{k}\left[L_j^{'}(l-2k) + L_j^{'}(l-2k-1)\right]A_{j,k} + \frac{1}{2}\sum_{k}\left[H_j^{'}(l-2k) + H_j^{'}(l-2k-1)\right]D_{j,k} \tag{8}$$

where $L_j^{'}$ and $H_j^{'}$ are the dual base of L_j and H_j respectively.

To evaluate the denoising effect of the reconstructed signal based on ND-SWT, the Root-Mean-Square Error (RMSE), Signal-to-Noise Ratio (SNR), Peak Signal-to-Noise Ratio (PSNR) and cross-correlation evaluation criteria have been used. The definitions with the mathematical expressions are stated as follows:

RMSE: It is used to compute the reconstruction error after denoising a signal which can be calculated by root of the ratio of the total number of samples of a given signal and the mean-square difference between original given signal $x(l)$ and denoised signal $\overline{x}\,(l)$. The RMSE is defined as follows:

$$RMSE = \sqrt{\frac{1}{n}\sum_{l=1}^{n}\left[x\,(l) - \overline{x}\,(l)\right]^2} \tag{9}$$

SNR: It is a measure that compares the level of a desired signal $\overline{x}\,(l)$ to the level of signal noise. The SNR is calculated as the ratio of mean signal power to the mean noise power and expressed as follows:

$$SNR = 10\log_{10}\left[\frac{\sum_{l=1}^{n} x^2\,(l)}{\sum_{l=1}^{n}\left[\overline{x}\,(l) - x\,(l)\right]^2}\right] \tag{10}$$

PSNR: It is a measure which can be computed by the ratio between the maximum possible power of a given signal $x(n)$ and the power of corrupting signal noise that affects the fidelity of its representation. The PSNR is defined as follows:

$$PSNR = 20\log_{10}\left[\frac{\max\,(x\,(n))}{RMSE}\right] \tag{11}$$

Cross-Correlation ($xcorr$): The similarity between two discrete time sequences is measured. If the value of cross-correlation $xcorr$ is close to 1, then the cleaned signal and the signal with noise closely resemble each other. The cross-correlation can be expressed as:

$$xcorr = \left[\frac{E\left(\overline{x}\left(n\right) - \mu_{\overline{x}}\right)\left(x\left(n\right) - \mu_{x}\right)}{\delta_{\overline{x}}\delta_{x}}\right] \tag{12}$$

where $\mu_{\overline{x}}$ and μ_x are the mean values of the denoised signal $\overline{x}\left(n\right)$ and the signal with noise $x(n)$ respectively and $\delta_{\overline{x}}$ and δ_x are denoted as the respective standard deviations of that two signals. The operator $E()$ is referred to the statistical expectation or mean function.

4. Results and Discussion

The description of the datasets, simulation environment, and used parameters are presented in this section. Moreover, the performance of the proposed work in denoising AE signals is analyzed based on several performance metrics.

4.1. Datasets and Simulation Setup

The experimental corrosion dataset acquired from hydrogen evolution on the carbon steel pipeline and synthetic datasets with the addition of different levels of Gaussian white noise were used to evaluate the performance of our proposed denoising method. The hydrogen evolution corrosion dataset was collected using four AE sensors deployed on the carbon steel pipeline at 12, 3, 6, and 9 o'clock positioning manner for a duration of approximately 146 s. There is a single type of measurement (AE signal amplitude in mV), which is recorded as waveforms in every microsecond interval, and each waveform duration is one millisecond. Our proposed method has been applied to the first 1000 waveforms of a single AE sensor and a total of 1,024,000 data points. On the other hand, the synthetic datasets were generated randomly with the addition of SNR 5 dB, 10 dB, 15 dB, 20 dB and 25 dB. Each sample was recorded every microsecond and a total of 1 millisecond duration. Apart from this, the simulations on the datasets were carried out using the Matlab environment to evaluate the performance of the proposed method. The three-layer ND-SWT method was adopted to decompose the given datasets based on Shannon entropy structural function. The four different mother wavelets (haar, db, sym and coif) were selected to compute the wavelet coefficients. The threshold function α was varied to remove the unnecessary wavelet coefficients to achieve noise reduction.

4.2. Denoising of Synthetic Datasets Added with Gaussian White Noise Based on ND SWT

The randomly generated sinusoidal clean signal and the addition of SNR 5 dB, 10 dB, 15 dB, 20 dB and 25 dB Gaussian white noise signals are presented in Figure 3. In the graph, the vertical axis displays the amplitude of a signal, whereas the horizontal axis represents the time. Here, the clean signal is used as a reference signal and our proposed denoising method is applied to the generated noisy signals to evaluate the performance in terms of accuracy in denoising without distorting the original properties of the reference signal. The synthetic clean signal and denoised signals by the ND-SWT are visualized in Figure 4 to observe the effectiveness of the ND-SWT method in denoising different levels of Gaussian white noise. It can be observed that even if there is a small effect on amplitude of the denoised signals only in the case of severe noisy signals, the rest of the properties of the denoised signals are almost similar to the clean signal. The performance comparison among the properties of the clean signal, the properties of the different levels of noisy signals, and the properties of the ND-SWT-based reconstructed denoised signals are presented in Table 3. It can be clearly seen from the table that all properties of the clean signal are influenced by the different levels of Gaussian white noise except the "Max Peak Frequency" property. However, the ND-SWT method suppresses the several degrees of noise effectively and the properties of the reconstructed denoised signals are almost similar to the properties of clean signal. Hence, it can be claimed that the ND-SWT is an effective method in denoising severe noisy signals without degrading the original properties of the signal.

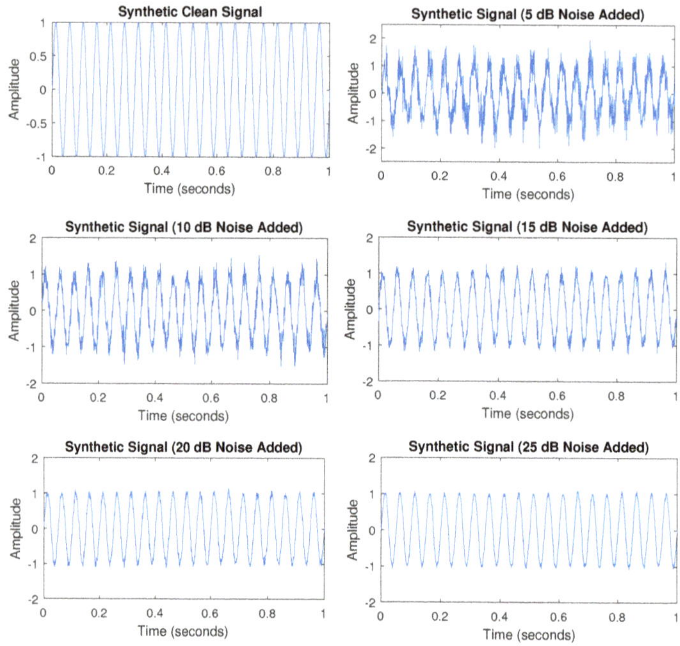

Figure 3. Synthetic clean signal and the signals with the addition of several degrees of Gaussian white noise.

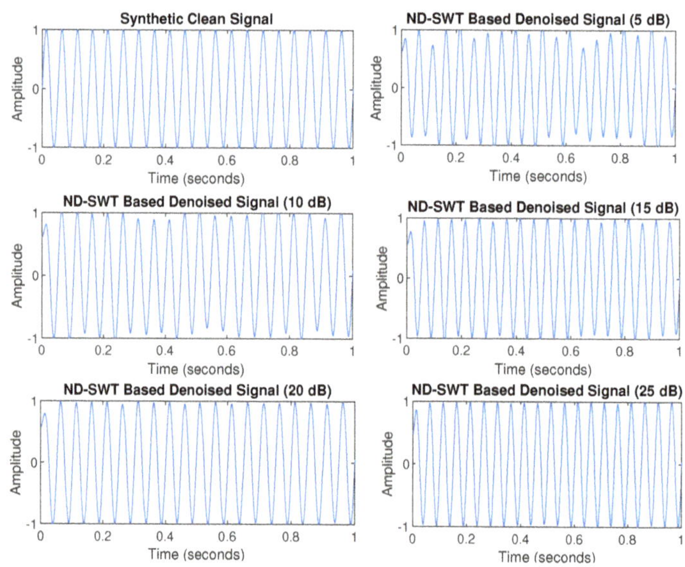

Figure 4. Synthetic clean signal and the ND-SWT-based denoised signals.

Table 3. Comparison among the properties of synthetic clean signal, different degrees of noisy signals and the ND-SWT-based denoised signals.

Properties	Clean Signal	25 dB	20 dB	15 dB	10 dB	5 dB
Number of Peaks	21	209	209	209	273	338
Max Peak Frequency (Hz)	19.53	19.53	19.53	19.53	19.53	19.53
Mean Frequency (Hz)	19.80	20.53	22.01	26.58	41.27	71.47
Angular Frequency (Hz)	125.54	1291.41	1291.41	1690.43	1970.83	2082.04
RMS Bandwidth (kHz)	0.87	60.24	60.24	110.20	107.47	101.97
Mean Frequency Power (dB)	−6.01	−5.97	−5.86	−5.80	−5.30	−4.26
RMSE	0.00	0.04	0.07	0.12	0.23	0.38
SNR (dBc)		24.49	20.91	15.66	9.29	6.04
xcorr (%)	100.00	99.84	99.52	98.48	95.17	87.76
	ND-SWT-Based Denoised Signals					
Number of Peaks	21	21	21	21	21	21
Max Peak Frequency (Hz)	19.53	19.53	19.53	19.53	19.53	19.53
Mean Frequency (Hz)	19.80	19.68	19.68	19.69	19.56	19.50
Angular Frequency (Hz)	125.54	125.79	125.79	125.79	125.79	125.92
RMS Bandwidth (kHz)	0.87	0.98	0.98	0.98	0.96	0.94
Mean Power (dB)	−6.01	−6.37	−6.37	−6.47	−6.55	−6.81
RMSE	0.00	0.03	0.04	0.05	0.07	0.11
SNR (dB)		50.89	55.84	54.78	48.29	54.96
xcorr (%)	100.00	99.91	99.83	99.74	99.56	98.91

4.3. Denoising of AE Signal Added with Friction Noise Using ND-SWT Method

In this section, the AE signals from frictional source measured by rubbing a plate of steel on the same test specimen and other parameters as mentioned in Section 3.2 to create an external noise during hydrogen evolution activity are extracted. Afterwards, The ND-SWT with coif3 wavelet method is used to suppress the frictional noise of acquired AE signal and evaluate the denoising performance. In Figure 5, a comparison of original frictional noisy AE signal, ND-SWT-based denoised signal and estimated noise are presented. It can be seen from the original AE signal that the AE frictional noise is proportional to the movement pattern and the force applied on the test specimen. The denoised signal in Figure 5 shows that the ND-SWT method effectively eliminates the frictional noise or overlapping frequency properties and retains the necessary hydrogen evolution AE features. Hence, the ND-SWT method can be used for the acquisition of friction noise-free AE signals which provide original frequency components for damage assessment in SHM applications.

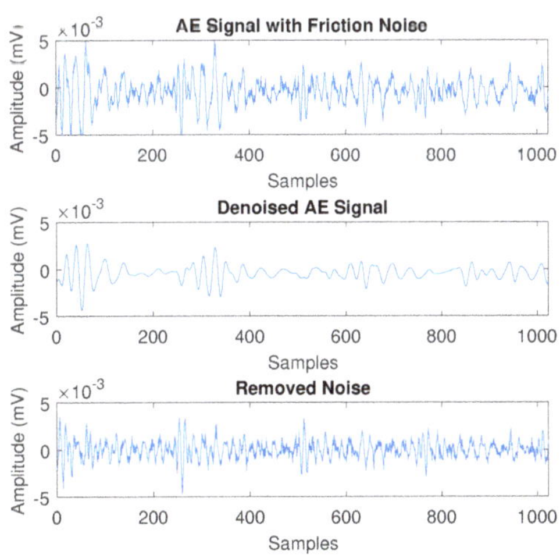

Figure 5. Comparison among frictional noisy AE signal generated from hydrogen evolution experiment, the signal after denoising and estimated noise.

4.4. Denoising of AE Signal Added with Friction and Vibration Noise Using ND-SWT Method

In this section, the AE signals measured on the same test specimen's other parameters are examined as discussed earlier in Section 3.2 by introducing both the external vibration and frictional and vibration noisy signal. The test specimen is rubbed and knocked randomly along the test specimen during a hydrogen evolution experiment for a certain duration to generate friction and vibration noise. The ND-SWT with coif3 wavelet method is applied to the acquired noisy AE signal for multi-scale decomposition to suppress both noises and evaluated for denoising performance. In Figure 6, a comparison of frictional and vibrational noisy AE signal, ND-SWT-based denoised signal and estimated noise are displayed. It can be seen that each cycle of the original AE signal has a burst type of emission wave which may contain the hydrogen evolution AE features whereas the rest of the signal is dominant with the continuous modulation due to knocking activity. The denoised signal in Figure 6 shows that the ND-SWT method effectively removes both the friction and vibration noise and retains the necessary AE information. Hence, the ND-SWT method can be used not only for removing friction noise but also eliminating vibrational noise of generated AE signals from SHM applications.

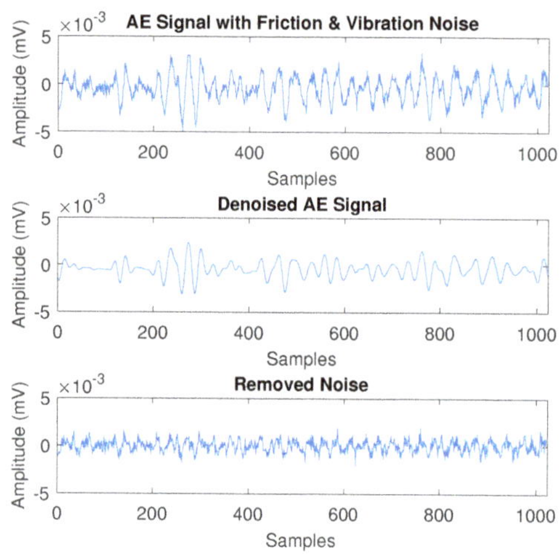

Figure 6. Comparison among frictional and vibrational noisy AE signal, the signal after denoising and estimated noise.

4.5. Frequency Spectrum of the Noisy AE Signals and ND-SWT-Based Denoised Signals

In this section, the hydrogen evolution-based normal AE features are explained. Based on the literature in [21], the normal frequency components of an AE signal concentrate the range between 30 kHz to 150 kHz during hydrogen evolution.

This range mainly depends on several bubbles generated during hydrogen evolution and the level of induced potential. Figure 7 shows a comparison of the frequency components for original frictional and vibrational noisy AE signals with the frequency components of the ND-SWT-based denoised AE signals. From Figure 7, it can be seen in the case of original frictional noise that it mainly influences the overlapping frequency band as well as having a small effect on the amplitude. The ND-SWT-based denoised signal contains the original frequency components by removing the overlapping frequency band and unnecessary amplitude. On the other hand, it can be observed that both friction and vibration noises concentrate on both amplitude overlapping frequency band features of AE signal which increases due to these noises. However, the denoised signal by the ND-SWT retains the original frequency components according to the literature by removing the unnecessary frequency bands.

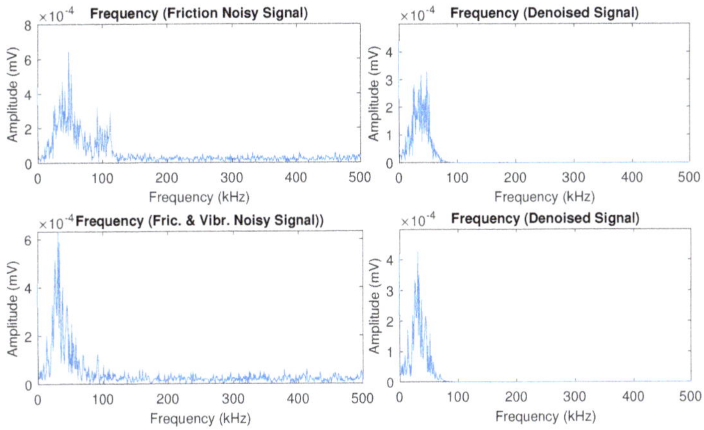

Figure 7. Comparison between the frequency components of the Original noisy AE signals and ND-SWT-based denoised signals.

4.6. Comparison of the Performance in Denoising AE Signal Based on Different Methods

In this section, the performance of various denoising methods on AE signals acquired from the hydrogen evolution experiments using different mother wavelets is analyzed. The performance of the proposed method in terms of denoising for "haar3", "db3", "sym3" and "coif3" wavelets is shown in Figure 8.

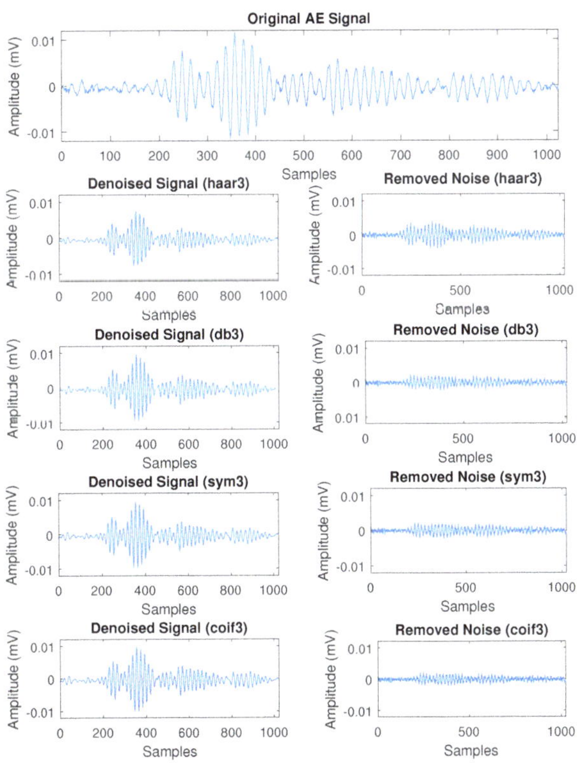

Figure 8. Noise reduction by ND-SWT using haar3, db3, sym3 and coif3 wavelets.

The plot illustrates the comparison between the original AE signal and the denoised AE signal along with the estimated suppressed noise for various types of wavelets. It can be observed that the proposed denoising method suppresses the highlighted friction noise generated from the mechanical or experimental environment and reconstructs the denoised signal with less distortion of the original signal for all types of wavelets. Please note that the "coif3" wavelet yields the lowest distortion after noise reduction of AE signal as compared to others. Moreover, Figures 9–11 represent the performance comparison of the most popular existing denoising methods between the original AE signal and the denoised AE signal along with the eliminated amount of noise for different wavelets. It can be observed that the existing methods yield poor performance in terms of denoising the AE signal as compared to the proposed methods. Even the "coif3" wavelet yields better performance in denoising for all existing methods, it still yields the poor noise suppression performance compared to the performance of the proposed method's "coif3" wavelet.

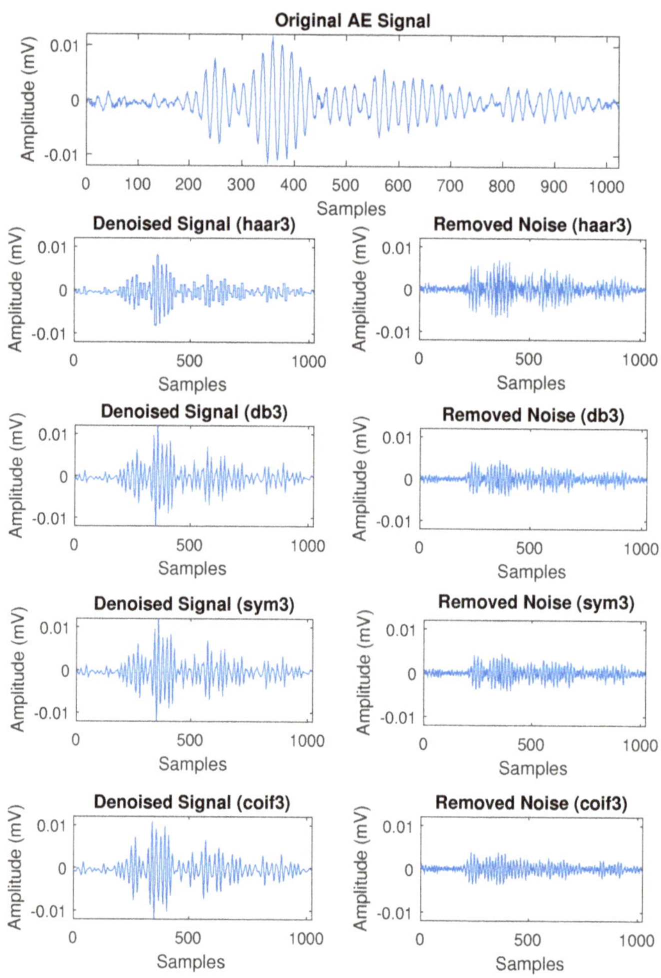

Figure 9. Noise reduction by DWT using haar3, db3, sym3 and coif3 wavelets.

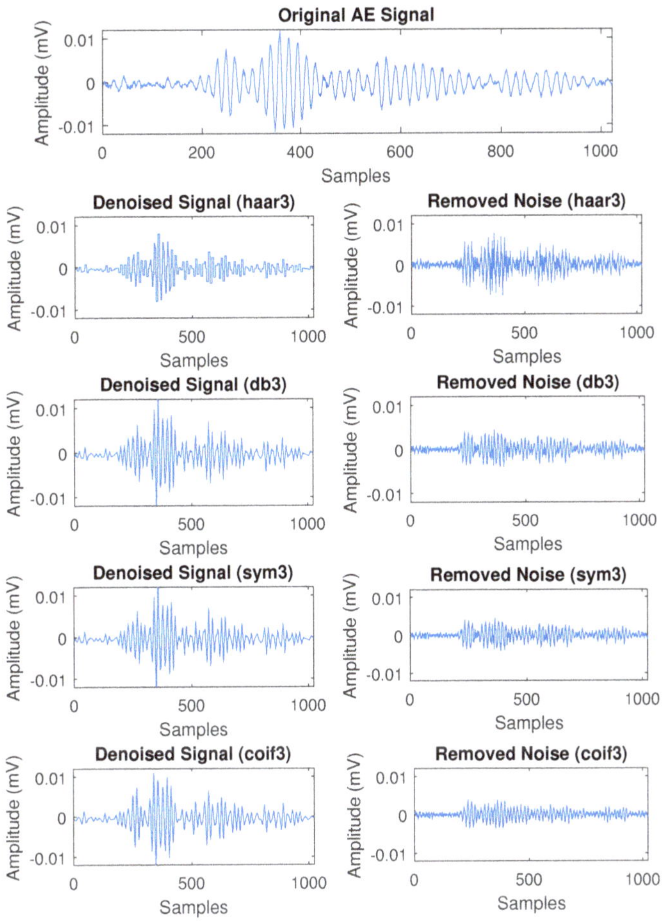

Figure 10. Noise reduction by WPT using haar3, db3, sym3 and coif3 wavelets.

4.7. Performance Comparison of Different Methods Using Various Performance Metrics

In this section, the performance of various denoising methods is evaluated using different performance metrics, including RMSE, SNR, PSNR, and cross-correlation. All performance metrics have been measured by comparing the properties of the original AE signals and noise-free reconstructed signals. Figures 12–15 present the performance of the proposed method on various metrics while denoising the AE signal with respect to the four different wavelets and compare it with the performances of the existing denoising methods. It can be clearly seen in the case of all performance metrics that the mother wavelet named "coif3" performs better in terms of denoising AE signals for all simulated methods than the performances of "haar3", "db3" and "sym3" wavelets. The main reason is that the "coiflets" uses high windows overlapping and six scaling wavelet function coefficients so that it increases adjacent samples in both averaging and differencing leads to a smoother wavelet. Hence, it is more capable of denoising AE signals. Apart from this, it can be observed from the graphs that the ND-SWT yields the optimum results in terms of accuracy in denoising AE signals, as it yields the lowest value of RMSE and highest values of SNR, PSNR, and cross-correlation as compared to the other existing DWT, WPT, and EMD-WPT methods. This is due to its functional characteristics, including revolving, multi-scale invariable, translation, translation invariability, and redundant properties.

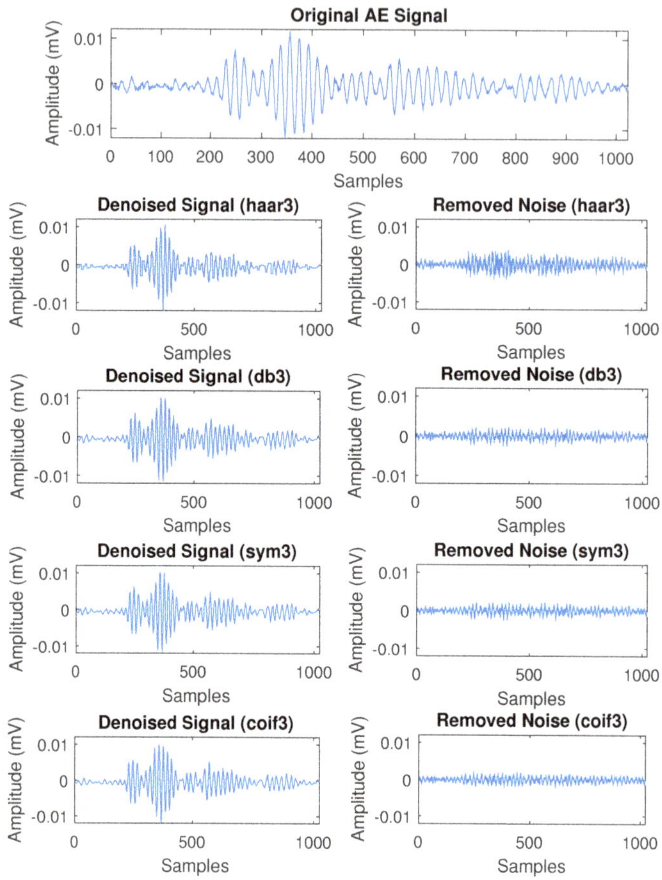

Figure 11. Noise reduction by EMD-WPT using haar3, db3, sym3 and coif3 wavelets.

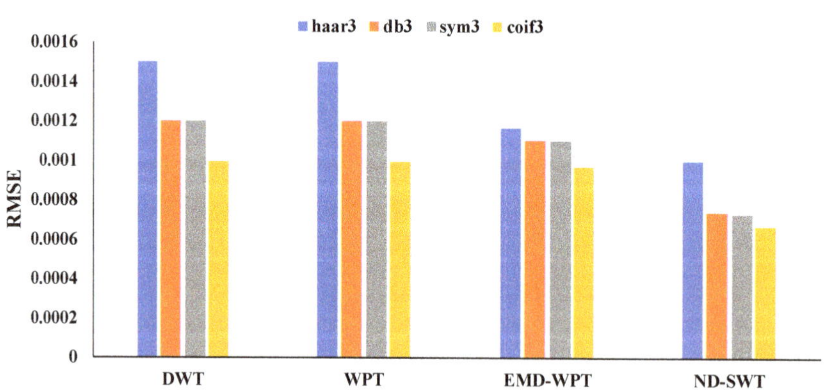

Figure 12. Comparison on RMSEs of the simulation AE signal with respect to four methods using four different wavelets.

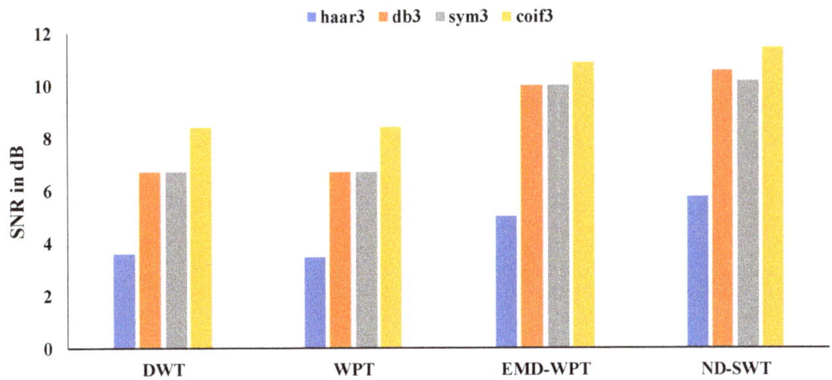

Figure 13. Comparison on SNRs of the simulation AE signal with respect to four methods using four different wavelets.

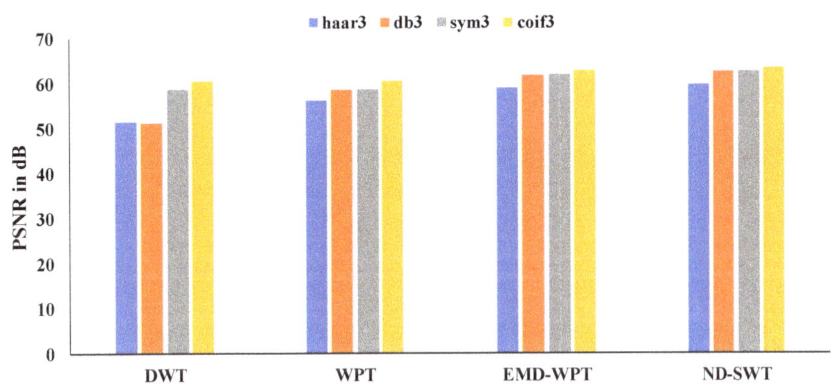

Figure 14. Comparison on PSNRs of the simulation AE signal with respect to four methods using four different wavelets.

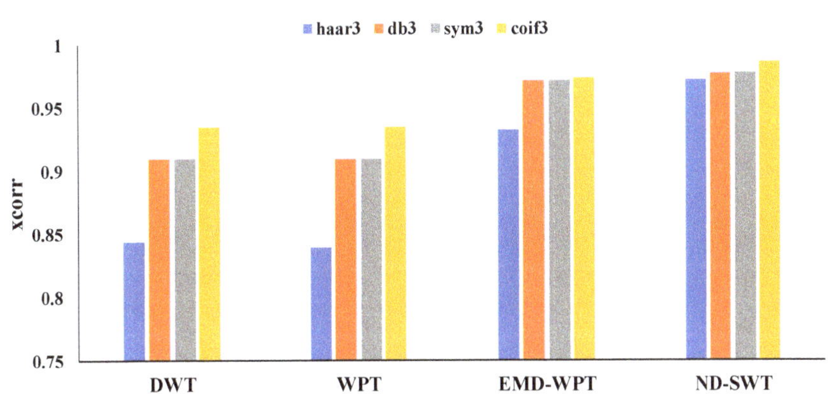

Figure 15. Comparison on cross-correlations of the simulation AE signal with respect to four methods using four different wavelets.

4.8. Frequency Spectrum Analysis of AE Signal after Noise Reduction

The frequency spectrum analysis explains the frequency of a signal or distributed frequency over a time-series. Figure 16 presents the frequency spectrum of the original AE signal, where the original signal has been transformed using FFT and displayed the magnitude in (mV) with respect to the frequency in (kHz) over time-series.

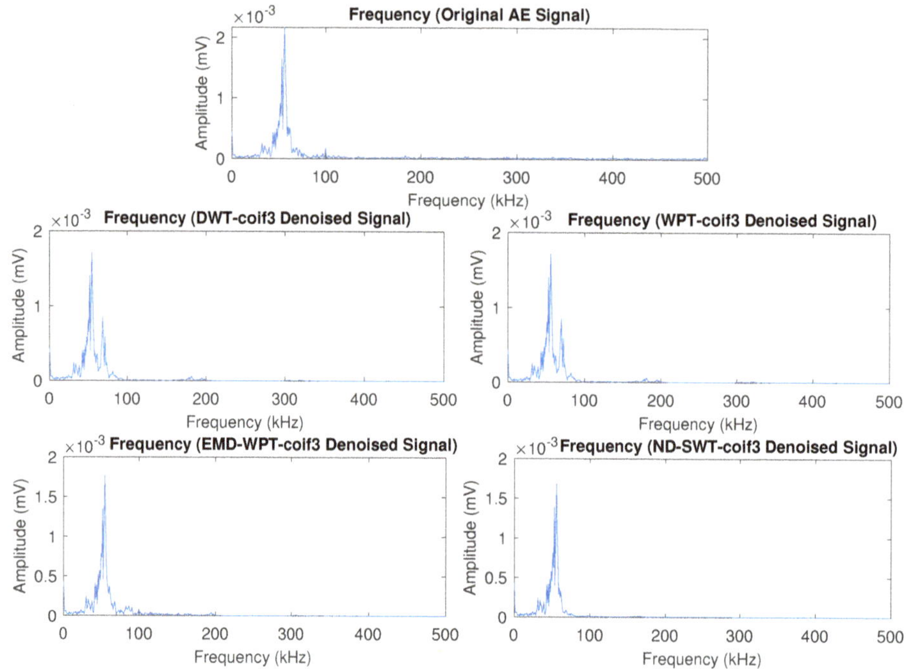

Figure 16. Frequency spectrum of AE signal after noise reduction by DWT, WPT and EMD-WPT and ND-SWT Methods.

The frequency spectrum illustrates a concentration of power at low frequencies, and it varies in amplitude and slope. Figure 16 also shows the frequency spectrum of the reconstructed denoised AE signal by the proposed and existing denoising methods. It can be clearly seen that the frequency spectrum of the ND-SWT-based denoised signal yields better originality in terms of the original frequency spectrum properties of the signal compared to the other existing methods. Thus, the ND-SWT method provides the accurate damage characterization facility in SHM using accurate frequency components of cleaned AE signals.

5. Concluding Remarks

Our findings and analysis confirm the fact that denoising method is very useful to clean noises from the AE signals in order to improve the quality of the AE signals and use them effectively for different applications in monitoring carbon steel structure, suffered from the hydrogen-induced damage reaction. The adopted ND-SWT provides efficient noise reduction of AE signals generated by hydrogen evolution experiments with the minimum reconstruction error hence, making it useful for large-scale noisy AE signals generated from various monitoring applications in SHM. The simulation results of the proposed method have proven that the SNR, PSNR, and cross-correlation performances are maintained high with the minimum RMSE compared to the performances of the recent existing denoising methods in SHM. The ND-SWT method can be applied to acquire AE clean signals during

SHM inspection in real environments. However, the preliminary analysis should be carried out in order to identify the setting parameters during acquisition including threshold value, length of wave, the accurate AE sensor and wave velocity for a specific structure. The selection of appropriate parameters can play a role of removing the reflected wave and other noise. The performance of the ND-SWT method in real practice can vary due to the influence of other external and internal factors such as temperature, humidity, load, pressure and so on which are unable to be considered in this work. The further study can be carried out on real-world SHM application including these factors to investigate the denoising performance of ND-SWT method.

Author Contributions: Review and final editing—Z.M., N.A.N.; data acquisition—M.S.M.; conceptualization and methodology—M.K.A.; data analysis—M.K.A., Z.M.; and original draft preparation—M.K.A. and N.A.A.R. All authors have read and agreed to the published version of the manuscript.

Funding: This research was funded by the Yayasan Universiti Teknologi PETRONAS-Fundamental Research Grant (YUTP-FRG) 2020 under Grant number 015LCO-187 and Mitsubishi Grant under Grant number 015LE0-001.

Acknowledgments: The authors acknowledge YUTP-FRG 2020 (Grant number 015LCO-187) and Mitsubishi Grant (Grant number 015LE0-001) for providing the financial support by the research grants to perform this research.

Conflicts of Interest: The authors declare no conflict of interest.

References

1. Joseph, R.; Giurgiutiu, V. Analytical and Experimental Study of Fatigue-Crack-Growth AE Signals in Thin Sheet Metals. *Sensors* **2020**, *20*, 5835. [CrossRef] [PubMed]
2. He, K.; Xia, Z.; Si, Y.; Lu, Q.; Peng, Y. Noise reduction of welding crack AE signal based on EMD and wavelet packet. *Sensors* **2020**, *20*, 761. [CrossRef] [PubMed]
3. Mei, H.; Haider, M.F.; Joseph, R.; Migot, A.; Giurgiutiu, V. Recent advances in piezoelectric wafer active sensors for structural health monitoring applications. *Sensors* **2019**, *19*, 383. [CrossRef] [PubMed]
4. Jack, T.A.; Pourazizi, R.; Ohaeri, E.; Szpunar, J.; Zhang, J.; Qu, J. Investigation of the hydrogen induced cracking behaviour of API 5L X65 pipeline steel. *Int. J. Hydrogen Energy* **2020**, *45*, 17671–17684. [CrossRef]
5. Wasim, M.; Ngo, T.D. Failure analysis of structural steel subjected to long term exposure of hydrogen. *Eng. Fail. Anal.* **2020**, *114*, 104606. [CrossRef]
6. Kang, Y.; Kim, M.; Kim, G.; Kim, N.; Song, S. Characteristics of Susceptible Microstructure for Hydrogen-Induced Cracking in the Coarse-Grained Heat-Affected Zone of Carbon Steel. *Metall. Mater. Trans. A* **2020**, *51*, 1–11. [CrossRef]
7. Nguyen, T.T.; Tak, N.; Park, J.; Nahm, S.H.; Beak, U.B. Hydrogen embrittlement susceptibility of X70 pipeline steel weld under a low partial hydrogen environment. *Int. J. Hydrogen Energy* **2020**, *45*, 23739–23753. [CrossRef]
8. Kharrat, M.; Ramasso, E.; Placet, V.; Boubakar, M. A signal processing approach for enhanced Acoustic Emission data analysis in high activity systems: Application to organic matrix composites. *Mech. Syst. Signal Proc.* **2016**, *70*, 1038–1055. [CrossRef]
9. Liu, X.L.; Liu, Z.; Li, X.B.; Rao, M.; Dong, L.J. Wavelet threshold de-noising of rock acoustic emission signals subjected to dynamic loads. *J. Geophys. Eng.* **2018**, *15*, 1160–1170. [CrossRef]
10. Khamedi, R.; Abdi, S.; Ghorbani, A.; Ghiami, A.; Erden, S. Damage characterization of carbon/epoxy composites using acoustic emission signals wavelet analysis. *Compos. Interfaces* **2019**, *27*, 111–124. [CrossRef]
11. Ji, J.; Li, Y.; Liu, C.; Wang, D.; Jing, H. Application of EMD Technology in Leakage Acoustic Characteristic Extraction of Gas-Liquid, Two-Phase Flow Pipelines. *Shock Vib.* **2018**, *2018*, 1529849 . [CrossRef]
12. Satour, A.; Montrésor, S.; Bentahar, M.; Elguerjouma, R.; Boubenider, F. Acoustic emission signal denoising to improve damage analysis in glass fibre-reinforced composites. *Nondestruct. Test. Eval.* **2014**, *29*, 65–79. [CrossRef]
13. Faisal, N.; Droubi, M.; Steel, J. Corrosion monitoring of offshore structures using acoustic emission sensors. *J. Inst. Corros.* **2017**, *139*, 14–16.
14. Calabrese, L.; Campanella, G.; Proverbio, E. Noise removal by cluster analysis after long time AE corrosion monitoring of steel reinforcement in concrete. *Constr. Build. Mater.* **2012**, *34*, 362–371. [CrossRef]

15. Hrairi, M. Statistical signal processing and sorting for acoustic emission monitoring of high-temperature pressure components. *Exp. Tech.* **2009**, *33*, 35–43. [CrossRef]
16. Huang, N.E. Introduction to the Hilbert–Huang transform and its related mathematical problems. In *Hilbert–Huang Transform and its Applications*; World Scientific: Singapore, 2014; pp. 1–26.
17. Schreiber, T. Interdisciplinary application of nonlinear time series methods. *Phys. Rep.* **1999**, *308*, 1–64. [CrossRef]
18. Ramos, R.; Valdez-Salas, B.; Zlatev, R.; Schorr Wiener, M.; Bastidas Rull, J.M. The discrete wavelet transform and its application for noise removal in localized corrosion measurements. *Int. J. Corros.* **2017**, *2017*, 7925404. [CrossRef]
19. Tan, C.; Wang, Y.; Zhou, X.; Wang, Z.; Zhang, L.; Liu, X. An integrated denoising method for sensor mixed noises based on wavelet packet transform and energy-correlation analysis. *J. Sens.* **2014**, *2014*, 650891 . [CrossRef]
20. Nason, G.P.; Silverman, B.W. The stationary wavelet transform and some statistical applications. In *Wavelets and Statistics*; Springer: Berlin/Heidelberg, Germany, 1995; pp. 281–299.
21. Wu, K.; Jung, W.S.; Byeon, J.W. Acoustic emission of hydrogen bubbles on the counter electrode during pitting corrosion of 304 stainless steel. *Mater. Trans.* **2015**, *56*, 587–592. [CrossRef]

processes

MDPI

Article

Firm's Sustainability and Societal Development from the Lens of Fishbone Eco-Innovation: A Moderating Role of ISO 14001-2015 Environmental Management System

Md. Abu Toha, Satirenjit Kaur Johl * and Parvez Alam Khan

Department of Management and Humanities, Universiti Teknologi PETRONAS, Seri Iskandar 32610, Perak, Malaysia; abutoha.jnu@gmail.com (M.A.T.); Parvez_17003460@utp.edu.my (P.A.K.)

* Correspondence: satire@utp.edu.my

Received: 22 June 2020; Accepted: 27 July 2020; Published: 15 September 2020

Abstract: Eco-innovation has gained considerable attention in academia as well as in industry due to its potential in mitigating environmental challenges and its positive correlation with firm performance. However, there are limited studies which have investigated the moderating relation of International Organization for Standardization (ISO) 14001:2015 between eco-innovation and firm sustainability in their contribution to societal development. This research is supported by a resource-based theory which explores the core-competencies of firms and challenges the resources creating the competitive advantage of the firm without compromising on the social responsibility aspect of the firm. This study proposes a fishbone eco-innovation business model, which includes production (product, process, and technology) and non-production (organization and marketing) business activities mapped with the 17 Sustainable Development Goals (SDGs) for societal development. This fishbone eco-innovation business model signals to the stakeholders about the organization's innovation in their green implementation, which goes beyond mere compliance. The contribution of the fishbone eco-innovation business model to societal development will create a unique competitive edge and green goodwill amongst the external stakeholders, which will attract sustainably responsible investors for investment. This article draws propositions and develops a conceptual model for future empirical research on eco-innovation and societal development.

Keywords: eco-innovation; firm sustainability; fishbone diagram; societal development; ISO 14001:2015; environment management system

1. Introduction

Weapons of mass destruction, life-threatening weather, natural catastrophes, climate change, and clean water crises are the topmost five risks that will have significant impacts in the next 10 years in the world [1]. Surprisingly, all five risks are directly and indirectly related to the environment and society, where these environmental risks are the central strategic issues for corporations and societal development. However, in the recent lockdown, steps taken by most countries in the worldwide Covid-19 pandemic ironically minimizes the environmental risk by improving the air quality and temperature around the world within a short period of time. This is primarily because carbon dioxide gas emissions into the atmosphere are reduced based on less industrial activities. This issue is noticeable but not enough for environmental sustainability [2–6]. The Covid-19 pandemic not only affects the air quality index but also affects social mobility during the lockdown because the lockdown demotivates social gatherings. Because of social gathering restrictions, societal expansions and growth are trapped in a cartage stage. As a result, business corporations face different issues regarding this pandemic in relation to sustainable societal development, such as health, supply chain, labor force, and cash flow [7,8], where most situations demand that environmental risk and societal growth issues be reconsidered post-COVID-19 pandemic in their sustainable business model to grab the full phase of sustainability for society.

Besides the COVID-19 pandemic, the imperativeness of environmental and societal development issues can also be understood by referring to the United Nations 17 sustainable development goals (UNSDG), in which the environmental loss is at US $2.2 trillion in the year 2018. However, the amount is still increasing due to traditional innovation approaches and GDP (Gross Domestic Product) investment and initiatives towards protecting the environment. The amount of global GDP financed in research and development (environment) increased from 1.52 percent to 1.68 percent from 2000 to 2016 [9], which does not match up to environmental losses. Hence, business corporations need to screen investment not only for sustainable business but also for innovative practices towards environmental challenges that lead to a competitive edge for the company and contribution towards societal development.

The motivations of this study are divided twofold. Firstly, due to the rapid boost in climate change and global warming, the whole world is undergoing a severe environmental problem, which creates a hindrance to social development. According to [10], 2.7 billion people are at risk due to climate change. Secondly, in the era of the Internet of Things, artificial intelligence, and the information age, organizations are still following traditional business models and end up emitting 36.83 billion tons of CO_2 and 7.70 million tons of solid waste, as well as seeing an increase in energy demand by 2.3% in the year 2018. Therefore, the current business model needs to be replaced with sustainable practices to boost the existing societal development velocity, which is another motivation for this research.

There is evidence that the fourth industrial revolution plays a crucial role in sustainability and societal development. It is claimed that the fourth industrial revolution has been initiated by vast scientific research and sustainable technological developments. These sustainable technological developments attempt to mitigate environmental risk and accelerate societal development through different techniques. A study of [11] stated that business corporations require environmentally friendly innovations to control environmental risks and promote societal development. On the other hand, the United Nations also supports environmentally friendly innovations through the Sustainable Development Goals (SDGs) 2030 plan to protect the environment and society through sustainability from current and future threats.

There is a growing body of studies that identify the importance of different forms of innovations relating to the environment and social development—namely, general innovation, green innovation, eco-innovation, and social innovation. Amongst all, the existing literature claims that eco-innovation has emerged as one of the utmost dynamic innovations for firm sustainability [12–14], whereas social innovation has emerged as a vital player in societal development [15]. The objective of this study is to investigate the effect of eco-innovation on societal development, including a firm's sustainability.

Eco-innovation depends on several determinants, such as green demand and supply, green regulation, green collaboration, green resources, and capabilities. These predeterminants accelerate the reduction in materials, improve energy efficiency, minimize emission, improve biodiversity, and raise living standards [16,17], as well as providing competitive advantages for the firm. These factors promote the three pillars of sustainability: sustainable competitive advantages, reduction in cost, and savings in energy. Improving emissions, biodiversity, and energy efficiency are related to the environmental pillar of sustainability, but the reduction in materials, improvement in energy efficiency, minimizing of emissions, and improvement of biodiversity are directly related to the social pillar of sustainability. Therefore, this study draws an argument that eco-innovation accelerates not only a firm's sustainability but also social development.

On the other hand, there are some other external factors that moderate eco-innovation. Among them, International Organization for Standardization (ISO) 14001:2015 is the most accepted standard which promotes environmental management. According to ISO's recent statistics, this is the second most popular business certification, and is implemented by approximately 200 countries all over the world. In addition to that, this business certification has improved 8% of the annual growth of companies around the globe [18–21]. This situation clearly demonstrates the certification's benefit, as it accelerates the firm's sustainable practices [22] and cost-saving through waste management, water management, the identification of environmental risks and opportunities, environmental compliance, employee engagement, and the promotion of safety practices. The said factors enhance trust amongst stakeholders, create green intentions, and help to portray the firms as responsible firms towards society. Therefore, this study claims that the ISO 14001:2015 certification moderates the nexus between eco-innovation and firm sustainability to enhance societal development. The following table gives a summary of the study of eco-innovation, firm sustainability, ISO 14001:2015, and societal development.

Table 1 highlights the gap from previous studies on eco-innovation concerning the business production process, which highlights the role of products, processes, technology, organization, and marketing with positive and negative correlations with sustainability. On the other hand, numerous studies have been conducted in the last few decades on firms' sustainable practices, innovation, social innovation, and environmental innovation in relation to stakeholders' contributions. Those researches argued the positive impact of eco-innovation on societal developments. However, there is still limited research being conducted on predicting the role of corporate eco-innovation practices on societal development.

Table 1. Summary of past studies on eco-innovation, firm sustainability, societal developments, and International Organization for Standardization (ISO) 14001:2015.

Reference	Eco-Innovation Dimension	Firm Sustainability Dimensions	Societal Development	ISO 14001:2015	Theory	Methodology	Findings
[23]	Technology	Environment	No focus	-	Resource-based view	Quantitative	Negative association
[24–28]	Marketing	Environmental, economic and social	Focus on social value addition	-	Resource-based view	Quantitative	Positive association
[29]	Organization	Social and environmental	Focus on employees' living standards	-	Resource-based view	Quantitative	Positive association
[30]	Product	Social and economic	Focus on product social value	-	Resource-based view	Quantitative	Negative association
[23,31–33]	Process	Environmental and economic	No focus	-	Resource-based view	Quantitative	Positive association
[34]	Technology	Environment, economic and social	Focus on sustainable social consumption patterns	-	-	Qualitative	Positive association
[35]	Technology	Environment, economic and social	Focus on societal natural resource allocations and utilization	-	-	Qualitative	Positive association
[34]	Organization	Social	Focus on polices and principles for societal transparency	-	-	Mixed	Positive association
[36]	Organization and technology	Social and environmental	Focus on social policy in the E.U. context	Considered as an independent variable in the framework	-	Quantitative	No association
[37]	Product and process	Social and environmental	Focus on the product social life cycle perspective	Considered as an independent variable in the framework	-	Quantitative	Positive association
[38]	-	Social	Focus on higher education institutions	Considered as a factor of sustainability practices	-	Qualitative	Negative association
[39]	Processes in eco-innovation	Social and environment	Focus on sustainable industrial process management and SDGs	Considered a sustainable effect on society	-	Qualitative	Positive association
[40]	Organizational eco-innovation	Social	Focus on employee sustainability through the social aspect	Considered as a factor that promotes social sustainability	-	Qualitative	Positive association

Table 1. *Cont.*

Reference	Eco-Innovation Dimension	Firm Sustainability Dimensions	Societal Development	ISO 14001:2015	Theory	Methodology	Findings
[17]	Overall eco-innovation	Environment, social and economic	Focus on economic performance, cost performance, resources savings, pollution prevention, and recycling	-	-	Quantitative	Positive association
[16]	Technology eco-innovation	Economic	Focus on lean management principles	Discussed it. Not considered in their study	-	Quantitative	Positive association
[41]	Overall eco-innovation	Organizational performance	Focus on eco-innovation performance	-	Resource-based and stakeholder perspectives	Qualitative	-

A limited number of studies have been published on the association between eco-innovation and firm sustainability that highlight the three pillars of sustainability. Additionally, there is a scarcity of studies on visual or graphical diagrams specifically related to eco-innovation [42,43], firm sustainability, and societal development. This study propounds the fishbone eco-innovation framework model with the moderating role of the environmental management system (ISO 14001:2015) on firms' sustainability and societal development [42,43].

The problem of the research is that most of the firms are still following traditional and conventional approaches to the environment and innovation, which results in an unsustainable environment that affects the society. On the other hand, there are few firms that concentrate on eco-innovation. However, there are also firms which are not aware of the cause and effect relationship of eco-innovation on societal development. This study also highlights the following questions: "Does eco-innovation affect firm's sustainability? How does the fishbone diagram associate with eco-innovation and firm's sustainability? How does eco-innovation affect sustainability and societal development? How does ISO 14001:2015 moderates the relationship between eco-innovation and firm's sustainability in the development of society?"

Eco-innovation has been gaining a considerable amount of attention from academicians as well as corporate sectors due to its potential in firms' sustainability [44] and societal development. At the micro-level, this research assists firms in taking precautionary steps in scaling up eco-innovation to formulate their business strategies and develop future plans. Besides that, this research can also assist top-level management to promote green and social goodwill, which will influence stakeholders' positive decisions, which could possibly lead the firm's community investment in social development. As a result, firms can build better relationship between internal and external stakeholders, thereby it is also expected to bring a positive nexus on the firm's bottom line.

The remaining section of this study will be discussed as follows: In the next section, the literature will be discussed. In Section 3, there will be a discussion on the proposed conceptual framework, followed by a discussion on future research directions. The conclusion is set out in the final section.

2. Literature Review

The definition of eco-innovation was first introduced by Fussler and James in 1996, where it is described as reducing negative environmental impacts while offering new products and processes to consumers and businesses. Eco-innovation also leads to sustainable protection and development by the introduction of innovative ideas, practices, and technologies. Eco-innovation can be described as a modern or substantially changed product, procedure, or business practice that seeks to minimize environmental costs, emissions, and the adverse consequences of resource usage instead of using existing approaches that do not take into consideration the environmental impact.

Many countries have acknowledged that eco-innovation is an impactful way to solve the ecological problems of today, including climate change and energy saving. Moreover, many countries perceive eco-innovation products and services in the market as a source of competitive advantage. Enterprises are seeking eco-innovation as an opportunity to curb negative impacts on the environment while building a decisive competitive edge and green goodwill in stakeholders' eyes. In addition to that, academic researchers believe that eco-innovation is one of the prime sources to achieve a firm's sustainability and societal development [24,45].

Sustainability can be defined as satisfying needs without compromising the resource capability for future generations' needs. Sustainability literature stands on three pillars—namely, economic, environmental, and social. Therefore, to attain sustainability, the firm needs to contribute on these three pillars. To explore firms' sustainability, this study conducts a detailed review of eco-innovation dimensions.

2.1. *Dimensions of Eco-Innovation and a Firm's Sustainability*

Research organizations such as [46,47] together with researchers—namely, [24,45]—identified five dimensions of eco-innovation, such as product, process, technology, organization, and marketing. The following section will highlight the five dimensions of eco-innovation and a firm's sustainability.

2.1.1. Product Eco-Innovation

Product eco-innovation is designing or improving a product that is new or beneficial in curbing environmental risk in order to enhance sustainability. Product eco-innovation refers to the materials used to produce products that can lower the impact on the environment [24]. Eco-innovation is used to improve the quality of products, services, and sustainability [24,46]. These goods and services emphasize eco-design to diminish the total effect on the environment through their production. [47] defined product eco-innovation by emphasizing its potential users. The introduction of goods or services which are new or significantly better quality concerning their intended uses is known as product eco-innovation [47,48] together with [24], claimed that product eco-innovation is positively associated with a sustainable firm's performance. The studies also pointed out seven specific indicators that are related to increasing a firm's value which focus on the sustainability issue. The indicators are: the usage of new cleaner material or a new input with a lower environmental impact, the usage of recycled materials, a reduction in the use of raw materials, a reduction in the number of product components, the elimination of dirty components, a product with a longer life cycle, and a product's ability to be recycled. It can be argued that product eco-innovation can accelerate competitive advantages and sustainability [48]. In contrast, a study from [30] stated that eco-product innovation negatively correlated with a substantial firm's performance due to the reactive approach of the firm.

This research advocates that product eco-innovation not only helps to curb the environmental risk but also promotes social development. [30] found that the environmental issue is correlated to social challenges. The improvement and reinforcement in the environment will positively minimize the social risk, which leads to achieving the societal development of the economy.

2.1.2. Process Eco-Innovation

According to [30], process eco-innovation is defined as the employment of new or significantly better-quality production or distribution methods which help to reduce environmental impact through significant variations. A study from [47] identified several traits, such as the reduction in material used, lower risk, and cost savings, which improve and strengthen the clean production of the firm.

Cleaner production in the manufacturing process and end of pipe technologies in the production line are the most commonly used terms in process eco-innovation, as claimed by [49]. In accordance with [50], process eco-innovation is no longer confined to explicit environmental performance, but additionally encompasses tacit environmental improvement and sustainable performance.

Several attempts from different researchers [30–33,47,49,51] have been made to identify the nexus of process eco-innovation and sustainable firm's performance. Among them [31–33] stated that process eco-innovation is positively related to a sustainable firm's value. [24] identified 11 specific factors that contribute to an increase in the firm's value. These factors are as follows: a reduction in chemical wastage, a reduction in water usage, a reduction in energy usage, the minimization of wastage, the recycling of components, the recycling of wastage, innovative environment-friendly technologies, renewable energy for the production process, R&D investment, the acquisition of machinery, and the acquisition of patents [52].

The adoption of eco-friendly factors in the process will maximize the resource efficiency based on the fact that firms and nations which are able to save their limited resources and conduct activities to save their resources will boost their societal development [53]. Therefore, this research advocates that eco-innovation has the potential for sustainable and societal development, which can be verified by the finding from [54] that innovation enhances and strengthens sustainable societal development.

2.1.3. Technological Eco-Innovation

In the ear of industrial revolution 4.0, the rapid development of technology changes the world constantly. Technological eco-innovation is defined as investments in green equipment and the installation of unconventional green production technologies in an organization. According to [55], technological eco-innovations play a crucial role in providing information to comprehensive material-saving plans and in managing credentials and statistics. The literature claims that firms require huge tangible resources and internal capabilities to implement technological eco-innovation. The study of [23] explained that the complex nature of technological eco-innovation necessitates firms to implement it through specific resources and green capabilities.

Several researchers, such as [44,56], have conducted studies on technological eco-innovation with the nexus of a firm's sustainability performance, with the findings that technological eco-innovation and a firm's sustainability are positively correlated. It is still debatable whether adopting eco-technology operational investments will enhance the efficiency of the production process, minimize the production time, or maximize the output with the same limited resources along with the health and safety of the operational employees. However, due to the capability of maximizing resource efficiency, this research believes that the invention and adoption of technological eco-innovation will help in societal development.

This research advocates that technological eco-innovation can play an enabling role in social development. There are various real-life examples where technologies play roles in social development, such as the case of mobile connectivity and the role of technology in the stock exchange. In a recent example, during the COVID-19 pandemic the whole world became dependent on technology in order to boost economic and social activity and effectuate governing activities. During the COVID-19 pandemic, social development is still progressing in the frozen and melting economy due to the role played by technology. Likewise, it has been evidenced by previous studies that technological eco-innovation not only plays an imperative role in enhancing sustainability but also in societal development as well.

2.1.4. Organizational Eco-Innovation

According to [46], organizational eco-innovation is the introduction of a new organizational system in a company's management processes, work-place of a organization, or external relations to delegate roles and decision-making to employees or manage the distribution of work within and amongst client activities. The organizational eco-innovation of the firm involves implementing several environmental policies and directions in motives to curb the environmental impact and boost sustainability.

In the existing literature on organizational eco-innovation, there are several supporting proofs [29,31–33,57,58] in support of the positive association between organizational eco-innovation and firm sustainability. There are several factors primarily responsible for a positive association between a sustainable firm's value and organizational eco-innovation—namely, green human resources, pollution prevention plans, environmental objectives, environmental audits, environmental advisories, investments in research, cooperation with stakeholders, new systems, and new markets [24]—which help firms to contribute to achieving sustainability and social development [29,57].

Organizational eco-innovation and social development have a direct relationship, as the organization establishes the innovation culture and continuously improves its innovation capability.

Apple is an example of a most innovative company, due to the organization's structure and innovation practices. For an example, Apple approved work from home concepts for its employees. This project saves both fuel energy as well as travelling time for employees, which improves societal living standards. Apple organization's structure, innovation, and products have changed the world of modern society, which leads to social development. This research aims to prove that organization eco-innovation is directly linked to societal development, without compromising the sustainability of the firm.

2.1.5. Marketing Eco-Innovation

According to [46], marketing eco-innovation is the employment of new marketing techniques relating to significant changes in product design or packaging, product placement, product promotion, or product pricing. These significant changes help to reduce harmful ecological impacts on society. In addition to that, these significant changes are also aligned with sustainability. To maintain this sustainability of firms, research organizations such as [47] attempted to identify significant marketing techniques that motivate consumers to purchase. According to them, marketing eco-innovation inclines to trace which marketing practices can be employed to inspire people. Hence, it can be in various forms, such as product design or packaging, product placement, product promotion, pricing, and eco-labelling. These forms of marketing eco-innovation are supported by a few researcher, such as [46,47].

In the existing literature on marketing eco-innovation, [24–28] argued that marketing eco-innovation received less consideration than any other dimensions of eco-innovation. However, several researchers such as [24–28] considered marketing eco-innovation in their studies and found a positive association between marketing eco-innovation and a firm's sustainability through the three pillars of sustainability to achieve societal development. According to [24,59], marketing eco-innovation is connected with three perspectives of sustainability, such as environmental, economic, and social. These studies claimed that there are three specific indicators, such as quality certifications, green design packaging, and reusable packaging. These three factors contribute towards sustainability, which is also responsible for the positive association between eco-innovation marketing and a firm's sustainability. In contrast, a study from [60] found a negative association between marketing eco-innovation and a firm's sustainability.

There are also researchers [60] that found a negative association between marketing eco-innovation and a firm's sustainability. On the other hand, there are contradicting arguments that find marketing eco-innovation improves the sustainability for the organization too [24,59]. In addition, marketing eco-innovation can be linked with societal developments. P&G is the perfect example of marketing eco-innovation, as they removed plastic packaging from their marketing concepts. They initiated a project called "alliance to end plastic waste" through their marketing research teams, and they promised society that their company's plastic packaging will be not found in the ocean in future. They contribute to societal developments as well as safeguarding sustainability.

2.2. *Three Pillars of Sustainability for Firm*

Sustainability stands on three pillars. The pillars are economic, environmental, and social. The three pillars of sustainability are also known as the Triple Bottom Line (TBL) framework in sustainability [61]. In addition to that, the three pillars of sustainability are associated with "three Ps." The three Ps stand for Profit for the economic pillar, Planet for the environment pillar, and People for the social pillar.

It was inferred from the previous studies of [34,61,62] that eco-innovation can affect the three pillars of sustainability. According to the study of [23,30–33], due to eco-innovation practices firms can gain economic profit from cost savings from the manufacturing process, which is connected to product eco-innovation and process eco-innovation. In addition to that, it was argued that environmentally friendly technology, which is linked to technology and the marketing dimensions of eco-innovation, would accelerate the environmental benefits of a firm by introducing new approaches to waste management and eco-marketing [23–28,44,56]. Furthermore, eco-innovation can accelerate the social pillar of sustainability through promoting green human resource management, which is associated with organizational eco-innovation [29,34]. Therefore, these three pillars of sustainability accelerate societal developments through firms' sustainable economic performance, firms' sustainable environmental performance, and firms' sustainable social performance. The following Figure 1 demonstrates this graphically, and is supported by the previous studies of [62–64].

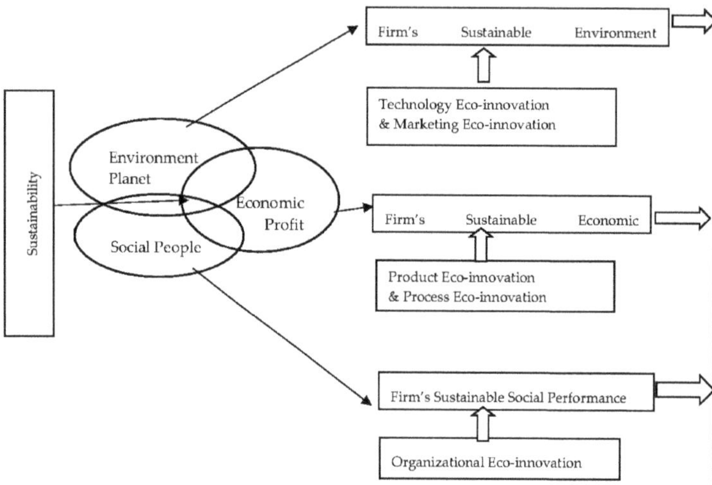

Figure 1. Sustainability and societal development through eco-innovation.

2.3. Societal Development

Social or societal development is the process of improving individual well-being within society. Societal development is linked with the three pillars of sustainability [40]. Societal development means investing in people [40]. It is also assumed that there would be no barriers to reach the dreams of every individual within the society with dignity and confidence [34,36,40]. In addition to that, a few studies [34,36,38] have claimed that societal development is important in the sustainable utilization and allocation of societal resources.

The sustainable utilization and allocation of limited resources within society largely depends on firms' management activities, as firms possess a significant portion of social resources [40]. The eco-innovation and sustainability practices of firms are part of modern firms' management activities. These activities, such as eco-innovation and sustainability practices, can be a great solution to accelerate societal development [34,36,38,40].

According to the studies of [29,34], societal development is linked with organizational eco-innovation. In addition to that, a study from [40] illustrated that worker's health, employee turnover, employee training, employee involvement in local community activity, and working conditions are the top five most favorable factors in the social dimensions of sustainability. These five factors represent more than 50% of the social dimensions of sustainability factors. Surprisingly, these five factors are also connected with organizational eco-innovation. Therefore, it has been established from previous literature that eco-innovation and firms' sustainability practices help to accelerate societal development.

2.4. The Moderating Role of ISO 14001:2015

The Environmental Management System (EMS) or ISO 14001:2015 is a set of rules, regulations, and guidance for businesses to be followed during operational and non-operational activities [65]. It has been adopted by businesses all over the world since the introduction of EMS. Recently, in the year

2015, EMS (1SO-14001) added risk and opportunity, which include the environmental dimensions of a product and the product development cycle [21]. The prime focus of EMS-14001:2015 is to improvise the environmental performance by implementing the sustainable and efficient use of resources and reducing waste. Various studies have revealed similar sets of research results, showing that ISO 14001:2015 enables a business corporation to reduce the environmental impact of their operational and non-operational activities [21,66,67].

The growth in the number of certified organizations around the world is increasing at the rate of 10 percent annually, which is a clear proof of the popularity of gaining the certification among investors and organizations. Several studies have identified that the ISO 14001:2015 standard has a positive impact on various aspects, such as the company's brand [68], which conform with laws and the minimization of pollution [69]. Despite that, there are also reports which have disputed that the standard provides beneficial effects on environmental sustainability [70], arguing that the implementation of ISO 14,001 does not contribute to substantial changes.

However, based on previous studies that have proved the fact that the implementation of the environmental management system affects a firm's eco-innovation, sustainability, and societal development, this research highlights that the moderating effect of ISO 14001:2015 is also directly in a nexus with societal development, as ISO 14001:2015 supports all five capital (environmental, social, human, technology, and finance) needs for the firm's operational and non-operational activities which contribute towards the social development of the nation.

3. Fishbone Chart of Eco-Innovation, Firm's Sustainability, and Societal Development: Conceptual Framework

This research proposes a conceptual framework in the eco-innovation fishbone model in order to enhance a firm's sustainability and the societal development of the economy. The fishbone chart in Figure 2 illustrates the five dimensions of eco-innovation (product, process, technology, organizational, and marketing). These dimensions are discussed as a critical analysis in the literature review inferred from previous studies that each dimension of eco-innovation is positively correlated with the firm's substantiality and societal development.

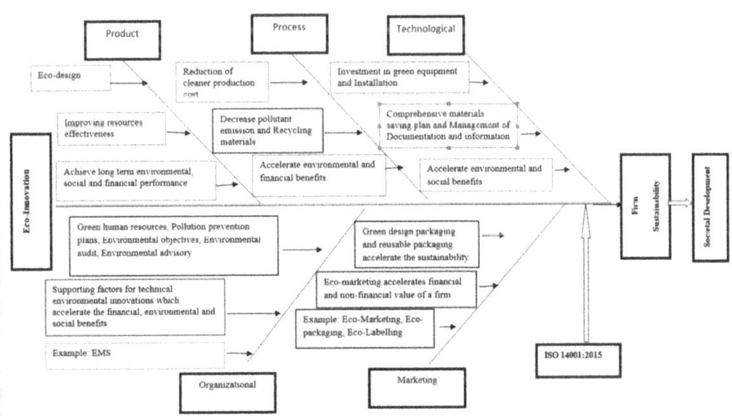

Figure 2. Conceptual framework on eco-innovation, firm sustainability, societal developments, and ISO 14001:2015 (authors' own elaboration).

Firstly, the product eco-innovation and eco-designed products can increase the effectiveness of resources, whereby product eco-innovation can be designed through highlighting the ISO certification, product disassembly/disposal, product lifecycle, and continuous improvement product. This parameter of eco-designed products contributes not only towards the environmental pillar but also towards two other sustainable pillars, which are the social and economic components of the firm.

The second variable of the eco-innovation fishbone model is "process", which advocates a clean process adoption that helps firms to monitor and curb the emission of pollution from the firm's production and service processes. The process of eco-innovation includes green material, process certification, renewable energy consumption, employee environmental training, waste recycle/disposal, and water eco-consumption.

The third variable of the fishbone model is "eco-innovative technology", which promotes the development or adoption of green technology in production and non-production activities to utilize limited resources effectively. Although several previous research findings have suggested that green technology is more expensive than normal technology [23], there are other researchers that have supported and provided evidence on its financial benefits to firms in the long run [44,56].

Fourthly, "organizational eco-innovation" is another element of the fishbone eco-innovation model, and highlights green human resources, pollution prevention plans, environmental audits, and environmental advisory boards in the firm.

Lastly, "marketing eco-innovation" is considered another potential cause of a firm's sustainability. Eco-marketing, eco-labeling, and eco-packaging are some of the examples of marketing eco-innovation. It was claimed that eco-marketing could accelerate the financial and non-financial value of a firm. These causes affect the sustainability of the firm. In addition to that, reusable packaging also helps the environment. As a result, it also improves the sustainability of the firm.

This fishbone research framework model of eco-innovation and the parameters of every element directly contribute towards building a green economy [71]. The eco-innovation fishbone model helps firms to move forward and achieve sustainability, which minimizes material wastage and utilizes resources effectively. The effective and efficient utilization of limited resources without emitting pollution shows the commitment of the individual firm, which then results in enabling conditions and strategies to meet the requirement of people of the nation without creating an extra burden for the future generation. The commitment of eco-innovation also minimizes the direct environmental effect not only on the public but also on other creatures of the world, which leads to a positive effect on social life. Therefore, this research presses the need to improve eco-innovation practices and strong commitments to ISO 14,001 by firms towards sustainability and societal development.

This proposed conceptual framework is backed by the resource-based view theory (RBV). Resource-based view theory is the most applicable theory to enlighten the association between eco-innovation and firms' sustainability [23]. According to the study of [72], RBV is based on four empirical indicators, which are valuable, rare, imperfectly imitable, and (non) substitutability. These four empirical indicators are used by firms to generate a higher rate of performance. The RBV theory also concentrates on managerial capability to identify the firm's internal resources, such as assets, unique capabilities, and core competencies, that can deliver higher performance and sustainable competitive advantages compared to the firm's rivals.

Likewise, eco-innovation is one of the firm's internal resources, capabilities, and competencies, and has four empirical indicators [23,72]. It accelerates higher rates of firms' sustainability performance. Moreover, a firm's sustainability stands on three pillars, named the economic, social, and environment pillars. Likewise, these three pillars of a firm's sustainability accelerate the high rate of societal development. RBV theory enlightens the nexus of the fishbone model of ISO 14001:2015 as an environmental management system that enhances a firm's sustainability and societal developments. The firm acquires ISO 14001:2015 certification by investing its internal resources, capabilities, and competences. This certificate accelerates the rate of a firm's sustainability, which also enhances societal development.

4. Discussion

This research highlights the increasing rate of environmental challenges and effects on society due to current business innovation practices. This study propounds the eco-innovation fishbone model to assist the enterprises in their sustainable practices towards societal development. The eco-innovation fishbone model also highlights the imperative role of ISO 14001:2015 in firm operation, leadership, planning, and improvement in business eco-innovation activities. This research advocates that the eco-innovation fishbone model be adopted in firms' operational and non-operational activities, enhancing firms' sustainability practices. Furthermore, the adoption of fishbone eco-innovation in business activities fulfills the firms' responsibility towards societal development. The following five paragraphs enlighten more on the results gained from each dimension of eco-innovation that are expected from the proposed conceptual model.

The eco-innovation of products highlights eco-design concepts which contribute to a firm's sustainability [43]. According to the study of [73], eco-design is a renovation approach towards product which considers the environmental impacts of its entire life cycle. Eco-design not only lowers environmental impact but also focuses on resource effectiveness [74]. Other major elements that need to be considered while developing or innovating the eco-product are reducing material intensity, renewable energy intensity, product climate change, product recyclability, and durability. These essential elements help a firm's product to be eco-labeled, which leads to achieving a positive effect on the firm's triple bottom line. Further to that, it also helps the firm in the effective utilization of limited resources, which boots the firm's contribution towards societal development.

Another element of the eco-innovation fishbone model is that the process of eco-innovation illustrates cleaner production during the manufacturing process without compromising sustainability and societal development [39,75]. Cleaner production is a strategy that combines various production factors, such as curbing greenhouse emission intensity, water intensity, waste intensity, and energy consumption. On top of that, it also promotes the recycling as well as the reusing of materials [39,75]. Other important elements in cleaner production are rainwater consumption and renewable energy increment. Self-generated renewable energy in the production process can minimize the production cost, which can affect the firm's triple bottom line. Apart from that, sustainability also contributes towards the 17 SDGs by modifying their production processes, which will lead to the firm's contribution towards societal development.

The third element of the fishbone eco-innovation model is eco-innovation technology, which refers to green technology innovation and adoption in the firm's operational and non-operational activities. The fishbone eco-innovation model uses eco-innovation technology due to the capability of green technology, which curbs material consumption, energy consumption, water consumption, waste, and production time, and provides a safer working environment for employees. The adoption or invention of green technology is also promoted by a European policy named the Environmental Technology Action Plan (ETAP) [76], with the objective of utilizing full green technology in industries to protect the environment as well as accelerating economic growth [76]. This technological eco-innovation adoption enhances the production process and contributes directly to the firm's triple bottom line. An example of this is Green Electric Cars, which is an alternative solution for minimizing energy usage [76]. Therefore, it can be argued that fishbone technology eco-innovation is able to enhance a firm's sustainability and societal development if the firm adopts green technology in their production process.

The non-operational element of fishbone-eco-innovation is organization eco-innovation, which highlights different environmental factors such as green human resources, pollution prevention plans, environmental objectives, climate change risks and opportunities, as well as environmental advisories. There are other operational supporting activities, such as green human resource management [40], which includes employee health, employee turnover, environmental employee training [40], employee community engagement, and incident-free working conditions [77,78]. These determinants are also enhanced by being certified with ISO 14001:2015, which is widely accepted for the environmental management system. The certification of ISO 14001:2015 and the implementation of non-operational elements enhance the firm's sustainability and fulfill the firm's responsibility for societal development [18–21].

The last element of the fishbone eco-innovation model is marketing eco-innovation, which embarks on green packaging, reusable packaging, green design marketing, eco-marketing, and eco-labeling, which highlights the eco-innovativeness of the firm to the company's stakeholder. The marketing eco-innovation raises awareness; attracts stakeholders; and creates green goodwill, which attracts future investment and revenue for the firm. This way, marketing eco-innovation manages balance the three bottom lines, namely the economic factor, without compromising on the environment and social well-being, which attract and retain sustainable consumers. As such, fishbone marketing eco-innovation fulfills the triple bottom line that will guide towards social and sustainable development.

There are several positive and negative findings of eco-innovation on a firm's performance (financial and non-financial). This research highlights the important role of ISO-14001:2015 in enhancing the fishbone eco-innovation business through green leadership, green planning, operation, performance evolution, and the improvement of business activities. The serious implementation of eco-innovation and ISO 14001:2015 will maximize resource efficiency and curb the waste generation, which will affect the firm's triple bottom line. The implementation of eco-innovation can also be enhanced by employing green mindset employees in business operational activities that promote green culture and resource efficiency. The effective utilization of resources and maximization of the output from business activities directly contribute to the United Nations' sustainable development goals. Therefore, it can be claimed that this model has practical applicability in the industry.

The practical applicability of the proposed conceptual fishbone eco-innovation model is applicable in different industries, such as oil and gas, consumer manufacturing goods, the chemical industry, and other industries. In the oil and gas industry, common challenges are oil spills, whereby to control the oil spill chemical engineers have proposed different traditional physical and biological techniques with limitations in expelling the oil totally. The emerging sorption method to expel the oil spill in the oil and gas industry is an exemplary technique of eco-innovation by employing biomass (leaves and husk) for cleaning the oil spill. These eco-innovation methods protect the firm from having environmental fines imposed that affect the firm's bottom line and sustainability [78].

The fishbone eco-innovation model is widely applicable to the consumer manufacturing goods industry, which involves product eco-innovation and process eco-innovation, which are directly allied with the utilization of green materials, waste recycling, and sustainable energy (air, wave, solar, biomass) to reduce the CO_2 emission intensity, which reduces the material consumption intensity (per-product), energy intensity, and water intensity without compromising the product quality [50].

These factors positively minimize the production cost and increase the revenue of the firm. The other non-operational element of the fishbone eco-innovation model (organizational and marketing eco-innovation) is to minimize further supporting production cost by fulfilling the environmental norms of the country as well as raising the awareness of responsible production and sustainable consumption amongst stakeholders. Therefore, the adoption of the fishbone eco-innovation model in the consumer manufacturing industry not only generates revenue but also saves the limited resources for future generation needs, leading to firm sustainability and societal development.

The fishbone eco-innovation model is also practically applicable to the chemical industry; eco solvents (ionic liquids) possess excellent eco-friendly properties, including that they are less toxic, easily renewable, can be used for regeneration, and are recyclable. There are wide applications in the production and purification processes in chemical industries, such as pharmaceuticals, polymers, food processing, and others. Another important application of eco-solvents is in the separation and recovery process from industrial waste effluents and curbing harmful emissions, such as CO_2, SO_2, as well as H_2S [78]. These eco-solvents help in achieving a low carbon society and social development by minimizing the environmental effects on human and aquatic life, which will be achieved with a well-structured firm production, process quality, and revenue generation for firm sustainability.

Lastly, this research intends to extend eco-innovation literature from the lens of fishbone eco-innovation from firm sustainability towards contributing to societal development. In addition to that, policymakers can formulate and amend their environmental regulation, as this study establishes the linkage between eco-innovation business capability and the societal development of the nation.

5. Future Research Direction

The future research of the proposed fishbone eco-innovation model can be validated by primary data with large sample size and a meta-analysis on the literature review, which will accelerate sustainability and promote business activities towards the United Nations 17 SDGs for societal development. The fishbone eco-innovation model can also be tested with different manufacturing processes, different service sectors, and different nations. The proposed model can also be used to conduct a comparative study of developed nations, developing nations, and emerging developing nations. The fishbone eco-innovation can also be tested on the role of top-level female employees in achieving firm sustainability and societal development using qualitative and quantitative approaches.

6. Conclusions

Eco-innovation has gained considerable attention among academics as well as corporate sectors due to its potential in mitigating environmental issues in addition to having a positive correlation with firm performance. However, there are limited studies that have monitored the moderating relation of ISO 14001:2015 between eco-innovation and firm sustainability, contributing to societal development. This research is supported by resource-based theory, which explores the core-competencies of enterprises and challenges resources in creating the competitive advantage of the firm without compromising the social responsibility of the firm. This study proposed a fishbone eco-innovation business model that includes production and non-production business activities towards the 17 SDGs for societal development.

This fishbone eco-innovation business model is expected to signal to the stakeholders about the organization's innovative ideas that go beyond mere compliance. The contribution of the fishbone eco-innovation business model towards societal development is to create the unique competitive edge of green intention amongst external stakeholders, which will attract investors for investment. Sustainably responsible investors will be inspired to invest by considering the practical applicability of this proposed model, such as energy efficiency, recycling materials, reduction in energy, reduction in carbon emission, green human resource management, zero waste from packaging, and reusable packaging. In addition to that, this proposed model will enhance more sustainable competitive advantages over the firm's rivals, which will also create an impact on the bottom-line performance of the firm. This article draws propositions and develops a conceptual model for further empirical research on eco-innovation and societal development.

Author Contributions: Conceptualization, M.A.T., P.A.K., and S.K.J.; literature review, M.A.T. and S.K.J.; writing—original draft, M.A.T.; writing—review and editing, M.A.T., P.A.K., and S.K.J. All authors have read and agreed to the published version of the manuscript.

Funding: This study is funded by FRGS (Fundamental Research Grand Scheme), Ministry of Higher Education, Malaysia, with the project title "To examine eco-innovation index for CO_2 emission reduction in the Malaysian energy sectors".

Acknowledgments: The authors want to give thanks to Ministry of Higher Education, Malaysia, for their FRGS grant in the project title "To examine eco-innovation index for CO_2 emission reduction in the Malaysian energy sectors". The authors want to also give thanks to the Department of Management and Humanities, Universiti Teknologi Petronas, Malaysia, for cooperating with all kinds of facilities throughout the research.

Conflicts of Interest: The authors declare no conflict of interest.

References

1. World Economic Forum (WEF). *The Global Risk Report 2018*; World Economic Forum: Geneva, Switzerland, 2018.
2. Suresh, A.; Chauhan, D.; Othmani, A.; Bhadauria, N.; Aswin, S.; Jose, J.; Mejjad, N. *Diagnostic Comparison of Changes in Air Quality over China before and during the COVID-19 Pandemic*; Research Square: Durham, NC, USA, 2020.
3. Baldasano, J. COVID-19 lockdown effects on air quality by NO_2 in the cities of Barcelona and Madrid (Spain). *Sci. Total Environ.* **2020**, *741*, 140353. [CrossRef]
4. Sharma, S.; Zhang, M.; Gao, J.; Zhang, H.; Kota, S.H. Effect of restricted emissions during COVID-19 on air quality in India. *Sci. Total Environ.* **2020**, *728*, 138878. [CrossRef] [PubMed]
5. Almond, D.; Du, X.; Zhang, S. *Did COVID-19 Improve Air Quality Near Hubei?* National Bureau of Economic Research: Cambridge, MA, USA, 2020.
6. Contini, D.; Costabile, F. *Does Air Pollution Influence COVID-19 Outbreaks?* Multidisciplinary Digital Publishing Institute: Basel, Switzerland, 2020.
7. Hakovirta, M.; Denuwara, N. *How COVID-19 Redefines the Concept of Sustainability*; Multidisciplinary Digital Publishing Institute: Basel, Switzerland, 2020.
8. Pirouz, B.; Shaffiee Haghshenas, S.; Shaffiee Haghshenas, S.; Piro, P. Investigating a serious challenge in the sustainable development process: Analysis of confirmed cases of COVID-19 (new type of coronavirus) through a binary classification using artificial intelligence and regression analysis. *Sustainability* **2020**, *12*, 2427. [CrossRef]
9. Secretary-General, U. *Special Edition: Progress Towards the Sustainable Development Goals*; Publication E/2019/68; United Nations Economic and Social Council: New York, NY, USA, 2019.
10. Forum, W.E. *Global Risk Report*; World Economic Forum: Geneva, Switzerland, 2016.
11. Fernando, Y.; Jabbour, C.J.C.; Wah, W.-X. Pursuing green growth in technology firms through the connections between environmental innovation and sustainable business performance: Does service capability matter? *Resour. Conserv. Recycl.* **2019**, *141*, 8–20. [CrossRef]
12. de Azevedo Rezende, L.; Bansi, A.C.; Alves, M.F.R.; Galina, S.V.R. Take your time: Examining when green innovation affects financial performance in multinationals. *J. Clean. Prod.* **2019**, *233*, 993–1003. [CrossRef]
13. Santos, D.F.L.; Lima, M.M.D.; Basso, L.F.C.; Kimura, H.; Sobreiro, V.A. Eco-innovation and financial performance at companies established in Brazil. *Int. J. Bus. Emerg. Mark.* **2017**, *9*, 68–89. [CrossRef]
14. Przychodzen, J.; Przychodzen, W. Corporate sustainability and shareholder wealth. *J. Environ. Plan. Manag.* **2013**, *56*, 474–493. [CrossRef]
15. Carrillo-Hermosilla, J.; Del Río, P.; Könnölä, T. Diversity of eco-innovations: Reflections from selected case studies. *J. Clean. Prod.* **2010**, *18*, 1073–1083. [CrossRef]
16. Leitão, J.; de Brito, S.; Cubico, S. Eco-Innovation Influencers: Unveiling the Role of Lean Management Principles Adoption. *Sustainability* **2019**, *11*, 2225. [CrossRef]
17. Yurdakul, M.; Kazan, H. Effects of Eco-Innovation on Economic and Environmental Performance: Evidence from Turkey's Manufacturing Companies. *Sustainability* **2020**, *12*, 3167. [CrossRef]
18. Kuhre, W.L. *ISO 14001 Certification: Environ. Manag. System*; Prentice Hall: Englewood Cliffs, NJ, USA, 2018.
19. Ikram, M.; Mahmoudi, A.; Shah, S.Z.A.; Mohsin, M. Forecasting number of ISO 14001 certifications of selected countries: Application of even GM (1, 1), DGM, and NDGM models. *Environ. Sci. Pollut. Res.* **2019**, *26*, 12505–12521. [CrossRef]
20. Iatridis, K.; Kesidou, E. What drives substantive versus symbolic implementation of ISO 14001 in a time of economic crisis? Insights from Greek manufacturing companies. *J. Bus. Ethics* **2018**, *148*, 859–877. [CrossRef]
21. Khan, P.A.; Johl, S.K. Nexus of comprehensive green innovation, environmental management system-14001-2015 and firm performance. *Cogent Bus. Manag.* **2019**, *6*, 1691833. [CrossRef]
22. Varadarajan, R. Innovating for sustainability: A framework for sustainable innovations and a model of sustainable innovations orientation. *J. Acad. Mark. Sci.* **2017**, *45*, 14–36. [CrossRef]
23. Ryszko, A. Proactive environmental strategy, technological eco-innovation and firm performance—Case of poland. *Sustainability* **2016**, *8*, 156. [CrossRef]
24. García-Granero, E.M.; Piedra-Muñoz, L.; Galdeano-Gómez, E. Eco-innovation measurement: A review of firm performance indicators. *J. Clean. Prod.* **2018**, *191*, 304–317. [CrossRef]

25. Lindh, H.; Williams, H.; Olsson, A.; Wikström, F. Elucidating the indirect contributions of packaging to sustainable development: A terminology of packaging functions and features. *Packag. Technol. Sci.* **2016**, *29*, 225–246. [CrossRef]
26. Wikström, F.; Williams, H.; Venkatesh, G. The influence of packaging attributes on recycling and food waste behaviour–an environmental comparison of two packaging alternatives. *J. Clean. Prod.* **2016**, *137*, 895–902. [CrossRef]
27. Chiarvesio, M.; Marchi, V.D.; Maria, E.D. Environmental innovations and internationalization: Theory and practices. *Bus. Strategy Environ.* **2015**, *24*, 790–801. [CrossRef]
28. Li, X.; Hamblin, D. Factors impacting on cleaner production: Case studies of Chinese pharmaceutical manufacturers in Tianjin, China. *J. Clean. Prod.* **2016**, *131*, 121–132. [CrossRef]
29. de Jesus Pacheco, D.A.; Carla, S.; Jung, C.F.; Ribeiro, J.L.D.; Navas, H.V.G.; Cruz-Machado, V.A. Eco-innovation determinants in manufacturing SMEs: Systematic review and research directions. *J. Clean. Prod.* **2017**, *142*, 2277–2287. [CrossRef]
30. Yao, Q.; Liu, J.; Sheng, S.; Fang, H. Does eco-innovation lift firm value? The contingent role of institutions in emerging markets. *J. Bus. Ind. Mark.* **2019**, *34*, 1763–1778. [CrossRef]
31. Marcon, A.; de Medeiros, J.F.; Ribeiro, J.L.D. Innovation and environmentally sustainable economy: Identifying the best practices developed by multinationals in Brazil. *J. Clean. Prod.* **2017**, *160*, 83–97. [CrossRef]
32. Aziz, N.; Wahab, D.A.; Ramli, R.; Azhari, C.H. Modelling and optimisation of upgradability in the design of multiple life cycle products: A critical review. *J. Clean. Prod.* **2016**, *112*, 282–290. [CrossRef]
33. Rodriguez, J.A.; Wiengarten, F. The role of process innovativeness in the development of environmental innovativeness capability. *J. Clean. Prod.* **2017**, *142*, 2423–2434. [CrossRef]
34. Kuo, T.-C.; Smith, S. A systematic review of technologies involving eco-innovation for enterprises moving towards sustainability. *J. Clean. Prod.* **2018**, *192*, 207–220. [CrossRef]
35. Song, M.; Cen, L.; Zheng, Z.; Fisher, R.; Liang, X.; Wang, Y.; Huisingh, D. How would big data support societal development and environmental sustainability? Insights and practices. *J. Clean. Prod.* **2017**, *142*, 489–500. [CrossRef]
36. Brogi, S.; Menichini, T. Do the ISO 14001 Environmental Management Systems Influence Eco-innovation Performance? Evidences from the EU Context. *Eur. J. Sustain. Dev.* **2019**, *8*, 292. [CrossRef]
37. Vattamparambil Nalan, V.; Satheesan, S. Life Cycle Perspective in Relation to ISO 14001:2015 and Associated Impacts of Design For Environment (DFE) and Sustainable Development. Master's Thesis, Halmstad University, Halmstad, Sweden, 2020.
38. Lan, Y.-C.; Lee, S.C. Development of an Environmental Management System Framework for Hong Kong Higher Education Institutions. In *Technologies and Eco-Innovation towards Sustainability II*; Springer: Singapore, 2019, pp. 25–38.
39. Abad Segura, E.; Morales, M.E.; Cortés García, F.J.; Belmonte Ureña, L.J. Industrial Processes Management for a Sustainable Society: Global Research Analysis. *Processes* **2020**, *8*, 631. [CrossRef]
40. Oželienė, D. Model of Company's Social Sustainability. *Soc. Tyrim.* **2018**, *41*, 89–100. [CrossRef]
41. Munodawafa, R.T.; Johl, S.K. A systematic review of eco-innovation and performance from the resource-based and stakeholder perspectives. *Sustainability* **2019**, *11*, 6067. [CrossRef]
42. Bossink, B. *Eco-Innovation and Sustainability Management*; Routledge: London, UK, 2013.
43. Hu, A.H.; Matsumoto, M.; Kuo, T.C.; Smith, S. *Technologies and Eco-Innovation towards Sustainability II*; Springer: Singapore, 2019.
44. Liao, Y.C.; Tsai, K.H. Innovation intensity, creativity enhancement, and eco-innovation strategy: T he roles of customer demand and environmental regulation. *Bus. Strategy Environ.* **2019**, *28*, 316–326. [CrossRef]
45. Sezen, B.; Cankaya, S.Y. Effects of green manufacturing and eco-innovation on sustainability performance. *Procedia-Soc. Behav. Sci.* **2013**, *99*, 154–163. [CrossRef]
46. Oecd, E. *Oslo Manual: Guidelines for Collecting and Interpreting Innovation Data*; European Comission: Paris, France, 2005; Volume 46.
47. Miedzinski, M.; Doranova, A.; Castel, J.; Roman, L.; Charter, M. *A Guide to Eco-Innovation for SMEs and Business Coaches*; European Comission: Brussels, Belgium, 2013.
48. Dong, Y.; Wang, X.; Jin, J.; Qiao, Y.; Shi, L. Effects of eco-innovation typology on its performance: Empirical evidence from Chinese enterprises. *J. Eng. Technol. Manag.* **2014**, *34*, 78–98. [CrossRef]

49. Rennings, K.; Ziegler, A.; Ankele, K.; Hoffmann, E. The influence of different characteristics of the EU environmental management and auditing scheme on technical environmental innovations and economic performance. *Ecol. Econ.* **2006**, *57*, 45–59. [CrossRef]
50. Salim, N.; Ab Rahman, M.N.; Wahab, D.A. A systematic literature review of internal capabilities for enhancing eco-innovation performance of manufacturing firms. *J. Clean. Prod.* **2019**, *209*, 1445–1460. [CrossRef]
51. Li, S.; Mirlekar, G.; Ruiz-Mercado, G.J.; Lima, F.V. Development of Chemical Process Design and Control for Sustainability. *Processes* **2016**, *4*, 23. [CrossRef]
52. Ianni, F.; Segoloni, E.; Blasi, F.; Di Maria, F. Low-Molecular-weight phenols recovery by eco-Friendly extraction from Quercus spp. wastes: An analytical and biomass-sustainability evaluation. *Processes* **2020**, *8*, 387. [CrossRef]
53. Xu, C.; Zhang, W.; Li, P.; Zhao, S.; Du, Y.; Jin, H.; Zhang, Y.; Wang, Z.; Zhang, J. High-performance aluminum-ion batteries based on AlCl 3/caprolactam electrolytes. *Sustain. Energy Fuels* **2020**, *4*, 121–127. [CrossRef]
54. Kantola, J.; Liu, Y.; Peura, P.; de Leeuw, T.; Zhang, Y.; Naaranoja, M.; Segev, A.; Huisingh, D. Innovative products and services for sustainable societal development: Current reality, future potential and challenges. *J. Clean. Prod.* **2017**, *162*, S1–S10. [CrossRef]
55. Tseng, M.-L.; Bui, T.-D. Identifying eco-innovation in industrial symbiosis under linguistic preferences: A novel hierarchical approach. *J. Clean. Prod.* **2017**, *140*, 1376–1389. [CrossRef]
56. Hojnik, J.; Ruzzier, M. The driving forces of process eco-innovation and its impact on performance: Insights from Slovenia. *J. Clean. Prod.* **2016**, *133*, 812–825. [CrossRef]
57. Peng, X.; Liu, Y. Behind eco-innovation: Managerial environmental awareness and external resource acquisition. *J. Clean. Prod.* **2016**, *139*, 347–360. [CrossRef]
58. Rajala, R.; Westerlund, M.; Lampikoski, T. Environmental sustainability in industrial manufacturing: Re-examining the greening of Interface's business model. *J. Clean. Prod.* **2016**, *115*, 52–61. [CrossRef]
59. Zhuo, Z.; Siyal, Z.A.; Shaikh, G.M.; Shah, S.A.A.; Solangi, Y.A. An Integrated Multi-Criteria Decision Support Framework for the Selection of Suppliers in Small and Medium Enterprises based on Green Innovation Ability. *Processes* **2020**, *8*, 418.
60. Zailani, S.; Jeyaraman, K.; Vengadasan, G.; Premkumar, R. Sustainable supply chain management (SSCM) in Malaysia: A survey. *Int. J. Prod. Econ.* **2012**, *140*, 330–340. [CrossRef]
61. Ghauri, M.R. How to Go Green as a Telecommunication Companyin a Global Market. Master's Thesis, Uppsala University, Uppsala, Sweden, 2013.
62. Parikh, R.; Bachwani, D.; Malek, M. An Analysis of Earn Value Management and Sustainability in Project Management in Construction Industry. *Stud. Indian Place Names* **2020**, *40*, 195–200.
63. Slaper, T.F.; Hall, T.J. The triple bottom line: What is it and how does it work. *Indiana Bus. Rev.* **2011**, *86*, 4–8.
64. Wilson, J.P. The triple bottom line. *Int. J. Retail Distrib. Manag.* **2015**, *43*, 432–447. [CrossRef]
65. Ferrón-Vílchez, V. Does symbolism benefit environmental and business performance in the adoption of ISO 14001? *J. Environ. Manag.* **2016**, *183*, 882–894. [CrossRef]
66. Sartor, M.; Orzes, G.; Touboulic, A.; Culot, G.; Nassimbeni, G. ISO 14001 standard: Literature review and theory-based research agenda. *Qual. Manag. J.* **2019**, *26*, 32–64. [CrossRef]
67. Castillo-Rojas, S.M.; Casadesús, M.; Karapetrovic, S.; Coromina, L.; Heras, I.; Martín, I. Is implementing multiple management system standards a hindrance to innovation? *Total Qual. Manag. Bus. Excell.* **2012**, *23*, 1075–1088. [CrossRef]
68. Spielmann, M.; de Haan, P.; Scholz, R.W. Environmental rebound effects of high-speed transport technologies: A case study of climate change rebound effects of a future underground maglev train system. *J. Clean. Prod.* **2008**, *16*, 1388–1398. [CrossRef]
69. Psomas, E.L.; Fotopoulos, C.V.; Kafetzopoulos, D.P. Motives, difficulties and benefits in implementing the ISO 14001 Environmental Management System. *Manag. Environ. Qual. Int. J.* **2011**, *22*, 502–521. [CrossRef]
70. Boiral, O.; Henri, J.-F. Modelling the impact of ISO 14001 on environmental performance: A comparative approach. *J. Environ. Manag.* **2012**, *99*, 84–97. [CrossRef]
71. Urbaniec, M. Towards sustainable development through eco-innovations: Drivers and barriers in Poland. *Econ. Sociol.* **2015**, *8*, 179. [CrossRef] [PubMed]
72. Barney, J. Special theory forum the resource-based model of the firm: Origins, implications, and prospects. *J. Manag.* **1991**, *17*, 97–98. [CrossRef]

73. Ivascu, L. Measuring the Implications of Sustainable Manufacturing in the Context of Industry 4.0. *Processes* **2020**, *8*, 585. [CrossRef]
74. Hemmati, S.; Elnegihi, M.M.; Lee, C.H.; Chong, D.Y.L.; Foo, D.C.; How, B.S.; Yoo, C. Synthesis of Large-Scale Bio-Hydrogen Network Using Waste Gas from Landfill and Anaerobic Digestion: A P-Graph Approach. *Processes* **2020**, *8*, 505. [CrossRef]
75. Bleischwitz, R.; Bahn-Walkowiak, B.; Irrek, W.; Schepelmann, P.; Schmidt-Bleek, F.; Giljum, S.; Lutter, S.; Bohunovski, L.; Hinterberger, F.; Hawkins, E. *Eco-Innovation-Putting the EU on the Path to a Resource and Energy Efficient Economy*; Wuppertal Spezial: Wuppertal, Germany, 2009.
76. Alonso-Fariñas, B.; Oliva, A.; Rodríguez-Galán, M.; Esposito, G.; García-Martín, J.F.; Rodríguez-Gutiérrez, G.; Serrano, A.; Fermoso, F.G. Environmental Assessment of Olive Mill Solid Waste Valorization via Anaerobic Digestion Versus Olive Pomace Oil Extraction. *Processes* **2020**, *8*, 626. [CrossRef]
77. Khan, H.W.; Moniruzzaman, M.; Nasef, M.M.E.; Bustam, M.A. Ionic liquid assisted cellulose aerogels for cleaning an oil spill. *Mater. Today Proc.* **2020**. [CrossRef]
78. Khan, H.W.; Reddy, A.V.B.; Nasef, M.M.E.; Bustam, M.A.; Goto, M.; Moniruzzaman, M. Screening of ionic liquids for the extraction of biologically active compounds using emulsion liquid membrane: COSMO-RS prediction and experiments. *J. Mol. Liq.* **2020**, *309*, 113122. [CrossRef]

processes

MDPI

Article

Heat and Mass Transfer during Lignocellulosic Biomass Torrefaction: Contributions from the Major Components—Cellulose, Hemicellulose, and Lignin

Ken-ichiro Tanoue [1,*], Kentaro Hikasa [1], Yuuki Hamaoka [1], Akihiro Yoshinaga [1], Tatsuo Nishimura [1], Yoshimitsu Uemura [2,3] and Akihiro Hideno [4]

1 Department of Mechanical Engineering, School of Sciences and Engineering for Innovation, Yamaguchi University, Tokiwadai 2-16-1, Ube, Yamaguchi 755-8611, Japan; kentaro.hikasa@shi-g.com (K.H.); y-hamaoka@nipponpapergroup.com (Y.H.); b065vd@yamaguchi-u.ac.jp (A.Y.); tnishimu@yamaguchi-u.ac.jp (T.N.)

2 NPO Kuramae Bioenergy, Minato-ku, Tokyo 108-0023, Japan; yuemura.my@gmail.com

3 Center for Biofuel and Biochemical Research, Universiti Teknologi PETRONAS, Seri Iskandar 32610, Malaysia

4 Paper industry innovation center of Ehime University, 127 Mendori-cho, Shkokuchuo 799-0113, Japan; a-hideno@agr.ehime-u.ac.jp

* Correspondence: tano@yamaguchi-u.ac.jp; Tel.: +81-836-85-9122

Received: 27 June 2020; Accepted: 6 August 2020; Published: 9 August 2020

Abstract: The torrefaction of three representative types of biomass—bamboo, and Douglas fir and its bark—was carried out in a cylindrical-shaped packed bed reactor under nitrogen flow at 573 K of the reactor wall temperature. As the thermal energy for the torrefaction was supplied from the top and the side of the bed, the propagation of the temperature profile of the bed is a crucial factor for discussing and improving the torrefaction reactor performance. Therefore, the temperature and gas flow rate (vector) profiles throughout the bed were calculated by model simulation so as to scrutinize this point. The measured temperature at a certain representative location (z = 30 mm and r = 38 mm) of the bed was well reproduced by the simulation. The volume faction of the bed at temperatures higher than 500 K at 75 min was 0.89, 0.85, and 0.99 for bamboo, and Douglas fir and its bark, respectively. It was found that the effective thermal conductivity is the determining factor for this difference. The heat of the reactions was found to be insignificant.

Keywords: biomass torrefaction; packed bed reactor; biomass major components; reaction enthalpy; numerical simulation

1. Introduction

Biomass is one of the representative renewable energy sources, and is one of the only energy sources that is tangible, as it consists of carbon, hydrogen, oxygen, and some minor atoms. Therefore, one of the near-future applications for biomass is the production of solid fuels. Torrefaction is a promising and simple technology for producing high-quality solid fuels from biomass [1–8]. Heat transfer within biomass is an important factor for determining the performance of torrefaction, as the size of biomass in the torrefaction reactor is larger than that for the other conversion technologies, such as gasification or liquefaction.

Kinetic studies of heat and mass transfer are also being conducted experimentally and numerically regarding how much production can be expected and how much heat is required by torrefaction. A two-step parallel successive reaction model for all major components during pyrolysis, when the temperature was higher than 400 °C, has been reported by Miller et al. [9]. Their model, which was suggested by Di Blasi (1994) [10] for cellulose pyrolysis, was extended to hemicellulose pyrolysis and lignin pyrolysis on the basis of previous experimental data. It has also been reported by Kawamoto that the cellulose of biomass is composed of crystalline (long block) and amorphous (short block) alternately, like a block co-polymer [11]. Decomposition starts at around 200 °C in the amorphous state, and the crystalline hardly decomposes at that temperature. At around 300 °C, amorphous decomposition is transmitted to the crystalline material, and the decomposition progresses rapidly.

The heat of the reactions during pyrolysis or torrefaction have been summarized in relation to the char formation [12–17]. In particular, the heat of the reactions during pyrolysis were measured directly using DSC (Differential Scanning Calorimetry), and it has been suggested that, firstly, an endothermic reaction will occur, and then an exothermic reaction will successively occur [12,14–16]. The heat of the pyrolyses of all of the major components have been reported qualitatively using DTA (differential thermal analysis) [18]. Although direct measurement by DSC for the heat of a reaction during cellulose pyrolysis was reported by Mok et al. [12], there are no data by DSC for the heat of reaction during hemicellulose pyrolysis and lignin pyrolysis. As heat and mass transfer during pyrolysis were reviewed by Di Blasi (2008) [19], it has been studied using a chemical reaction model suitable for the heat and mass transfer experimental results [20–26], and an analysis using the heat of chemical reaction commensurate with temperature change has also been conducted [20–23].

In addition, the effect of inorganic materials in biomass feedstock on torrefaction has been investigated by S. Zhang et al. [27]. They found that when the torrefaction temperature was 270 °C, the product yields for raw rice husk were 55% in char, 23% in liquid, and 21% in gas, while the product yields for rice husk with a reduced potassium concentration of less than 1 wt% of all inorganic compounds were 60% in char, 23% in liquid, and 17% in gas.

There are many varieties of biomass, but few reports have been generalized and analyzed for the thermochemical reaction of the thermal decomposition behavior, and it is not clear about heat and mass transfer in low temperature regions, like torrefaction.

In this study, we aimed to generalize the biomass torrefaction behavior using the concentrations of the major constituents of cellulose, hemicellulose, and lignin. Specifically, as shown in Figure 1, the heat of the reaction for the major components of the biomass during pyrolysis was measured. We also experimentally investigated the biomass torrefaction process on a packed bed of small particles of biomass, because a big chunk of a biomass slab is difficult to handle and has no assurance for spatial uniformity. Furthermore, the heat and mass transfer simulation during torrefaction in the biomass packed bed were performed using the heat of the chemical reaction and the pyrolysis model by Miller et al. [9]. The validity of the numerical simulation model was compared with that of the biomass torrefaction process experiment.

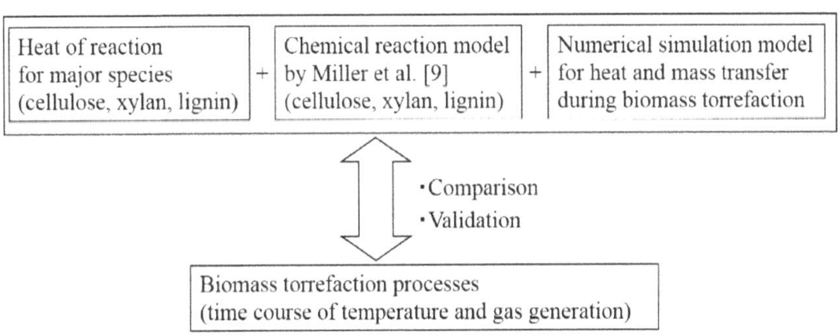

Figure 1. Concept of this work to investigate the biomass torrefaction processes.

2. Materials and Methods

2.1. Sample Preparation

Bamboo powder, Douglas fir powder, and the bark powder of Douglas fir were selected as representative biomass species. They were pulverized and sieved into #3.35 mm, #1.68 mm, #1.00 mm, and #500 μm using a motorized sieve (ANF-30, Nitto Kagaku Co., Ltd., Nagoya, Aichi, Japan). The size of about 400 particles for 500 μm or less was measured using a microscope. The constituent sugars and lignin in the bamboo and Douglas fir samples were analyzed using a modified method based on the technical report NREL/TP-510-42618 [28]. Table 1 shows the mass percentages in the major components of biomass.

Table 1. Mass percentage of major components in the biomass powder.

Biomass Species	Component (wt%)					
	Cellulose	Hemicellulose			Lignin	Others
		Xylan	Arabinan	Mannan		
Bamboo	34.1	25.2	2.3	0.2	24	14.2
Douglas fir	42.5	2.8	1.1	14	22	17.6
Bark of Douglas fir	25.4	3.3	0.9	3.9	51	15.5

Bamboo has three major components of cellulose, xylan in hemicellulose, and lignin, while Douglas fir consists of cellulose, mannan in hemicellulose, and lignin. On the other hand, there are only two major components of cellulose and lignin for the bark of the Douglas fir.

The effect of the potassium concentration in the biomass on torrefaction was also investigated, but no significant difference was found in the product yield [27]. Therefore, in this study, we investigated the effect of the cellulose, hemicellulose, and lignin concentrations on the heat and mass transfer during the biomass torrefaction.

2.2. Thermogravimetric (TG) Analysis and Differential Scanning Calorimetry (DSC)

In order to measure the heat of the reaction during the torrefaction of biomass, thermogravimetry analysis and differential scanning calorimetry analysis of pure cellulose powder (CAS RN 9004-34-6, Sigma Aldrich, Meguro-ku, Tokyo, Japan), pure xylan powder (CAS RN 9004-34-6, FUJIFILM Wako Pure Chemical Corporation, Osaka, Osaka, Japan) and pure dealkaline lignin powder (CAS RN 8068-05-1, Tokyo kasei, Chuo-ku, Tokyo, Japan) were conducted. For DSC (DSC3100s, MAC science, Chuo-ku, Tokyo, Japan) and TG (TG/DTA6300, Hitachi High-Tech Science Corporation, Minato-ku, Tokyo, Japan), the heating rate, nitrogen gas flow rate, and input mass were 10 K/min, 0.5 L/min, and 5 ± 0.25 mg, respectively. The sample sizes of the cellulose, xylan, and lignin powder were 30, 45, and 64 μm, respectively. Figure 2 shows the mass decrease profile from 300 K to 700 K from the TG analysis, as well as the relationship between the heat flow and temperature obtained by the DSC analysis for (a) cellulose, (b) xylan, and (c) lignin. The heat of reaction for these components was obtained by the following procedure.

(1) The time or temperature of the reaction start point $t_{\mathrm{RS},i}$ ($T_{\mathrm{RS},i}$) and reaction end point $t_{\mathrm{RE},i}$ ($T_{\mathrm{RE},i}$) were decided using the trend of the DSC and TG curves.
(2) The base line between $t_{\mathrm{RS},i}$ and $t_{\mathrm{RE},i}$ was linearly drawn in the DSC diagram.
(3) The cross point $t_{\mathrm{shift},i}$ ($T_{\mathrm{shift},i}$) between the DSC curve and the base line was defined as the shifted point from the endothermic chemical reaction to the exothermic chemical reaction.
(4) The endothermic and exothermic heats of reactions were given by Equations (1) and (2).

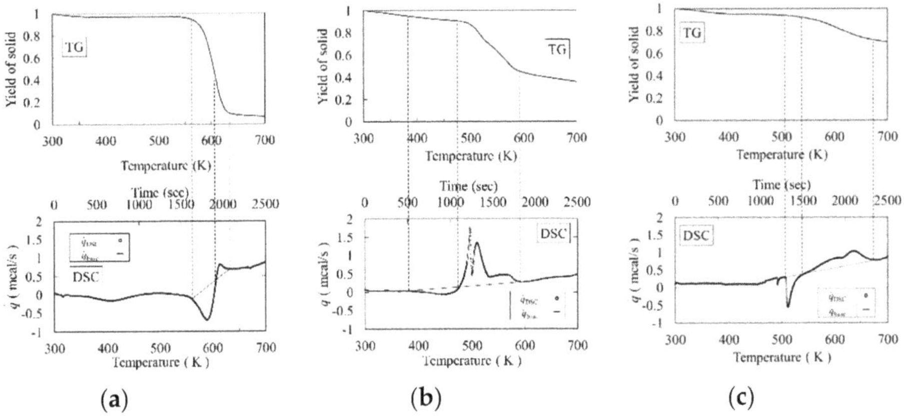

Figure 2. Results of thermogravimetory (TG) and differential scanning calorimetry (DSC) experiments. (**a**) Cellulose powder; (**b**) xylan powder; (**c**) lignin powder.

$$-\Delta H_{\mathrm{R,endo.},i} = \frac{1}{\mathrm{m}_0}\int_{t_{\mathrm{RS},i}}^{t_{\mathrm{shift},i}} \left(q_{\mathrm{base},i} - q_{\mathrm{DSC},i}\right)\mathrm{d}t \tag{1}$$

$$-\Delta H_{\mathrm{R,exo},i} = \frac{1}{\mathrm{m}_0}\int_{t_{\mathrm{shift},i}}^{t_{\mathrm{RE},i}} \left(q_{\mathrm{DSC},i} - q_{\mathrm{base},i}\right)\mathrm{d}t \tag{2}$$

Table 2 shows the obtained heat of the reaction for cellulose pyrolysis, xylan pyrolysis, and lignin pyrolysis. From Table 2, the absolute value of the endothermic heat of the reaction was higher than that of the exothermic heat of reaction during cellulose pyrolysis, while the absolute value of the exothermic heat of reaction was higher than that of the endothermic heat of reaction during xylan pyrolysis and lignin pyrolysis.

Table 2. Results for heats of reaction during pyrolysis of major components of biomass by DSC and TG.

Component i	$T_{\mathrm{RS},i}$ (K)	$T_{\mathrm{RE},i}$ (K)	$T_{\mathrm{shift},i}$ (K)	$-\Delta H_{\mathrm{R,endo.},i}$ (kJ/kg)	$-\Delta H_{\mathrm{R,exo.},i}$ (kJ/kg)
Cellulose	560	638	607	−125.8	22.6
Xylan	371	602	477	−56.4	245.0
Lignin	491	682	549	−62.9	127.1

2.3. Experimental Apparatus and Procedure

The experimental apparatus is shown in Figure 3. The apparatus consisted of anitrogen gas supply, a tubular reactor, furnace, data logger, thermocouples, cold trap for tar and water, and a wet-type gas flow meter. First, 150 g of biomass powder was placed into the reactor, of which the diameter and height were 108 mm and 230 mm, respectively. N_2 gas of 0.5 standard liter per minute (SLM) was fed into the reactor so as to avoid oxidation. In addition, the temperature in the packed bed and temperature at the reactor wall were measured throughout the experiment using twelve thermocouples, and were recorded in the hard disc of a computer through a data logger. During pyrolysis, the generated moisture and tar were condensed thorough the stainless-steel bend pipe with a silicon tube for water cooling by the aspilator, and were trapped by the egg-plant shaped flask. The amount of generated gas during torrefaction was measured by the wet type gas meter and was recorded by the video camera. The temperature at the reactor wall was set at 573 K. Table 3 shows the experimental conditions for the biomass packed bed. Although the particles were pulverized and sieved using four sieves (#3.35mm, #1.68 mm, #1.0 mm, and #0.5 mm), the particle size differed depending on the biomass species. In our previous research [20], we investigated the effect of the particle size on gas generation, and the maximum gas generation error was 2.4% at D_p = 0.74 mm and D_p = 0.34 mm. Therefore, for these biomass samples, this work proceeded on the assumption that the particle size dependence of the gas generation amount could be small. The biomass particles that were used were those that had been dried at 110 °C for 12 h. The bulk density depended on the type of biomass. So, the heights of the packed bed HB were also different. In consideration of the reproducibility of the total generated gas volume during the biomass torrefaction, experiments were conducted twice or more. The experimental coefficient of the validation for the total gas volume during biomass torrefaction was less than 2%.

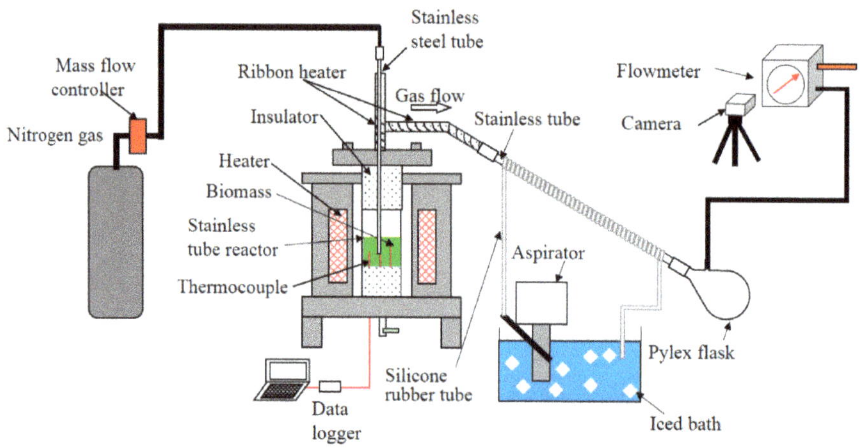

Figure 3. Experimental apparatus for measuring heat and mass transfer during the torrefaction of biomass.

Table 3. Experimental conditions of the biomass packed bed.

Biomass Species	Particle Size (μm)	Bulk Density (kg/m^3)	Height of Biomass Packed Bed, H_B (mm)
Bamboo	196	254	85
Douglas fir	317	311	68
Bark of Douglas fir	222	122	170

2.4. Numerical Simulation

The numerical simulation during pyrolysis was conducted in a two-dimensional cylindrical coordinate. Table 4 shows the governing equations for the numerical simulation. Figure 4 shows the calculation domain and boundary conditions. Although the cellulose of biomass is composed of crystalline (long block) and amorphous (short block) alternately, like a block co-polymer, as has been reported by Kawamoto [11], the co-polymer's effect of cellulose on kinetics during pyrolysis has not been investigated. In this paper, Miller's chemical reaction model and kinetic parameters during pyrolysis [9] were adapted to take into account the effect of the major components of the biomass. The pressure equation was derived from the mass balances, Ideal gas law, and Darcy's equation.

These governing equations and boundary conditions were discretized by the control volume method. In order to stabilize the numerical simulation, a hybrid scheme was adopted for the convection terms in the heat transfer and mass balances. The temperature and pressure were solved by the Euler implicit method. The material balances were solved by the fourth-order Runge–Kutta method.

The dependence of all of the physical properties on temperature was described in our previous report [24]. The change of the porosity [24–26] during torrefaction was adopted. Grid sensitivity on the time course of the temperature and generated gas flow rate were investigated using (N_r, N_z) = (68, 102), (86, 128), (100, 149), and (114, 170), where N_r and N_z show the mesh number along the r component and along the z component, respectively. The maximum relative error of the growth rate distributions between (N_r, N_z) = (100, 149) and (N_r, N_z) = (114, 170) was about 1.6%. Therefore, (N_r, N_z) = (100, 140) was adopted for the bamboo torrefaction. As the heights of the packed beds were also different from the species of the biomass, (N_r, N_z) = (100, 119) and (N_r, N_z) = (100, 342) were adopted for the Douglas fir torrefaction and bark torrefaction, respectively.

Table 4. Chemical reaction model and governing equations of the numerical simulation during the biomass torrefaction.

Chemical Reaction Model [9]	Major component i of biomass $\xrightarrow{k_{1,i}}$ Intermediate material of component i $\xrightarrow{k_{2,i}}$ Tar_i; $\xrightarrow{k_{3,i}}$ β_i $Char_i$ + (1- β_i) Gas_i
Energy balance	$\frac{\partial\{((\rho C)_S+\varepsilon(\rho C)_G)T\}}{\partial t} + \frac{1}{r}\frac{\partial}{\partial r}\left(r(\rho C)_G UT\right) + \frac{\partial}{\partial z}\left((\rho C)_G WT\right)$ $= \frac{1}{r}\frac{\partial}{\partial r}\left(r\lambda_{eff}\frac{\partial T}{\partial r}\right) + \frac{\partial}{\partial z}\left(\lambda_{eff}\frac{\partial T}{\partial z}\right) + \sum_{i=1}^{3}\sum_{j=1}^{3}\left(R_{j,i}\right)(-\Delta H_i)$ where, $(\rho C)_S = \sum_{i=1}^{3}\left(\rho_i C_i + \rho_{IM,i} C_{IM,i} + \rho_{Char,i} C_{Char,i}\right)$ $(\rho C)_G = \rho_{N_2} C_{N_2} + \sum_{i=1}^{3}\left(\rho_{Tar,i} C_{Tar,i} + \rho_{Gas,i} C_{Gas,i}\right)$ $R_{1,i} = k_{1,i}\,\rho_i,\ R_{2,i} = k_{2,i}\,\rho_{IM,i},\ R_{3,i} = k_{3,i}\,\rho_{IM,i}$
Mass balance for component i	$\frac{d\rho_i}{dt} = -k_{1,i}\,\rho_i$
Mass balance for intermediate material of component i	$\frac{d\rho_{IM,i}}{dt} = k_{1,i}\,\rho_i - k_{2,i}\,\rho_{IM,i} - k_{3,i}\,\rho_{IM,i}$
Mass balance for char i	$\frac{d\rho_{Char,i}}{dt} = \beta_i\, k_{3,i}\,\rho_{IM,i}$
Mass balance for tar i	$\frac{\partial(\varepsilon\rho_{Tar,i})}{\partial t} + \frac{1}{r}\frac{\partial}{\partial r}\left(r\rho_{Tar,i}U\right) + \frac{\partial}{\partial z}\left(\rho_{Tar,i}W\right) = k_{2,i}\,\rho_{IM,i} = S_{Tar,i}$
Mass balance for gas i	$\frac{\partial(\varepsilon\rho_{Gas,i})}{\partial t} + \frac{1}{r}\frac{\partial}{\partial r}\left(r\rho_{Gas,i}U\right) + \frac{\partial}{\partial z}\left(\rho_{Gas,i}W\right) = (1-\beta_i)k_{3,i}\,\rho_{IM,i} = S_{Gas,i}$
Pressure equation	$\frac{\partial(\varepsilon\frac{P}{T})}{\partial t} + \frac{1}{r}\frac{\partial}{\partial r}\left(rU\frac{P}{T}\right) + \frac{\partial}{\partial z}\left(W\frac{P}{T}\right) = R_0\left(\frac{1}{M_{Tar}}\sum_{i=1}^{3} S_{Tar,i} + \frac{1}{M_{Gas}}\sum_{i=1}^{3} S_{Gas,i}\right)$
Darcy's law	$U = -\frac{\kappa}{\mu}\frac{\partial P}{\partial r},\ W = -\frac{\kappa}{\mu}\frac{\partial P}{\partial z}$

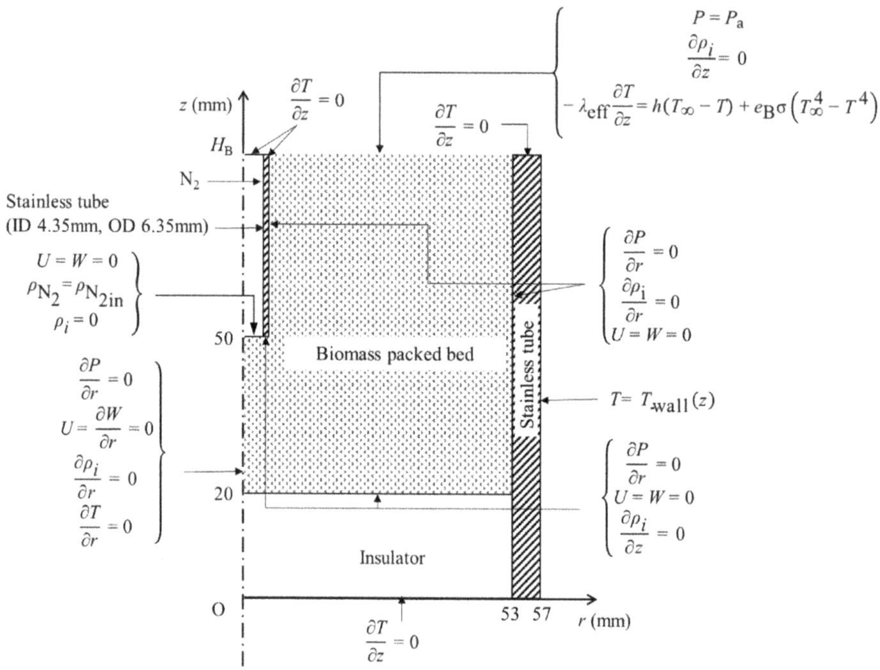

Figure 4. Boundary conditions of the numerical simulation during the biomass torefaction.

3. Results and Discussion

3.1. Time Course of Temperature and Gas Flow Rate during Torrefaction

Figure 5 shows the time course of the temperature at $r = 38$ mm and $z = 30$ mm in the biomass packed bed and the generated gas flow rate during the biomass torrefaction. The keys and lines show the experimental results and calculation results, respectively. In Figure 5a, for the Douglas fir powder, the temperature at $r = 38$ mm and $z = 30$ mm in the bed gradually increased with time. After that, the temperature approached a constant temperature of about 573 K. The calculation temperature agreed well with the experimental one. The gas started generating at about 30 min. At $t > 40$ min, the generated gas flow rate increased adequately and had a maximum at $t = 50$ min. At this time, the temperature was $T = 450$ K, where the gas generated could be from the thermal decomposition of the mannan. After that, the gas flow rate decreased gradually with time. Although the calculation gas flow rate of the black solid line was higher than that of the experimental one, the tendency of the time course agreed quantitatively. Firstly, gas generation could be started by the hemicellulose decomposition of the green line, and had the maximum gas generation. Then, the decomposition of the lignin and the cellulose occurred. For the bamboo powder in Figure 5b, the experimental results for the time course of the temperature also agree well with the calculation results. The gas generation during the bamboo powder torrefaction was higher than that during the Douglas fir torrefaction, and could be started by the hemicellulose decomposition. Then, as the decomposition of the lignin and cellulose occurred, the gas generation had a quasi-state value and decreased with time. For the bark of the Douglas fir powder in Figure 5c, the experimental results for the time course of the temperature also agree well with the calculation results. The maximum gas generation during the bark of the Douglas fir powder torrefaction was higher than that during the Douglas fir torrefaction, and could be started by the lignin decomposition. Then, as the decomposition of the hemicellulose and cellulose occurred successively, the gas generation had a quasi-state value and decreased with time. The total generated gas of the bark of the Douglas fir torrefaction was lower than that of the bamboo powder torrefaction. Therefore, it was found that the time course of the temperature and gas generation during the torrefaction of the biomass depended strongly on the mass percentage of the major components.

From Figure 5, the numerical calculation result of the gas generation flow rate during torrefaction was higher than the experimental one for all of the biomass species. Although the Miller model has been compared with the experimental results of several biomasses in a temperature range of 400 °C or higher [9], there are no data using Miller's reaction model in the low temperature range of 300 °C. It is also necessary to study the heat transfer and gas generation behavior of cellulose, lignin, and hemicellulose, which are the major components of biomass at around 300 °C. Furthermore, it is necessary to reexamine the pyrolysis model with reference to Di Blasi's experiment [29], regarding the reaction rate constants k_2 and k_3 of the second step in the Miller model [9].

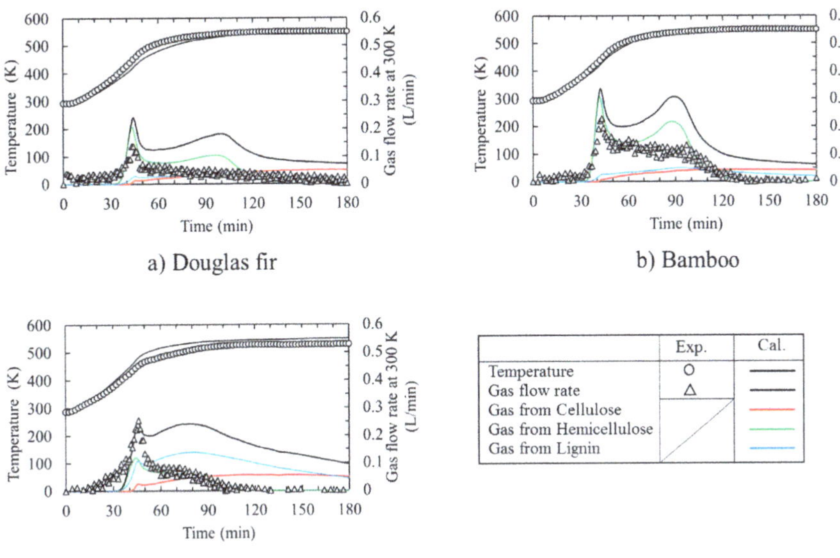

Figure 5. Time course of the temperature and generated gas flow rate during the biomass torrefaction.

3.2. Heat and Mass Transfer during Torrefaction of Biomass

Figure 6 shows the calculation results for the special profile of the temperature, gas flow velocity vector (left), and the solid density (SD, right) at different reaction times. The rectangular blank at the top center in the right figure represents a stainless-steel pipe. The temperature of the bed rose from the top and left, which means the side wall, with the torrefaction time. The zone for temperatures higher than 540 K started prevailing (volume faction = 0.83) at 180 min. Together with the high temperature zone propagation, the local SD of the bed became smaller with a similar profile. Surprisingly, the decrease in SD propagation was not as significant as the temperature propagation in the bed. In other words, the decrease in SD propagation showed a certain time delay of 60 to 90 min in comparison with the temperature profile propagation. This may be related to the sweep gas flow profile, time required for completing the reaction, or the difference between primary decomposition (k_1) and secondary decomposition (k_2 and k_3) in the bed. In order to clarify this point of "what caused this delay?", further investigation is required. At t = 30 min, the heat transfer due to the thermal conduction occurred from the reactor wall, and at the top surface of the backed bed, the temperature at the center and bottom region was less than 400 K. No torrefaction occurred at 30 min. At t = 60 min, as the temperature at the wall and top surface of the packed bed was higher than 500 K, decomposition started from the corner of the top surface. With the passing of time, the temperature near the region of the packed bed was higher than 500 K, the decomposition was propagated. At t = 180 min, the Douglas fir packed bed at the region of the top and near the wall decomposed to about 70% of the initial packed bed, while the Douglas fir packed bed at the other region could not be pyrolyzed.

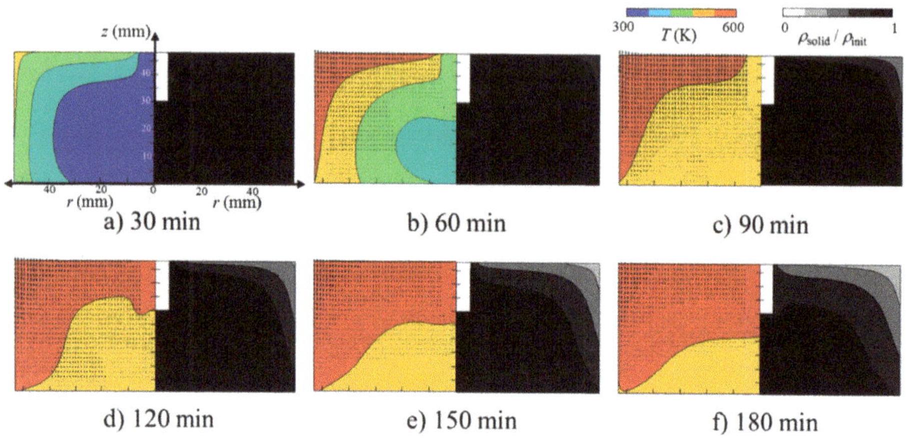

Figure 6. Calculation results for the time course of the temperature, velocity vectors, and total solid distribution during Douglas fir powder torrefaction. ρ_{init} = 353 kg/m^3.

Figure 7 shows the calculation results for the heat and mass transfer at 75 min during various biomass torrefactions. As shown in Figure 6b for the bamboo powder torrefaction, the temperature in the packed bed was higher than that for the Douglas fir torrefaction, due to the bulk density. Furthermore, as the highest mass percentage of the major component in the bamboo was xylan, the decomposition rate in the bamboo packed bed was higher than that in the Douglas fir packed bed. As shown in Figure 6c for the bark of the Douglas fir powder torrefaction, the temperature in the packed bed was also higher than that for the Douglas fir torrefaction, because of the bulk density. However, as the most greatest percentage of the major component in the bark of Douglas fir was lignin, the decomposition rate was higher than that in the Douglas fir packed bed. The volume faction of the bed at temperatures higher than 500 K at 75 min was 0.89, 0.85, and 0.99 for bamboo, and Douglas fir and its bark, respectively. It was found that the effective thermal conductivity was the determining factor for this difference, because the effective thermal conductivity of the bed at temperatures higher than 540 K at 75 min was 0.0254, 0.0252, and 0.0303 W/(m^2 K) for bamboo, and Douglas fir and its bark, respectively. Overall, the bamboo torrefaction was the highest in this study.

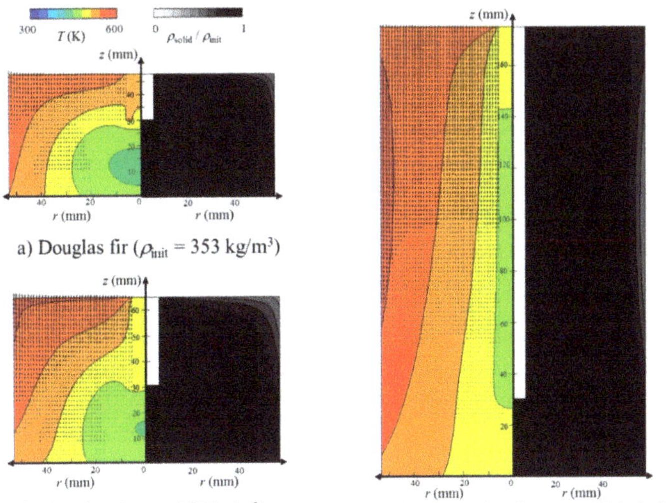

a) Douglas fir (ρ_{init} = 353 kg/m^3)

b) Bamboo (ρ_{init} = 261 kg/m^3)

c) Bark of Douglas fir (ρ_{init} = 97 kg/m^3)

Figure 7. Calculation results for temperature, velocity vectors and solid distributions at 75 min during torrefaction for various kinds of biomass.

4. Conclusions

In this study, the torrefaction of three representative types of biomass, namely, bamboo, and Douglas fir and its bark, was carried out experimentally and numerically, paying attention to the concentrations of the major constituents of cellulose, hemicellulose, and lignin in a cylindrical-shaped packed bed reactor under a nitrogen flow at 573 K of the reactor wall temperature. The following conclusions were obtained.

(1) From DSC experiments, the cellulose pyrolysis was progressed mainly by endothermic reaction (s) while xylan pyrolysis and lignin pyrolysis were proceeded mainly by exothermic reaction (s).

(2) Hemicellulose decomposition could be occurred firstly during Douglas fir torrefaction and bamboo torrefaction. And then lignin and cellulose decomposition would be occurred. So, a quasi-state gas flow rate could be observed. On the other hand, bark of Douglas fir torrefaction depends strongly on the lignin decomposition.

(3) The time course of the temperature in the packed bed agreed well with that of the calculation by taking into account the heat of reaction for not only the Douglas fir, but also for bamboo and the bark of Douglas fir. On the other hand, the numerical calculation result of the gas generation flow rate during torrefaction was higher than the experimental one for all of the biomass species, because there are no data using Miller's reaction model [9] in the low temperature range of 300 °C. It is also necessary to study the heat transfer and gas generation behavior of cellulose, lignin, and hemicellulose, which are the major components of biomass at around 300 °C. Furthermore, it is necessary to reexamine the pyrolysis model with reference to Di Blasi's experiment [29] regarding the reaction rate constants k_2 and k_3 of the second step in the Miller model [9].

(4) The zone at temperatures higher than 540 K prevailed (volume faction = 0.83) at 180 min. Together with the high temperature zone propagation, the local SD of the bed became smaller with a similar profile. Surprisingly, the decreased solid density (SD) propagation was not as significant as the temperature propagation in the bed. In other words, the SD decrease propagation showed a certain time delay of 60 to 90 min in comparison with the temperature profile propagation. This may be related to the sweep gas flow profile, time required for completing the reaction, or the difference between primary decomposition (k_1) and secondary decomposition (k_2 and k_3)

in the bed. In order to clarify this point of "what caused this delay?", further investigation is required. As the temperature at the wall and top surface of the packed bed was higher than 500 K, the decomposition started from the corner of the top surface. With the elapsed time, the temperature near the region of the packed bed was higher than 500 K, and the decomposition was propagated. At t = 180 min, the Douglas fir packed bed at the region of the top and near the wall decomposed to about 70% of the initial packed bed, while the Douglas fir packed bed at the other region could not be pyrolyzed.

(5) For the bamboo powder torrefaction, the temperature in the packed bed was higher than that for the Douglas fir torrefaction because of the bulk density. Furthermore, as the greatest mass percentage of the major component in the bamboo was xylan, the decomposition rate in the bamboo packed bed was higher than that in the Douglas fir packed bed. As the greatest mass percentage of the major component in the bark of Douglas fir was lignin, the decomposition rate was higher than that in the Douglas fir packed bed. The volume faction of the bed at temperatures higher than 500 K at 75 min was 0.89, 0.85, and 0.99 for bamboo, and Douglas fir and its bark, respectively. It was found that the effective thermal conductivity was the determining factor for this difference, because the effective thermal conductivity of the bed at temperatures higher than 540 K at 75 min was 0.0254, 0.0252, and 0.0303 W/(m^2 K) for bamboo, and Douglas fir and its bark, respectively. Overall, the bamboo torrefaction was the highest in this study.

Author Contributions: Conceptualization, K.-i.T. and K.H.; methodology, K.-i.T. and Y.H.; software, K.-i.T.; validation, A.Y. and Y.U.; formal analysis, K.-i.T., K.H., and A.H.; investigation, K.H., Y.H., A.Y., and A.H.; resources, A.H.; data curation, K.H., Y.H., A.Y., and A.H.; writing (original draft preparation), K.-i.T.; writing (review and editing), K.-i.T. and Y.U.; visualization, A.Y.; supervision, T.N. and Y.U.; project administration, K.-i.T.; funding acquisition, T.N. All authors have read and agreed to the published version of the manuscript.

Funding: This work was supported, in part, by a Grant-in Aid for Scientific Research C (no. 17K06196) from the Japan Society for the Promotion of Science.

Acknowledgments: Special thanks to Toshihide Irii and Hamza Bin Rahim for preparing the experimental equipment.

Conflicts of Interest: The authors declare no conflict of interest.

Nomenclature

C	Heat capacity (J/(kg K))
e	Emissivity
H_B	Height of biomass packed bed (m)
h	Heat transfer coefficient between biomass and gas (W/(m^{2}K))
$\Delta H_{R,endo.,i}$	Endothermic heat of reaction during pyrolysis of biomass major component i (J/kg)
$\Delta H_{R,exo.,i}$	Exothermic heat of reaction during pyrolysis of biomass major component i (J/kg)
i	Major component i of biomass (i = cellulose, hemicellulose, and lignin)
$k_{1,i}, k_{2,i}, k_{3,i}$	Reaction rate constant in Miller's chemical reaction model for component i [9] (1/s)
M_{Tar}	Molecular weight of tar (= 0.11 kg/mol [22])
M_{Gas}	Molecular weight of gas (= 0.38 kg/mol [22])
m_0	Input mass of TG and DSC experiments (kg)
N	Maximum grid number
P	Pressure (Pa)
q_{base}	Heat flow of the base line from the endothermic chemical reaction to the exothermic chemical reaction (J/s)
q_{DSC}	Heat flow in DSC curve (J/s)
R_i	Reaction rates in Miller's chemical reaction model for component i [9] (kg/(m^{3}s))
R_0	Universal gas constant (= 8.314 J/(mol K)) (J/(mol K))
r	r coordinate in the packed bed (mm)
S_{Tar}	Reaction rate of tar for component i (kg/(m^{3}s))
S_{Gas}	Reaction rate of gas for component i (kg/(m^{3}s))

T	Temperature (K)
$T_{RS,i}$	Temperature at reaction start time in TG curve for component i (K)
$T_{RE,i}$	Temperature at reaction end time in TG curve for component i (K)
$T_{shift,i}$	Temperature at the shifted time from the endothermic chemical reaction to the exothermic chemical reaction in TG curve for component i (K)
$t_{RS,i}$	Reaction start time in TG curve for component i (s)
$t_{RE,i}$	Reaction end time in TG curve for component i (s)
$t_{shift,i}$	Shifted time from the endothermic chemical reaction to the exothermic chemical reaction in the TG curve for component i (s)
U	Volume averaged Darcy's velocity along the r-axis in the packed bed (m/s)
W	Volume averaged Darcy's velocity along the z-axis in the packed bed (m/s)
z	z coordinate in the packed bed (mm)
Greek symbol	
βi	kinetic parameter in Miller's chemical reaction model for component i [9]
ε	Porosity in the packed bed
κ	Permeability in the packed bed (m^2)
λ	Thermal conductivity (W/(mK))
μ	Viscosity (Pa s)
ρ	Density (kg/m^3)
σ	Stefan–Boltzman constant (= 5.669×10^{-8} W/m^2K^4) (W/m^2K^4)
Subscript	
a	Atmosphere
Char	Char
eff	Effective
Gas	Gas
im	Intermediate material
init	Initial value
N_2	Nitrogen
r	r component
s	Solid
Tar	Tar
v	Volatile
wall	Wall of the stainless-steel tube
z	z component
∞	Environmental condition on the wall of the stainless-steel tube

References

1. Hisam, S.M.; Uemura, Y.; Tazli, A.M. Effects of Temperature and Concentration of Oxygen on Torrefaction of Empty Fruit Bunches. *J. Jpn. Inst. Energy* **2016**, *95*, 1110–1114.
2. Yoshida, T.; Kubojima, Y.; Kamikawa, D.; Kiguchi, M.; Tanaka, K.; Miyago, M.; Masui, M.; Ohyabu, Y.; Kobayashi, A.; Igarashi, H. Production Test of Torrefied Woody Biomass Solid Fuel in an Original Small-scale Plant. *J. Jpn. Inst. Energy* **2018**, *97*, 231–235. [CrossRef]
3. Chen, W.H.; Chen, J.; Peng, J.; Bi, X.T. A state-of-the-art review of biomass torrefaction, densification and applications. *Renew. Sustain. Energy Rev.* **2015**, *44*, 847–866. [CrossRef]
4. Ribeiro, J.M.C.; Godina, R.; Matias, J.C.D.O.; Nunes, L.J.R. Future perspectives of biomass torrefaction: Review of the current state-of-the-art and research development. *Sustainability* **2018**, *10*, 2323. [CrossRef]
5. Rodrigues, A.; Loureiro, L.; Nunes, L.J.R. Torrefaction of woody biomasses from poplar SRC and Portuguese roundwood: Properties of torrefied products. *Biomass Bioenergy* **2018**, *108*, 55–65. [CrossRef]
6. Nunes, L.J.; Matias, J.C. Biomass Torrefaction as a Key Driver for the Sustainable Development and Decarbonization of Energy Production. *Sustainability* **2020**, *12*, 922. [CrossRef]
7. Loureiro, L.M.; Nunes, L.J.; Rodrigues, A.M. Woody biomass torrefaction: Fundamentals and potential for Portugal. *Silva Lusit.* **2017**, *25*, 35–63.

8. Filipe dos Santos Viana, H.; Martins Rodrigues, A.; Godina, R.; Carlos de Oliveira Matias, J.; Jorge Ribeiro Nunes, L. Evaluation of the physical, chemical and thermal properties of Portuguese maritime pine biomass. *Sustainability* **2018**, *10*, 2877. [CrossRef]
9. Miller, R.S.; Bellan, J. A Generalized Biomass Pyrolysis Model Based on Superimposed Cellulose, Hemicellulose and Liqnin Kinetics. *J. Comb. Sci. Tech.* **1997**, *126*, 97–137. [CrossRef]
10. Di Blasi, C. Numerical simulation of cellulose pyrolysis. *Biomass Bioenergy* **1994**, *7*, 87–98. [CrossRef]
11. Kawamoto, H. Reactions and molecular mechanics of cellulose pyrolysis. *Mokuzai Gakkaishi* **2015**, *61-1*, 1–24. [CrossRef]
12. Mok, W.S.L.; Antal, M.J., Jr. Effects of pressure on biomass pyrolysis. II. Heats of reaction of cellulose pyrolysis. *Themochimica Acta* **1983**, *68*, 165–186. [CrossRef]
13. Milosavljevic, I.; Oja, V.; Suuberg, E.M. Thermal Effects in Cellulose Pyrolysis: Relationship to Char Formation Processes. *Ind. Eng. Chem. Res.* **1996**, *35*, 653–662. [CrossRef]
14. Ratha, J.; Wolfingera, M.G.; Steinera, G.; Krammera, G.; Barontinib, F.; Cozzanib, V. Heat of wood pyrolysis. *Fuel* **2003**, *82*, 81–91. [CrossRef]
15. Gomez, C.; Velo, E.; Barontini, F.; Cozzani, V. Influence of Secondary Reactions on the Heat of Pyrolysis of Biomass. *Ind. Eng. Chem. Res.* **2009**, *48*, 10222–10233. [CrossRef]
16. Joungmo, C.; Jeffrey, M.D.; George, W.H. The Intrinsic Kinetics and Heats of Reactions for Cellulose Pyrolysis and Char Formation. *ChemSusChem* **2010**, *3*, 1162–1165.
17. Richard, B.B.; Ahmed, F.G. Biomass torrefaction: Modeling of reaction thermochemistry. *Bioresour. Technol.* **2013**, *134*, 331–340.
18. Yang, H.; Yan, R.; Chen, H.; Lee, D.H.; Zheng, C. Characteristics of hemicellulose, cellulose and lignin pyrolysis. *Fuel* **2007**, *86*, 1781–1788. [CrossRef]
19. Di Blasi, C. Modeling chemical and physical processes of wood and biomass pyrolysis. *Prog. Ener. Comb. Sci.* **2008**, *34*, 47–90. [CrossRef]
20. Pyle, D.L.; Zaror, C.A. Heat transfer and kinetics in the low temperature pyrolysis of solids. *Chem. Eng. Sci.* **1984**, *39-1*, 147–158. [CrossRef]
21. Koufopanos, C.A.; Papayannakos, N. Modelling of the Pyrolysis of Biomass Particles. Studies on Kinetics, Thermal and Heat Transfer Effects. *Can. J. Chem. Eng.* **1991**, *69*, 908–915. [CrossRef]
22. Won, C.P.; Arvind, A.; Howard, R.B. Experimental and theoretical investigation of heat and mass transfer processes during wood pyrolysis. *Combust. Flame* **2010**, *157*, 481–494.
23. Okekunle, P.O.; Pattanotai, T.; Watanabe, H.; Okazaki, K. Numerical and Experimental Investigation of Intra-Particle Heat Transfer and Tar Decomposition during Pyrolysis of Wood Biomass. *J. Therm. Sci. Technol.* **2011**, *6*, 360–375. [CrossRef]
24. Tanoue, K.; Suetomi, T.; Nishimura, T.; Taniguchi, M.; Sasauchi, K. Thermal conduction and gas generation undergoing pyrolysis in the packed bed of woody biomass. *J. Jpn. Inst. Energ.* **2012**, *91*, 976–984. (In Japanese) [CrossRef]
25. Tanoue, K.; Suetomi, T.; Uemura, Y.; Nishimura, T.; Taniguchi, M.; Sasauchi, K. Effect of Tar Decomposition On Gas Generation During Pyrolysis in Packed Bed of Woody Biomass. *Int. J. Biomass Renew.* **2013**, *2*, 1–6.
26. Tanoue, K.; Hamaoka, Y.; Nishimura, T.; Taniguchi, M.; Sasauchi, K. Influence of Volume Shrinkage and Water Evaporation on Heat Transfer and Chemical Reactions During the Pyrolysis of A Cellulose-Powder-Packed Bed. *Int. J. Biomass Renew.* **2018**, *7*, 1–16.
27. Zhang, S.; Su, Y.; Ding, K.; Zhu, S.; Zhang, H.; Liu, X.; Xiong, Y. Effect of inorganic species on torrefaction process and product properties of rice husk. *Bioresour. Technol.* **2018**, *265*, 450–455. [CrossRef]
28. Sluiter, A.; Hames, B.; Scarlata, R.C.; Sluiter, J.; Templeton, D.; Crocker, D. Determination of structural carbohydrates and lignin in biomass. *Lab. Anal. Proced.* **2008**, *1617*, 1–16.
29. Di Blasi, C.; Lanzetta, M. Intrinsic kinetics of isothermal xylan degradation in inert atmosphere. *J. Anal. Appl. Pyrolysis* **1997**, *40*, 287–303. [CrossRef]

processes

MDPI

Article

Modelling of a Single Passage Air PV/T Solar Collector: Experimental and Simulation Design

Noran Nur Wahida Khalili [1], Mahmod Othman [1,*], Mohd Nazari Abu Bakar [2] and Lazim Abdullah [3]

1 Fundamental and Applied Sciences Department, Universiti Teknologi PETRONAS, Seri Iskandar 32610, Malaysia; noran_17005800@utp.edu.my
2 Faculty of Applied Sciences, Universiti Teknologi MARA, Arau 02600, Malaysia; mohdnazari@uitm.edu.my
3 School of Informatics and Applied Mathematics, Universiti Malaysia Terengganu, Kuala Terengganu 21030, Malaysia; lazim_m@umt.edu.my
* Correspondence: mahmod.othman@utp.edu.my

Received: 1 May 2020; Accepted: 17 June 2020; Published: 29 June 2020

Abstract: The hybrid photovoltaic/thermal solar collector has attracted research attention for more than five decades. Its capability to produce thermal energy simultaneously with electrical energy is considered attractive since it provides higher total efficiency than stand-alone photovoltaic or thermal systems separately. This paper describes theoretical and experimental studies of a finned single pass air-type photovoltaic/thermal (PV/T) solar collector. The performance of the system is calculated based on one dimensional (1D) steady-state analysis using one dimensional energy balance equations, where simulation was carried out using MATLAB. Experiments were carried out to observe the performance of the solar collector under changes in air mass flow rate. Experimental values on photovoltaic panel temperature and air temperature on both air inlet and outlet, together with the ambient temperature and solar radiation were measured. The simulation results were validated against the results obtained from experiments using the error analysis method, Root Mean Square Error. At a solar irradiance level of 800 to 900 W/m^2, the thermal efficiency increases to 20.32% while the electrical efficiency increases to 12.01% when the air mass flow rate increases from 0.00015 kg/s to 0.01 kg/s. The error analysis shows that both experimental and simulation results are in good agreement.

Keywords: photovoltaic/thermal; solar collector; single pass; 1D steady-state

1. Introduction

The photovoltaic/thermal PV/T solar collector is a system of heat exchanger that is capable to harness not only electrical energy, but also thermal energy from the solar radiation absorbed with no noise, pollution or moving parts. PV/T is used to describe a solar thermal collector with integrated PV cells that can produce electrical and thermal energy at the same time [1]. Solar energy could be produced more effectively compared to operating individually, as the simultaneous operation of PV and solar thermal collector offers higher solar conversion rates [2]. By combining both systems in the same frames and brackets, PV/T also offers a smaller size collector that requires a smaller area and lower equipment cost than having two collectors with different systems [3]. Therefore, researchers have shown increasing interest in PV/T technology, wherein solar energy is predicted to be deeply exploited and utilized.

Heat and electrical energy collection is the main consideration in the related factors that affect the overall performance of the PV/T system [4]. The process of harnessing solar energy produces thermal energy, which increase the temperature of the solar cells. This is because, in a conventional PV solar collector, only photons with energy greater than or equal to the band-gap energy of the cells will create electron-hole pairs and produce electricity [5]. The rest will be wasted in the form of heat.

Many studies showed that the open circuit voltage of a PV system is reduced if there is increase in temperature, which could produce lower electrical efficiency. Therefore, it has been proposed and proved that the overall PV system efficiency can be improved by the addition of a cooler, which could reduce the temperature of the PV panel [6]. Working fluid, such as water and air, are usually chosen as the medium of heat removal [7]. The role of the fluid is crucial to help reduce the temperature of the PV module and increase its photovoltaic efficiency by removing excess heat [8]. Studies carried out in previous years have investigated the effect of heat removing fluid. Water-typed PV/T collectors have common parts. What makes them different is that while the flat plat collector has the transparent glass on top, it is replaced with a plank PV for the water PV/T collector. Since the demand for clean water, as well as thermal and electrical energy, increases day by day, solar distillation is often chosen to convert contaminated water to clean water [9].

The studies in theoretical and experimental analysis of photovoltaic/thermal (PV/T) solar collectors have been conducted by researchers since the mid-1970s and have evolved overtime [3]. Wolf [10] carried out some of the earliest analysis of the performance of combined solar photovoltaic and heating systems for a single residence, while Florschuetz [11] extended the famous Hottel–Whillier model to analyze the performance of the PV/T flat plate solar collector. The technology of air-based PV/T innovation (PV plus thermal) is widely used and its known general operation efficiencies range from 20–40%. Since the efficiency of crystalline silicon cells range from 10–12%, the thermal part provides the rest of the efficiency [12]. There are several designs of air passage for a solar PV/T that have been used previously. Hegazy [13] performed a further study of the thermal, electrical, hydraulic and overall performances of flat plate photovoltaic/thermal (PV/T) air collectors. Four of the most preferred designs are considered in this study. The models represent where the air is flowing, either over the absorber (Model I) or under it (Model II) and on both sides of the absorber in a single pass (Model III) or in a double pass fashion (Model IV). The effects of air-specific flow rate and the selectivity of the absorber plate and PV cells on the performances have been examined. It has been found that Model III has the highest efficiency in higher radiation, while Model I has the lowest performance. Saygin et al. [14] proposed a modified photovoltaic/thermal (PV/T) solar collector where, in the study, air passes through a slot at the middle of the glass cover in the solar collector. They investigated the PV system with no cooling and compared the PV/T collector with different distances between the panel and glass cover (3 cm, 5 cm and 7 cm). In their research, they found that when the distance between PV module and cover was 3 cm, the highest thermal performance was obtained while a 5 cm gap between them gives out the highest electrical efficiency. Kumar and Rosen [15] investigated a double pass photovoltaic/thermal (PV/T) solar air heater attached with vertical fins in the lower channel. Their findings showed that the solar cell temperature is reduced significantly from 82 °C to 66 °C. It is also stated that one of the crucial parameters in designing PV/T collectors is the packing factor. Dubey et al. [16] analyzed two types of PV module, glass-to-glass and glass-to-tedlar, with and without duct (air pass way) and found that the first PV module type with duct produces higher efficiency than second type of PV module. This occurred because the radiation that falls on the non-packing area of the glass-to-glass PV module is transmitted through the glass cover, while in the other type of module, all radiation is absorbed by the tedlar. The heat transfer occurs through conduction, causing higher temperature of solar cells in glass to the tedlar-type, resulting in lower efficiency.

The researchers also show their interest in investigating the operation of water-type PV/T solar collectors. Kiran and Devadiga [17] studied the performance of a PV/T system with water as a cooling agent. It is found that without cooling, the electrical efficiency of the system is lower as compared to when cooling is added. The total efficiency of PV/T system was found to be higher than the individual stand-alone PV panel and currents solar water collector. Azad [18] has conducted a study comparing the experimental analysis of two heat pipe solar collectors with different numbers of heat pipes and a flow-through collector. Three collectors with six pipes, 12 pipes and one-flow through type have been constructed and tested simultaneously. It is concluded that to increase the efficiency of heat pipe collectors, the number of heat pipes and the effective absorber area need to be increased by the proper design of the heat pipe condenser. However, as water-type PV/T collector is costly and hard to maintain, the air-type PV/T collector is often preferred.

It is known that the PV cell temperature change affects the electrical efficiency of the PV module. Therefore, it is crucial to note the parameters that could affect the PV cells temperature; most importantly, the PV cell encapsulation and material properties. The effects of external factors such as the solar irradiance, ambient temperature and wind speed must also be considered [19]. A study has been completed that proved the accumulation of dust, shading, and bird fouling has a significant effect on PV current and voltage, and consequently, the harvested PV energy [20]. The changes in surrounding conditions or atmospheric phenomena could be considered as the most affecting factors contributing to the performance of a PV/T collector. An analytic expression for solar radiation can be defined on a clear sunny day, however it is not possible during cloudy days. Hence, there has been a study on a statistical tool to detect abnormal operating conditions that affect the performance of the PV panel effectively for PV plants that are not equipped with a weather station [21].

The common method used to simulate a performance of a PV/T solar collector is a one-dimensional (1D) energy balance equation representing the heat transfer occurring in the collector. Sathyamurthy et al. [22] studied a one-dimensional energy balance equation and wrote a review on the enhancement techniques of solar still, including flat plate collectors, pulsating heat pipes, concentrating collectors, evacuated tube collectors, parabolic trough, thermoelectric effect, solar water heater and PV/T still. Zondag et al. [23] used this method to investigate and compare the performance of seven different design types of PV/T collectors to study the electrical and thermal efficiency of collectors of each design which can be grouped into four; sheet-and-tube PV/T collector, channel PV/T collector, free flow PV/T collector and two-absorber PV/T collector. It is concluded that the highest efficiency is produced by channel-below-transparent-PV design. In [5], the researchers improved the PV/T solar collector design which integrated a PV panel with a water heating component; a serpentine-shaped copper tube and air heating component which is a single pass air channel. The simulations using 1D energy balance equations showed that when both fluids are operated alone, the total thermal and electrical performances are only satisfactory, and higher when operated simultaneously. Tiwari and Sodha [24] evaluated the overall performance of hybrid PV/thermal (PV/T) air collectors with different configurations; unglazed and glazed PV/T air heaters, with and without tedlar. It is found that in the unglazed PV/T module, there is no difference in the solar cell temperature of with and without tedlar, except there is a slight increase in outlet air temperature in the model due to absence of tedlar. Sarhaddi et al. [25] presented a detailed thermal and electrical model developed to calculate the thermal and electrical parameters of a typical PV/T air collector which includes the parameters such as solar cell, temperature, back surface temperature, outlet air temperature, open-circuit voltage, short-circuit current, maximum power point, voltage, maximum power point current, etc.

In this study, a 1D mathematical model of energy balance equations was uniquely developed representing the heat transfer in the solar collector system built. The simulation is used to represent the solar collector system to better understand the effect of the parameter changes on the its performance. Since the operation of the PV/T solar collector involves uncertainty of the weather changes, the simulation is extensively used to carry out a comprehensive study. The unique design of the solar collector that is used to develop the model in this study is a single pass air PV/T solar collector attached with fins. The objective of this paper is to present a comprehensive mathematical model with a focus on the simulation description details. The results of the simulation are then verified with actual data collected from the experiment carried out. This paper is structured as follows. The design of the solar collector fabricated specifically for this study is explained in Section 2. The mathematical modelling representing the solar collector system is discussed in Theoretical Analysis in Section 3 and its solution in Section 4; while in Section 5, the experimental results obtained are presented as a validation to the simulation results. The results and discussion are presented in Section 6, and the last section is dedicated for the conclusion of the study.

2. Design of Solar Collector

Based on the literature review, a design of a solar collector consists of a PV module and a single pass air channel below the module as shown in Figure 1. This configuration has been widely used by previous researchers, however, the system built in this study focuses on a compact design of an air PV/T solar collector. The PV panel used is a commercial 50 W monocrystalline silicon CG-M050 panel of dimension 69 cm × 54 cm × 3 cm. The PV panel is a three-layer panel with a tempered glass on top and an absorber plate at the bottom, comprised of solar cells encapsulated by a thin adhesive layer called ethylene-vinyl acetate (EVA). Fins are added parallel to the air flow direction as a heat transfer rate enhancement factor by providing augmented heat transfer area. The gap between the absorber plate and the back plate created an air passage for the heat transfer. The system operated with air is forced in and out of the collector by two 12 V DC fans to extract the excess heat from the PV module to be converted into useful heat in other energy applications. Aluminium is selected to fabricate the backplate and the fins attached on it by taking into account its ability absorb and transfer heat from solar cell, as well as the density of the material and its economic viability [26]. The fabrication of the PV panel into a solar collector was carried out by a collaboration with a local manufacturer. All data from the experiment are collected through a data logger connected to a PC. The temperature, air mass flow rate and solar irradiance are measured using temperature sensors, flowmeters and a pyranometer respectively.

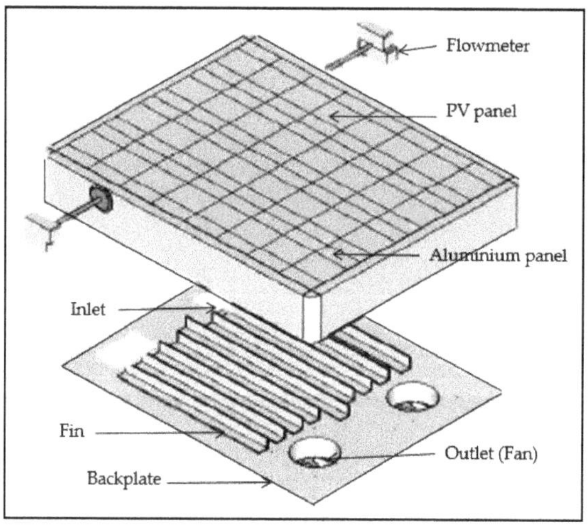

Figure 1. Exploded view of the photovoltaic/thermal (PV/T) solar collector.

Figure 1 shows the exploded view of the designed air PV/T solar collector where the set of fins can be seen attached to the backplate of the collector. The aluminium panel acts as a frame to support the collector system, while Figure 2 shows the cross-sectional view of the collector.

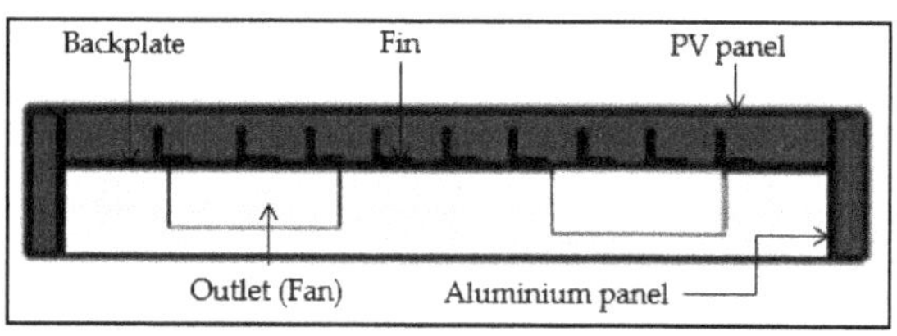

Figure 2. Cross sectional view of the PV/T solar collector.

3. Theoretical Analysis

In this study, a steady-state one dimensional analysis is obtained. The governing equations are obtained which involve steady-state energy balance equations at various points of the solar collector. The thermal schematic model of the collector is shown in Figure 3.

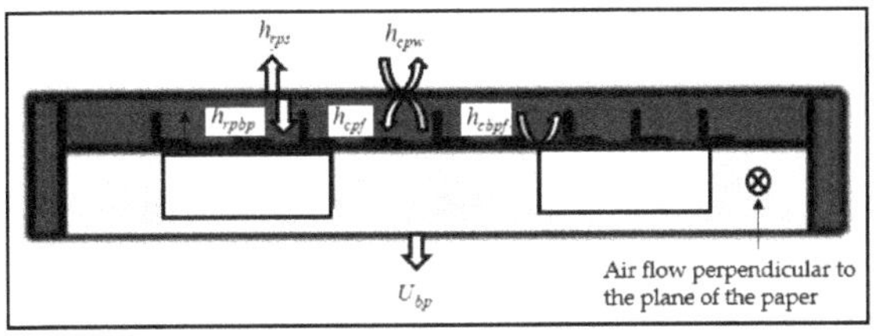

Figure 3. The thermal schematic model of a finned single pass PV/T air solar collector.

To simplify the analysis, the following assumptions were made:

1. Steady state of energy transfer was achieved.
2. Heat capacity effects of the solar cells and back plate were neglected.
3. The temperatures of the glass cover, solar cells and plates vary only in the direction of air flow.
4. The side losses from the system are negligible.

Following the work by [23,27,28], the energy balance equations for each temperature nodes can be written as follows:

1. For the nodes of the solar cells of the PV module;

$$\overbrace{\tau_g\alpha_p(1-PF)G}^{1}+\overbrace{\tau_g\alpha_{pv}(PF)\left(1-\eta_{pv}\right)G}^{2}=\overbrace{h_{cpf}\left(T_p-T_f\right)}^{3}+\overbrace{h_{rpbp}\frac{A_c}{A_{ab}}\left(T_p-T_{bp}\right)}^{4}$$
$$+\overbrace{h_{rps}\left(T_p-T_s\right)}^{5}+\overbrace{h_{cpw}\left(T_p-T_w\right)}^{6} \quad (1)$$

2. For the air temperature nodes;

$$\frac{\dot{m}C_f}{W}\frac{dT_f}{dx}=\overbrace{h_{cpf}\left(T_p-T_f\right)}^{3}+\overbrace{h_{cbpf}\frac{A_{ab}}{A_c}\eta_p\left(T_{bp}-T_f\right)}^{7} \quad (2)$$

where fin effectiveness, $\eta_p = 1-\frac{A_{fin}}{A_{ab}}\left(1-\eta_{fin}\right)$ fin efficiency, $\eta_{fin} = \frac{\tan hmh_{fin}}{mh_{fin}}$ and $m=\left(\frac{2h_{cbpf}}{k_{fin}w_{fin}}\right)^{\frac{1}{2}}$.

3. For the nodes of the surface of back plate with fins;

$$\overbrace{U_{bp}\left(T_{bp}-T_a\right)}^{8}+\overbrace{h_{cbpf}\frac{A_c}{A_{ab}}\eta_p\left(T_{bp}-T_f\right)}^{7}+\overbrace{h_{rpbp}\frac{A_{ab}}{A_c}\left(T_{bp}-T_p\right)}^{4}=0 \quad (3)$$

The heat transfer terms involved in the equations above are defined as follows:

1. The rate of the solar energy absorbed by the absorber plate of the PV module per unit area.
2. The rate of the solar energy received by solar cells of the PV module after transmission and the rate of electrical energy available per unit area.

3. The rate of heat transfer to the air flow per unit area.
4. The rate of energy radiated to the surface of the back plate and fins per unit area.
5. The rate of energy radiated to the sky per unit area.
6. The rate of heat transfer to the wind per unit area.
7. The rate of heat transfer to the air flow per unit area.
8. The rate of heat lost to ambient through back plate per unit area.

Equation (1) can be rearranged into:

$$X_1 = A\left(T_p - T_f\right) + B\left(\frac{1}{\gamma}\right)\left(T_p - T_{bp}\right) + C\left(T_p - T_s\right) + D\left(T_p - T_w\right) \tag{4}$$

The heat transfer coefficients are defined as

$$A = h_{cpf},\ B = h_{rpbp},\ C = h_{rps},\ D = h_{cpw}$$

$$X_1 = \tau_g \alpha_p (1 - PF)G + \tau_g \alpha_{pv}(PF)\left(1 - \eta_{pv}\right)G,$$

$$\gamma = \frac{A_{ab}}{A_c}$$

Equation (2) can be rearranged into:

$$X_2 \frac{dT_f}{dx} = A\left(T_p - T_f\right) + E\gamma\eta_p\left(T_{bp} - T_f\right), \tag{5}$$

where the heat transfer coefficients are defined as

$$E = h_{cbpf},$$

$$X_2 = \frac{\dot{m}C_f}{W}$$

Equation (3) can be rearranged into:

$$U_{bp}\left(T_{bp} - T_a\right) = E\gamma\eta_p\left(T_f - T_{bp}\right) + B\gamma\left(T_{bp} - T_p\right). \tag{6}$$

From Equations (4) and (6), the expression for temperature of absorber plate and backplate can be written as Equations (7) and (8), respectively:

$$T_p = \frac{H_{17} - T_f H_{15}}{H_{16}} \tag{7}$$

$$T_{bp} = \frac{T_p H_{12} - T_f h_{cpf} - H_{11}}{h_{rpbp}} \tag{8}$$

The variables T_p and T_{bp} can be eliminated from Equation (5) by substituting Equations (4) and (6) into it. The following linear first order differential equation with boundary condition is obtained.

$$\frac{dT_f}{dx} = A_0 - A_1 T_f \tag{9}$$

$$T_f\big|_{x=0} = T_{in} = T_a$$

A_0 and A_1 are the constants obtained through algebraic manipulations where

$$A_0 = \frac{H_{17}H_{18} - H_{11}H_{16}h_{cbpf}\gamma^2\eta_p}{\frac{\dot{m}C_f}{W}H_{16}h_{rpbp}}$$

$$A_1 = \frac{H_{15}H_{18} + H_{16}H_{19}}{\frac{\dot{m}C_f}{W}H_{16}h_{rpbp}}$$

$$H_{11} = \tau_g\alpha_p(1 - PF)G + \tau_g\alpha_{pv}(PF)\left(1 - \eta_{pv}\right)G + h_{rps}T_s + h_{cpw}T_w$$

$$H_{12} = h_{cpf} + \frac{h_{rpbp}}{\gamma} + h_{rps} + h_{cpw}$$

$$H_{14} = U_{bp} + h_{cbpf}\gamma\eta_p + \frac{h_{rpbp}}{\gamma}$$

$$H_{15} = h_{cpf}H_{14} + h_{cbpf}h_{rpbp}\eta_p$$

$$H_{16} = \left(\frac{h_{rpbp}}{\gamma}\right)^2 - H_{12}H_{14}$$

$$H_{17} = -H_{11}H_{14} - T_aU_{bp}\frac{h_{rpbp}}{\gamma}$$

$$H_{18} = h_{cpf}h_{rpbp} + h_{cbpf}\gamma^2\eta_pH_{12}$$

$$H_{19} = h_{rpbp}H_{13} + h_{cpf}h_{cbpf}\gamma^2\eta_p$$

Solving the first order linear differential equation in Equation (9) gives the solution as expressed in Equation (10) below:

$$T_f(x) = \frac{A_0}{A_1} + T_ae^{-A_1x} - \frac{A_0}{A_1}e^{-A_1x} \tag{10}$$

The average temperature of the PV is calculated using numerical integration rule as in Equation (11) to obtain Equation (12).

$$\overline{T}_{pv} = \overline{T}_p = \frac{1}{L_c}\int_0^{L_c} T_pdx \tag{11}$$

$$\overline{T}_{pv} = \overline{T}_p = \frac{1}{L_c}\int_0^{L_c} T_pdx = \left[\frac{H_{17}}{H_{16}}\right] - \left[\frac{H_{15}}{H_{16}L_c}\right]\left[-\frac{T_a}{A_1}e^{-A_1l_c} + \frac{A_0}{A_1}L_c + \frac{A_0}{(A_1)^2}e^{-A_1l_c} + \frac{T_a}{A_1} - \frac{A_0}{(A_1)^2}\right] \tag{12}$$

Following are the heat transfer coefficients employed in the above equations. The radiative heat transfer coefficient between the absorber plate of the PV panel and the sky is given by Equation (13) [25].

$$h_{rps} = \sigma\varepsilon_g\left(T_p + T_s\right)\left[T_p^2 + T_s^2\right] \tag{13}$$

where T_s is equivalent sky temperature represented by Swimbank's formula [29] given by Equation (14).

$$T_s = 0.0552\left(T_a^{1.5}\right) \tag{14}$$

The radiative heat transfer coefficient between the absorber plate and the backplate is obtained from Equation (15).

$$h_{rpbp} = \frac{\sigma\varepsilon_g\left(T_p + T_s\right)\left[T_p^2 + T_s^2\right]}{\left(\frac{1-\varepsilon_{pv}}{\varepsilon_{pv}}\right) + \frac{1}{A_{ab}/A_c} + \left(\frac{1-\varepsilon_{bp}}{\varepsilon_{bp}}\right)\frac{A_{ab}}{A_c}} \tag{15}$$

The wind convective heat transfer coefficient is calculated by Equation (16) (in Mac Adam (1954) as cited in [30]).

$$h_w = 5.7(3.8v) \tag{16}$$

The convective heat transfer coefficient between air and absorber plate is evaluated by Equation (17), and between air and backplate by Equation (18).

$$h_{cpf} = \frac{Nuk_f}{D_h} \tag{17}$$

$$h_{cbpf} = \frac{Nuk_f}{D_h} \tag{18}$$

in which D_h is the cross-section area of the collector and k_f is thermal heat conductivity in the air channel given by Equation (19).

$$k_f = \left[1.0057 + 0.000066\left(\frac{T_p + T_{bp}}{2} - 300\right)\right]1000 \tag{19}$$

Nu, the Nusselt number is determined based on the flow of the air (laminar, transition and turbulent) as determined in Equations (20)–(22) (Heaton (1964) as cited in [31]).

For $Re < 2300$ (Laminar flow),

$$Nu_{lam} = Nu_{ref} + \frac{a\left((Re)(Pr)\frac{D_h}{L_c}\right)^m}{1 + b\left((Re)(Pr)\frac{D_h}{L_c}\right)^n} \tag{20}$$

where

$$Nu_{ref} = 5.3,\ a = 0.00190,\ b = 0.00563,\ m = 1.71 \text{ and } n = 1.17$$

For $2300 < Re < 6000$ (Transition flow) (Hausen (1943) as cited in [32]),

$$Nu_{tran} = 0.116\left(Re^{\frac{2}{3}} - 125\right)Pr^{\frac{1}{3}}\left[1 + \left(\frac{D_h}{L_c}\right)^{\frac{2}{3}}\right]\left(\frac{vis_f}{vis_{pv}}\right)^{0.14} \tag{21}$$

For $Re > 6000$ (Turbulent flow) [33],

$$Nu_{turb} = 0.018Re^{0.8}Pr^{0.4} \tag{22}$$

4. Solution to the Equations

The model explained in Section 3 is used to perform simulation by using MATLAB to analyze the performance of the solar collector. The proposed algorithm built for the simulation is shown in the flowchart in Figure 4. The program is initiated with the setting up of the values of parameters (Table 1) and estimated temperatures of the absorber plate, T_p and the temperature of the back plate, T_{bp} which are then used to calculate the temperature of air and thermophysical properties of the fluids. Then, the simulation runs to calculate the values of the temperature of air, T_f to be used to compute the new value of T_p and T_{bp}. The process continues by calculating the difference between the estimated temperature and the computed temperature. If the difference calculated is less than 0.01 °C, the process stops iterating, and the new computed values of the temperature will replace the old ones.

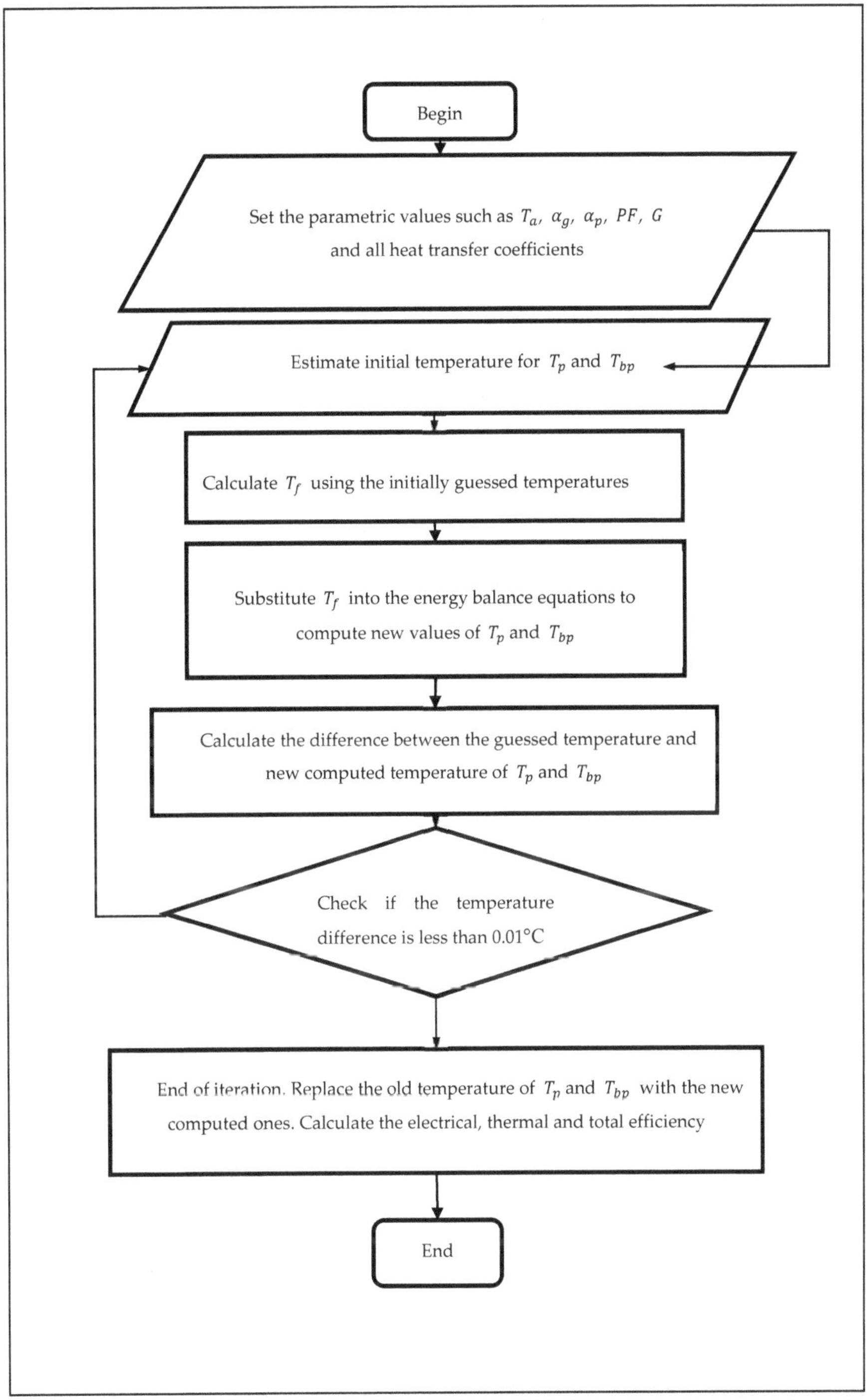

Figure 4. Flowchart for the MATLAB algorithm.

Table 1. Values of coefficients and ambient conditions used in the simulations.

Parameter		Value	Parameter		Value
Width of collector	W_c	0.54	Height of fin	h_{fin}	0.01905
Length of collector	L_c	0.69	Width of fin	w_{fin}	0.001
Depth of collector	D_c	0.02	Distance between fins	d_{fin}	0.05
Absorptivity of PV	α_{pv}	0.938	Number of fins	n_{fin}	9
Absorptivity of absorber plate	α_p	0.951	Ambient temperature (K)	T_a	303
Absorptivity of glass	α_g	0.06	Wind speed	v	2.0
Emissivity of PV	ε_{pv}	0.8	Irradiance	G	800
Emissivity of glass	ε_g	0.94	The solar cell temperature at reference condition	T_{ref}	298
Emissivity of back plate	ε_{bp}	0.96	PV panel power temperature coefficient	β_{ref}	0.005444
Electrical conductivity	α	5.67×10^{-8}	Electrical efficiency at the reference temperature	η_{ref}	0.14

Energy Analysis

The thermal efficiency is calculated by simplifying the equations by [23]:

$$\eta_{th} = \frac{\dot{m}C_f(T_0 - T_i)}{A_c G} \tag{23}$$

The efficiency of crystalline silicon cells decreases with increasing temperature, since the open circuit voltage and fill factor decreases, but the short circuit current slightly increases. The electrical efficiency of the collector written as a function of temperature based on [23,34] is as follows:

$$\eta_{ele} = \eta_{ref}\left(1 - \beta_{ref}\left(\overline{T}_{pv} - T_{ref}\right)\right) \tag{24}$$

Since there is difference in the nature of the electrical and thermal energy, the total overall thermal equivalent efficiency of a PV/T solar collector, as referred to [11] is calculated as:

$$\eta_{total} = \eta_{th} + \eta_{ele} \tag{25}$$

5. Experimental Validation

The PV/T solar collector is set up (see Figure 5) for data collection in the compound of Universiti Teknologi PETRONAS (UTP). The set-up is based on the following conditions. To simplify the analysis, the following assumptions were made:

1. Mounting Location—PV modules can be mounted on ground, where a power source is available, while heat capacity effects of the solar cells and back plate were neglected.
2. Shading—Photovoltaic arrays are adversely affected by shading. A well-designed PV system needs clear and unobstruct access to the sun's rays from 9 a.m. to 3 p.m. Even small shadows, such as the shadow of a single branch of a leafless tree can significantly reduce the power output of a solar module. Keep in mind that an area may be unshaded during one part of the day but shaded at another part of the day.
3. Orientation—PV modules are ideally oriented towards true south.
4. Tilt—The PV modules need to be installed according to the latitude angle of the location where it is placed. For the compound of UTP where the collector system is installed, the tilt angle is approximately 4.3590°.

Figure 5. Experiment set-up for data collection of the PV/T solar collector.

Validation methods were performed by comparing results obtained experimentally and theoretically based on the trend shown on the related graphs. In this study, the error analysis performed was Root Mean Square Error (RMSE) [25];

$$RMSE = \sqrt{\frac{1}{n}\sum_{i=1}^{n}\left(x_{sim,i} - x_{exp,i}\right)^2} \quad (26)$$

6. Results and Discussion

The mathematical model that was developed is used to carry out the simulation in MATLAB and the data collected from the experiment were used to calculate the performance of the solar collector. The algorithm in Figure 4 is used in predicting the values of parameters in the simulation, which are validated using the experimental results. The developed model is said to be valid if the results of both theoretical and experimental curves show good agreement. The results obtained from the simulation and experiment carried out are shown graphically in the figures below.

As the mass flow rate is set higher, the PV and outlet temperature drops. This can be seen in the graph of temperature rise (difference between inlet air temperature and outlet air temperature) against the air mass flow rate (Figure 6). The temperature rise decreases with the increase in air mass flow rate since heat was transferred away by the moving air.

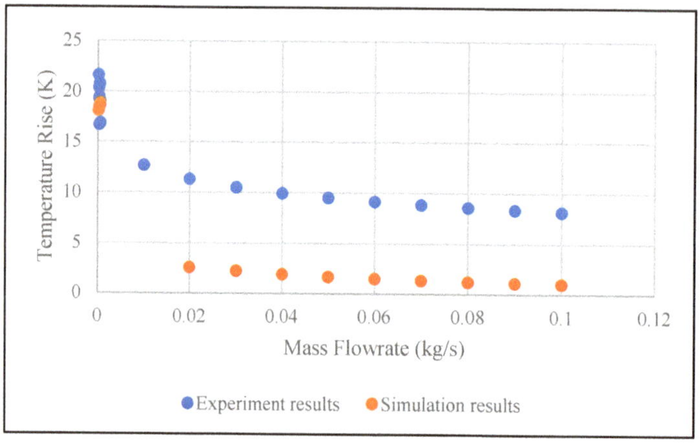

Figure 6. Comparison between the theoretical and experimental results on the effect of air mass flow rate to the temperature at G = 800–900 Wm^{-2} and $T_{a,avg}$ = 310 K.

Figures 7 and 8 show the effects of an increase in air mass flow rate on the collector efficiencies; electrically and thermally. At a solar irradiance level of 800 to 900 W/m^2, the thermal efficiency increases to 20.32% while electrical efficiency increases to 12.01% when the air mass flow rate increases from 0.00015 kg/s to 0.01 kg/s. It can be concluded that the solar collector system achieved an optimum thermal efficiency when the air mass flow rate approached 0.01 kg/s, while optimum electrical efficiency was achieved at an air mass flow rate of 0.02 kg/s. In connection with Figure 6, it can be seen that when mass flow rate increases, the thermal efficiency increases since the average temperature of the collector drops, until a certain point. The increase in mass flow rate increases the heat transfer coefficient between the air and the air channel; which leads to a decrease in photovoltaic cells. This caused the electrical efficiency to increase, although the increase is comparatively small to the increase in mass flow rate. In spite of that, it is shown in Figure 5 that the electrical efficiency increases with the mass flow rate. In the simulation carried out, the environmental parameters are set such that ambient temperature is at 310 K, solar irradiance at 850 W/m^2, and wind speed at 2 m/s. While the high flow rate produced thermal efficiency for the collector, it also increased the unnecessary electricity demand to operate the fans. However, if the air mass flow rate is too low, the collector temperature increases which leads to lower electrical efficiency. Hence, it will only be beneficial if the mass flow rate is kept at an optimum point because the significant increase in efficiencies is only at a low flow rate and approaches a plateau at higher flow rates. While thermal efficiency showed good agreement between simulation and experimental results, the slight difference between the simulated value and actual electrical efficiency is due to the fact that the data are collected under natural Malaysian weather conditions. The uncertainty of weather is a challenge faced during the data collection process since any atmospheric changes could affect the performance of the solar collector. However, due to the chaotic nature of weather to be simulated, the model is highly sensitive to the initial conditions which could lead to a very different realization of the simulated weather [35]. Therefore, since the results obtained in this study are of a similar trend presented in previous studies, it could be said that the simulation and experimental results are in good agreement.

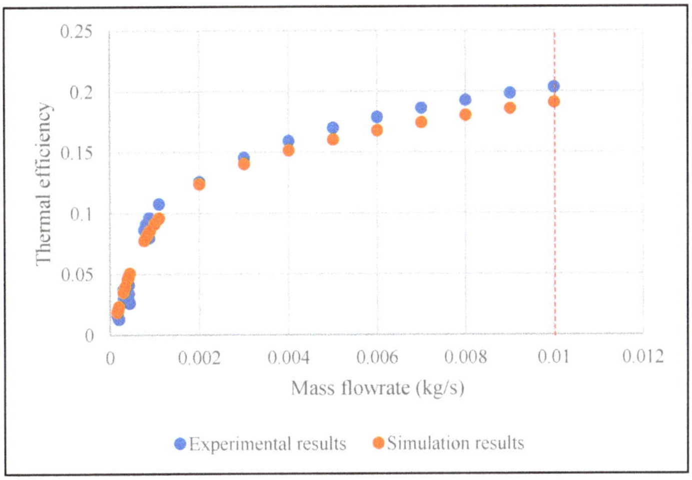

Figure 7. Comparison between the theoretical and experimental results on the effect of air mass flow rate to the thermal efficiency at G = 800–900 Wm^{-2} and $T_{a,avg}$ = 307.5 K.

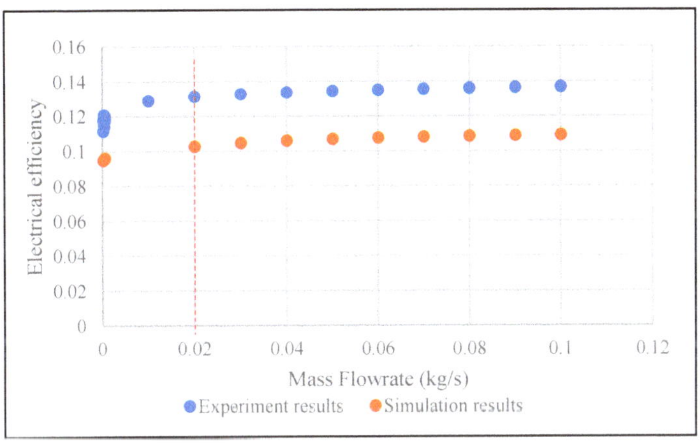

Figure 8. Comparison between the theoretical and experimental results on the effect of air mass flow rate to the electrical efficiency at G = 800–900 Wm^{-2} and $T_{a,avg}$ = 310 K.

Figure 9 shows the relationship between current production and solar irradiance. Both are directly proportional and it is shown in the figure that the higher the solar radiation, the higher the current produced. In a silicon solar panel, the electron band, E_g, decreases with increasing temperature and it enabled the absorption of additional photons [36]. These additional photons consequently slightly increase the short-circuit current, I_{sc} (photocurrent), produced. The graph in Figure 10 depicts a bell-curved shape which shows the solar radiation reading captured by the pyranometer throughout a day of the experiment carried out. The radiation increases when it approaches noon, and decreases when the sun sets. The ambient temperature that was recorded during the day increases from 30.64 °C in the morning to 38.85 °C during noon and later decreases to 33.53 °C in the evening.

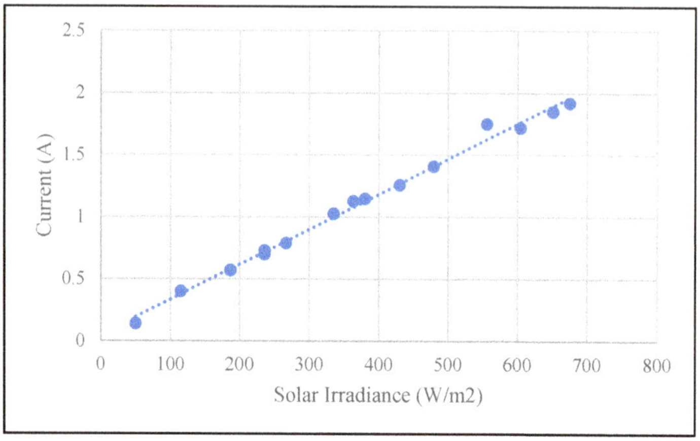

Figure 9. Direct relationship between current produced by the PV panel and the solar irradiance.

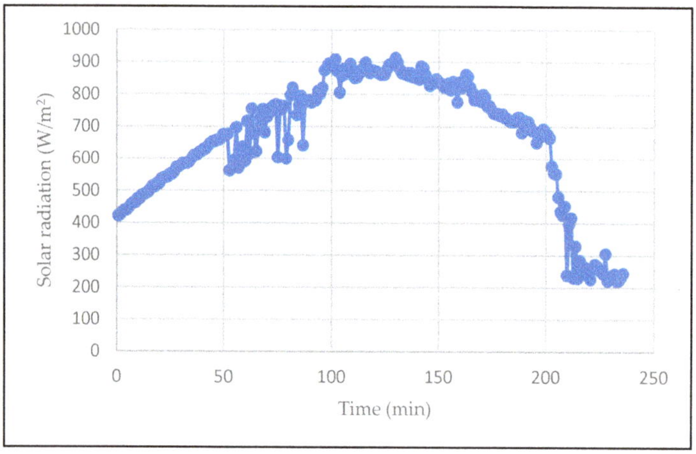

Figure 10. Reading of solar irradiance throughout a day (6 November 2019) during experiment.

The results of the error analysis performed on the data collected experimentally and theoretically are summarized in Table 2. When the ambient temperature is at 301–304 K, the values of RMSE for the thermal efficiency, η_{th} when solar irradiance are 100–200 W/m^2, 300–400 W/m^2 and 500–700 W/m^2 are 0.07%, 0.10% and 0.02% respectively. Meanwhile, for electrical efficiency, η_{ele}, when ambient temperature is at 301–304 K, the RMSE values are 0.02%, 0.03% and 0.02%. On average, the values of RMSE for T_{pv}, T_f, η_{th}, η_{ele} and T_{rise} are 5.99%, 7.74%, 0.07%, 0.02% and 5.38% respectively.

Table 2. Values of coefficients and ambient conditions used in the simulations.

Ambient Temp. (K)	Solar Irradiance (W/m^2)	Root Mean Square Error (RMSE) (%)				
		T_{pv}	T_f	η_{th}	η_{ele}	T_{rise}
301–304	100–200	5.22	1.56	0.07	0.02	5.46
	300–400	5.31	4.00	0.10	0.03	6.68
	500–700	4.69	9.96	0.02	0.02	2.93
305–306	100–200	5.40	0.98	0.18	0.02	6.15
	300–400	3.26	5.15	0.08	0.02	5.16
	500–600	3.86	9.74	0.02	0.02	3.34
	700–800	5.68	14.01	0.02	0.02	3.62
307–308	100–200	8.09	0.92	0.17	0.02	8.66
	300–400	5.54	5.67	0.08	0.04	6.92
	500–700	4.69	12.03	0.02	0.02	3.92
	800–900	7.20	16.72	0.01	0.02	3.96
309–311	100–400	9.12	3.10	0.09	0.03	11.17
	800–900	9.83	16.76	0.00	0.02	1.98

7. Conclusions

A PV/T solar collector integrating a PV module and a single pass air channel with a compact size design is discussed in this paper. As an enhancing factor of the heat transfer, a set of fins parallel to the air flow was attached to the back plate. A set of one-dimensional heat transfer equations representing the collector system was developed by constructing the temperature nodes at various points; on the PV module, back plate, air inlet and air outlet. The developed model was used to simulate the performance of the PV/T solar collector in MATLAB. A focus on comprehensive mathematical modelling and a detailed description of the simulation process was presented to provide a clear understanding of the model developed. From the experiment that was carried out, the following conclusions are made in this study; the developed model portrayed the performance of the solar collector since the results of both simulation and experiment are in good agreement. Hence, the model developed for the solar collector could contribute to the further analysis on the performance of the PV/T solar collector with different configurations, under different parameters.

Author Contributions: All authors contributed for this study. Conceptualization, N.N.W.K., M.O., M.N.A.B. and L.A.; methodology, N.N.W.K., M.O. and M.N.A.B.; validation, M.O., M.N.A.B. and L.A.; formal analysis, N.N.W.K., M.O. and M.N.A.B.; resources, N.N.W.K. and M.N.A.B.; software, N.N.W.K., M.N.A.B.; supervision, M.O., M.N.A.B. and L.A.; writing—original draft preparation, N.N.W.K. and M.O.; writing—review and editing, M.O., M.N.A.B. and L.A.; funding acquisition, M.O. All authors have read and agreed to the published version of the manuscript.

Funding: This research was funded by Yayasan Universiti Teknologi PETRONAS, grant number 0153AA-H27.

Acknowledgments: This research was supported by Universiti Teknologi PETRONAS, Malaysia. The authors would like to convey their gratitude to the reviewers for their significant remarks and suggestion for this paper.

Conflicts of Interest: The authors declare no conflict of interest.

Nomenclature

A_{ab}	total exposed surface area (m^2)	bp	back plate
A_c	solar collector surface area (m^2)	c	convection
C_f	specific heat capacity of fluid (J/(kg K))	f	fluid
h	heat transfer coefficient	g	PV panel tempered glass
$\dot{m}$	air mass flow rate (kg/s)	i	input
T	temperature (K)	o	output
U	overall heat loss coefficient (W/(m K^2))	pv	PV
vis_f	fluid viscosity	r	radiative
w_{fin}	fin thickness (m)	s	sky
		w	wind
Greek Letters			
η	efficiency	Abbreviations	
		ele	electrical
Subscripts		PF	packing factor
a	ambient	th	thermal

References

1. Othman, M.Y.H.; Yatim, B.; Sopian, K.; Abu Bakar, M.N. Performance analysis of a double-pass photovoltaic/thermal (PV/T) solar collector with CPC and fins. *Renew. Energy* **2005**, *30*, 716–722. [CrossRef]
2. Kumar, A.; Baredar, P.; Qureshi, U. Historical and recent development of photovoltaic thermal (PVT) technologies. *Renew. Sustain. Energy Rev.* **2015**, *42*, 1428–1436. [CrossRef]
3. Chow, T.T.; Pei, G.; Fong, K.F.; Lin, Z.; Chan, A.L.S.; Ji, J. Energy and exergy analysis of photovoltaic–thermal collector with and without glass cover. *Appl. Energy* **2009**, *86*, 310–316. [CrossRef]
4. Lee, J.H.; Hwang, S.G.; Lee, G.H. Efficiency Improvement of a Photovoltaic Thermal (PVT) System Using Nanofluids. *Energies* **2019**, *12*, 3063. [CrossRef]
5. Abu Bakar, M.N.; Othman, M.; Din, M.H.; Manaf, N.A.; Jarimi, H. Design concept and mathematical model of a bi-fluid photovoltaic/thermal (PV/T) solar collector. *Renew. Energy* **2014**, *67*, 153–164. [CrossRef]
6. Al-Khazzar, A.A.A. Behavior of Four Solar PV Modules with Temperature Variation. *Int. J. Renew. Energy Res.* **2016**, *6*, 1091–1099.
7. Othman, M.Y.; Hamid, S.A.; Tabook, M.A.S.; Sopian, K.; Roslan, M.H.; Ibarahim, Z. Performance analysis of PV/T Combi with water and air heating system: An experimental study. *Renew. Energy* **2016**, *86*, 716–722. [CrossRef]
8. Zhou, C.; Liang, R.; Zhang, J. Optimization design method and experimental validation of a solar PVT cogeneration system based on building energy demand. *Energies* **2017**, *10*, 1281. [CrossRef]
9. Diwania, S.; Agrawal, S.; Siddiqui, A.S.; Singh, S. Photovoltaic–thermal (PV/T) technology: A comprehensive review on applications and its advancement. *Int. J. Energy Eviron. Eng.* **2020**, *11*, 33–54. [CrossRef]
10. Wolf, M. Performance analyses of combined heating and photovoltaic power systems for residences. *Energy Convers.* **1976**, *16*, 79–90. [CrossRef]
11. Florschuetz, L.W. Extension of the Hottel-Whillier model to the analysis of combined photovoltaic/thermal flat plate collectors. *Sol. Energy* **1979**, *22*, 361–366. [CrossRef]
12. Hazami, M.; Riahi, A.; Mehdaoui, F.; Nouicer, O.; Farhat, A. Energetic and exergetic performances analysis of a PV/T (photovoltaic thermal) solar system tested and simulated under to Tunisian (North Africa) climatic conditions. *Energy* **2016**, *107*, 78–94. [CrossRef]
13. Hegazy, A.A. Comparative study of the performances of four photovoltaic/thermal solar air collectors. *Energy Convers. Manag.* **2000**, *41*, 861–881. [CrossRef]
14. Saygin, H.; Nowzari, R.; Mirzaei, N.; Aldabbagh, L.B.Y. Performance evaluation of a modified PV/T solar collector: A case study in design and analysis of experiment. *Sol. Energy* **2017**, *141*, 210–221. [CrossRef]
15. Kumar, R.; Rosen, M.A. Performance evaluation of a double pass PV/T solar air heater with and without fins. *Appl. Therm. Eng.* **2011**, *31*, 1402–1410. [CrossRef]
16. Dubey, S.; Sandhu, G.S.; Tiwari, G.N. Analytical expression for electrical efficiency of PV/T hybrid air collector. *Appl. Energy* **2009**, *86*, 697–705. [CrossRef]

17. Kiran, S.; Devadiga, U. Performance analysis of hybrid PV/Thermal systems. *Int. J. Emerg. Technol. Adv. Eng.* **2014**, *4*, 80–86.
18. Azad, E. Experimental analysis of thermal performance of solar collectors with different numbers of heat pipes versus a flow-through solar collector. *Renew. Sustain. Energy Rev.* **2018**, *82*, 4320–4325. [CrossRef]
19. Michael, J.J.; Selvarasan, I.; Goic, R. Fabrication, experimental study and testing of a novel photovoltaic module for photovoltaic thermal applications. *Renew. Energy* **2016**, *90*, 95–104. [CrossRef]
20. Mustafa, R.J.; Gomaa, M.R.; Al-Dhaifallah, M. Environmental impacts on the performance of solar photovoltaic systems. *Sustainability* **2020**, *12*, 608. [CrossRef]
21. Vergura, S. A statistical tool to detect and locate abnormal operating conditions in photovoltaic systems. *Sustainability* **2018**, *10*, 608. [CrossRef]
22. Sathyamurthy, R.; El-Agouz, S.A.; Nagarajan, P.K.; Subramani, J.; Arunkumar, T.; Mageshbabu, D.; Prakash, N. A review of integrating solar collectors to solar still. *Renew. Sustain. Energy Rev.* **2017**, *77*, 1069–1097. [CrossRef]
23. Zondag, H.A.; De Vries, D.W.; Van Helden, W.G.J.; Van Zolingen, R.J.C.; Van Steenhoven, A.A. The yield of different combined PV-thermal collector designs. *Sol. Energy* **2003**, *74*, 253–269. [CrossRef]
24. Tiwari, A.; Sodha, M.S. Parametric study of various configurations of hybrid PV/thermal air collector: Experimental validation of theoretical model. *Sol. Energy Mater. Sol. Cells* **2007**, *91*, 17–28. [CrossRef]
25. Sarhaddi, F.; Farahat, S.; Ajam, H.; Behzadmehr, A.M.I.N.; Adeli, M.M. An improved thermal and electrical model for a solar photovoltaic thermal (PV/T) air collector. *Appl. Energy* **2010**, *87*, 2328–2339. [CrossRef]
26. Chabane, F.; Moummi, N.; Benramache, S.; Bensahal, D.; Belahssen, O. Collector efficiency by single pass of solar air heaters with and without using fins. *Eng. J.* **2013**, *17*, 43–55. [CrossRef]
27. Jarimi, H.; Abu Bakar, M.N.; Othman, M.; Din, M.H. Bi-fluid photovoltaic/thermal (PV/T) solar collector: Experimental validation of a 2-D theoretical model. *Renew. Energy* **2016**, *85*, 1052–1067. [CrossRef]
28. Othman, M.Y.; Yatim, B.; Sopian, K.; Abu Bakar, M.N. Performance studies on a finned double-pass photovoltaic-thermal (PV/T) solar collector. *Desalination* **2007**, *209*, 43–49. [CrossRef]
29. Nowak, H. The sky temperature in net radiant heat loss calculations from low-sloped roofs. *Infrared Phys.* **1989**, *29*, 231–232. [CrossRef]
30. Biwole, P.H.; Woloszyn, M.; Pompeo, C. Heat transfers in a double-skin roof ventilated by natural convection in summer time. *Energy Build.* **2008**, *40*, 1487–1497. [CrossRef]
31. Duffie, J.A.; Beckman, W.A.; Blair, N. *Solar Engineering of Thermal Processes, Photovoltaics and Wind*; John Wiley & Sons: Hoboken, NJ, USA, 2020.
32. Bejan, A.; Kraus, A.D. *Heat Transfer Handbook*; John Wiley & Sons: Hoboken, NJ, USA, 2003; Volume 1.
33. Tan, H.M.; Charters, W.W.S. An experimental investigation of forced-convective heat transfer for fully-developed turbulent flow in a rectangular duct with asymmetric heating. *Sol. Energy* **1970**, *13*, 121–125. [CrossRef]
34. Michael, J.J.; Iniyan, S.; Goic, R. Flat plate solar photovoltaic–thermal (PV/T) systems: A reference guide. *Renew. Sustain. Energy Rev.* **2015**, *51*, 62–88. [CrossRef]
35. Gelfand, A.E.; Fuentes, M.; Hoeting, J.A.; Smith, R.L. *Handbook of Environmental and Ecological Statistics*; CRC Press: Boca Raton, FL, USA, 2017.
36. Radziemska, E.; Klugmann, E. Thermally affected parameters of the current–voltage characteristics of silicon photocell. *Energy Convers. Manag.* **2002**, *43*, 1889–1900. [CrossRef]

processes

MDPI

Article

Thermophysical Properties of Newly Synthesized Ammonium-Based Protic Ionic Liquids: Effect of Temperature, Anion and Alkyl Chain Length

Nur Hidayah Zulaikha Othman Zailani [1], Normawati M. Yunus [1,*], Asyraf Hanim Ab Rahim [1] and Mohamad Azmi Bustam [2]

1 Department of Fundamental and Applied Sciences, Center of Research in Ionic Liquids (CORIL), Institute of Contaminant Management for Oil and Gas, Universiti Teknologi PETRONAS, Seri Iskandar 32610, Malaysia; zulaikha95.nhz@gmail.com (N.H.Z.O.Z.); asyrafhanim92@gmail.com (A.H.A.R.)

2 Department of Chemical Engineering, Center of Research in Ionic Liquids (CORIL), Institute of Contaminant Management for Oil and Gas, Universiti Teknologi PETRONAS, Seri Iskandar 32610, Malaysia; azmibustam@utp.edu.my

* Correspondence: normaw@utp.edu.my; Tel.: +605-368-7689

Received: 17 May 2020; Accepted: 21 June 2020; Published: 25 June 2020

Abstract: Ionic liquids which are often classified as low melting point salts have received significant attention from research groups and industries to be used in a wide range of applications. Many of these applications require thorough knowledge on the thermophysical properties of the ionic liquids before utilizing their full potentials in various fields. In this work, a series of alkylammonium cation and carboxylate anion-based room temperature protic ionic liquids (PILs) were synthesized by varying length of alkyl chain of the cation from diethyl to dibutyl combined with pentanoate, hexanoate and heptanoate anions. These ammonium-based PILs named as diethylammonium pentanoate [DEA][C5], diethylammonium hexanoate [DEA][C6], diethylammonium heptanoate [DEA][C7], dibutylammonium pentanoate [DBA][C5], dibutylammonium hexanoate [DBA][C6] and dibutylammonium heptanoate [DBA][C7] were characterized using Nuclear Magnetic Resonance (NMR) spectroscopy. The thermophysical properties of the PILs namely density, dynamic viscosity and refractive index were measured and analyzed. Density, ρ and dynamic viscosity, η were determined at T = (293.15 to 363.15) K and refractive index, n_D was measured at T = (293.15 to 333.15) K. The fitting parameters are proposed for the empirical correlations of density, dynamic viscosity and refractive index. The values of thermal expansion coefficient, α_p, molecular volume, V_m, standard entropy, S° and lattice potential energy, U_{pot} also have been calculated by using the specified equations. The thermal decomposition temperature, T_d was also determined using a thermogravimetric analyzer (TGA) while the differential scanning calorimetry (DSC) technique provided the glass transition, T_g, melting point, T_m and crystallization, T_c temperatures of the PILs. The experimental results revealed that the dependency of the experimental values namely the ρ, η, n_D, and T_d on the alkyl chain of the anion, size of the cations and the temperature of measurement.

Keywords: protic ionic liquids; density; viscosity; refractive index; phase transition; thermal expansion coefficient; standard entropy; lattice potential energy

1. Introduction

Ionic liquids (ILs) are molten salts generally made from large organic cations and organic or inorganic anions. These ions interact poorly with each other which causes the salts to exist in liquid forms at room temperature or below 100 °C. ILs possess unique properties such as low flammability, high specific solubility, wide liquid range, high electrical conductivity and good thermal and chemical stability [1–5]. These attractive properties have driven researchers to utilize ILs, particularly imidazolium-based ILs, as alternative solvents in various applications such as in membrane separation, gas absorption and organic reactions [6–10]. Recently, protic ionic liquids (PILs) have emerged as promising candidates of alternative solvents due of their low cost of production and a simple one-step synthesis procedure. Besides that, over the past few years, PILs have attracted greater attention from researchers and quite a few works have been done to reveal several potential applications using PILs. Since PILs can be synthesized by the transfer of protons from Brønsted acids to Brønsted bases, PILs can be useful solvents for acid catalyzed reactions such as Diels–Alder reaction [11–13]. Besides that, the presence of active protons in PILs will allow PILs to be suitable electrolytes in fuel cells [14–16]. In addition, PILs are generally more hydrophilic and able to dissolve metals to larger extents than counter aprotic ionic liquids, AILs [14,17]. Ohno and co-workers provided the description of the modern type of protic ionic liquids [18] while Greaves and Drummond wrote an extensive review on the properties and applications of protic ionic liquids [19]. Furthermore, the ionicity of protic ionic liquids is of particular interest due to the possibility of incomplete proton transfer between the acid and base which may contribute to the presence of a neutral acid and base mixture [17,19,20]. A limit of 1% neutral species presence in an ionic liquid was proposed for the acid-base neutralization product to be called 'pure ionic liquid' [21]. Nevertheless, in recent years, ammonium-based PILs have been widely synthesized and used in a wide range of research's applications such as in organic synthesis, oxygen reduction reaction, polymer dissolution, CO_2 absorption and separation processes [22–27]. In view of this, proper designing of new PILs, whether by tuning the combination or structures of cations and anions or introducing some special functional groups, is very essential to ensure the PILs meet the standard requirement and physiochemical properties needed by the applications. Furthermore, the measurement of relevant physical properties within the ILs area is one of the most outstanding fields of study as understanding the thermophysical properties of pure ILs and their mixtures in the wide pressure and temperature ranges is essential to determine their potential applications [28]. For example, experimental density, viscosity, and their derived property such as molar volume data of ILs may be of great information in mass and heat transfer processes of working fluids [29]. Therefore, there is a need for systematic thermodynamic and thermophysical measurements for ILs to highlight their availability for use at the industrial processes level. Furthermore, these experimental thermophysical properties of ILs would allow the improvement of overall quality and reliability of the ILs in any applications [30].

This work is a continuation of our previous work to produce ammonium-based PILs as solvents for CO_2 absorption. The interest rises from the fact that this type of PILs has demonstrated promising ability to absorb CO_2 under experimental conditions, in addition to their simple synthesis procedure i.e., a one-step neutralization reaction [22,31,32]. On different note, this type of PILs may find applications in the areas of lubrication and lead acid battery modification [33,34]. Previously, we have reported the synthesis, characterization, thermophysical properties and CO_2 absorption of ammonium-based PILs utilizing bis (2-ethylhexyl) ammonium, tributylammonium and ethanolammonium cations with acetate and butyrate anions [22]. The results of the work have motivated us to further explore the synthesis and CO_2 absorption of ammonium-based PILs by using systematic combinations of simple amines and carboxylic acids. In this work, we report the synthesis of six new ammonium-based PILs, namely diethylammonium pentanoate [DEA][C5], diethylammonium hexanoate [DEA][C6], diethylammonium heptanoate [DEA][C7], dibutylammonium pentanoate [DBA][C5], dibutylammonium hexanoate [DBA][C6] and dibutylammonium heptanoate [DBA][C7] using one-step neutralization reaction as available in literature [35,36]. The purity of all these new ammonium-based PILs were checked by ^1H and ^{13}C NMR. The water content of these PILs was determined prior studying various thermophysical properties. The thermophysical properties namely density, viscosity and refractive index were measured and discussed in this study. The thermal decomposition, glass transition, crystallization and melting point temperatures of the new ammonium-based PILs were determined. The important parameters such as molecular volume, standard molar entropy and lattice energy were estimated by empirical equations. Further, the experimental values of density, viscosity and refractive index analysis were fitted using several empirical equations.

2. Materials and Methods

2.1. Chemicals

All six ammonium-based PILs were synthesized using two amines and three organic acids from Merck, Darmstadt, Germany. All chemicals with analytical grade were used in this work. The amines and acids CAS numbers, abbreviations and grade percentage are as follows: diethylamine (109-89-7, 99.0%), dibutylamine (111-92-2, 99.0%), pentanoic acid (109-52-4, 98.0%), hexanoic acid (142-62-1, 98.0%) and heptanoic acid (111-14-8, 99.0%).

2.2. Synthesis of PILs

The PILs were synthesized using a 1-step neutralization reaction. An equimolar amount of acid was added to an equimolar amount of base in 1:1 mol ratio. The organic acids were added dropwise to the flask under stirring with a magnetic bar. The mixtures were constantly stirred for 24 h at room temperature. In order to remove traces of water that might be coming from surrounding atmosphere or from the starting reagents, the synthesized PILs were heated at 70 °C under vacuum approximately for 8 h upon the completion of the reaction [35]. The dried samples were sealed and kept until further analysis. There was no noticeable solid crystal or precipitation formed when the PILs were purified after synthesis [36]. The proton transfer reaction results in carboxylate salt namely pentanoate, hexanoate and heptanoate of diethyl or dibutylamine. Generally, the reaction can be expressed as follows:

$$(R_X)_2NH + HOOC(R_Y) \rightarrow (R_X)_2NH_2^{+}\,{}^{-}OOC(R_Y)$$

where R_X is the alkyl substitutions into the amine compound (ethyl or butyl). Meanwhile, R_Y is the alkyl substitution into the acid compound (pentyl, hexyl or heptyl). For example, diethylammonium pentanoate is formed when R_X is ethyl and R_Y is pentyl. In this work, diethylamine [DEA] and dibutylamine [DBA] were used as the sources of the cations while carboxylate acids with alkyl chain length of pentyl, [C5], hexyl, [C6] and heptyl [C7] were utilized to provide the anions of the PILs. The combinations of the acids and the amines resulted in six PILs as tabulated in Table 1.

2.3. Structural Characterization and Water Content

The structural confirmation of the PILs was done by using Nuclear Magnetic Resonance (NMR) spectroscopy (Bruker, Billerica, MA, USA). In this work, 100 μL sample was dissolved in 600 μL solvent ($CDCl_3$) for each analysis. Both ^1H and ^{13}C spectra were recorded using Bruker Ascend TM 500 from Bruker, Billerica, MA, USA. The spectra are reported in parts per million and the multiplicities are abbreviated as *s* (singlet), *d* (doublet), *t* (triplet) and *m* (multiplet). Meanwhile, Volumetric Karl Fisher and Stromboli Oven (Model: V30 Mettler Toledo, Columbus, OH, USA) was used to analyze the water content in the synthesized PILs.

2.4. Thermophysical Properties Characterization

The density and viscosity of PILs were measured simultaneously via scanning preset at temperatures from 293.15 to 363.15 K using Anton Paar Stabinger Viscometer SVM3000 (Graz, Austria). The measurements were done in three replications and the average value was taken for further study. The viscometer was calibrated using a standard fluid provided by the supplier and a validation test was also conducted using a commercial imidazolium IL with known density and viscosity values.

The refractive index values of the PILs were determined using ATAGO RX-5000 Alpha Digital Refractometer from Tokyo, Japan in the temperature range of 293.15 to 333.15 K. Several standard organic solvents provided by the supplier were used to calibrate the instrument. A validation test was also conducted using a commercial imidazolium IL and the result was compared with the values available from literatures.

The decomposition temperatures of the PILs were determined by using thermogravimetric analyzer (TGA) (Simultaneous Thermal Analyer (STA) 6000 from Perkin Elmer, Walthan, Massachusetts, United States). In this study, Simultaneous Thermal Analyzer (STA) 6000 from Perkin Elmer (Waltham, MA, USA) was used to complete the thermal stability analysis. An approximate 10 mg of sample was weighted in a crucible pan and the analysis was performed in the temperature range of 30–650 °C under 20 mL/min nitrogen flow with a heating rate of 10 °C·min^{-1}.

The phase transition and melting point analyses were carried out by using Differential Scanning Calorimeter (DSC) 1 Star system, Mettler Toledo (Columbus, OH, USA). About 10 mg of samples was weighted in aluminum pans and sealed. The reaction was conducted in the temperature range of 80 to −150 °C with a heating rate of 10 °C·min^{-1}. The data were collected in the second cooling plot.

Table 1. List of acids, bases and new ammonium-based protic ionic liquids (PILs) synthesized in this work.

Base	Acid	Name of Synthesized PILs	Abbreviation
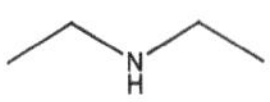 Diethylamine	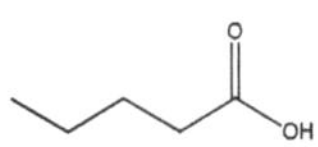 Pentanoic acid	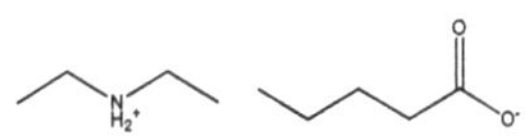 Diethylammonium pentanoate	[DEA][C5]
	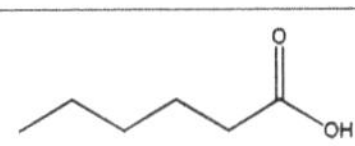 Hexanoic acid	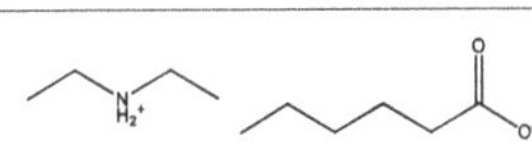 Diethylammonium hexanonate	[DEA][C6]
	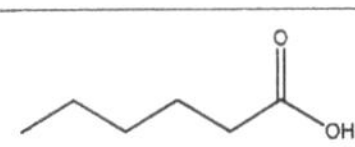 Heptanoic acid	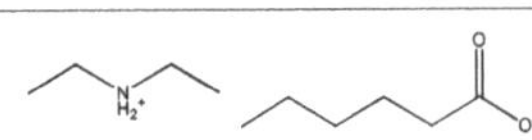 Diethylammonium heptanoate	[DEA][C7]
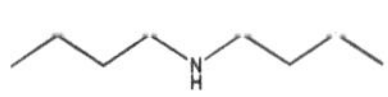 Dibutylamine	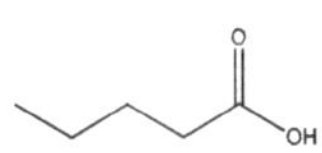 Pentanoic acid	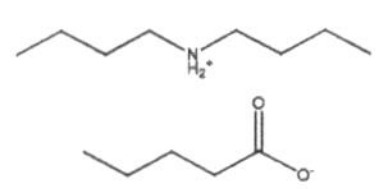 Dibutylammonium pentanoate	[DBA][C5]
	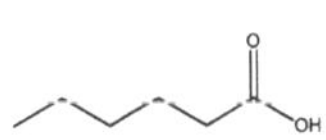 Hexanoic acid	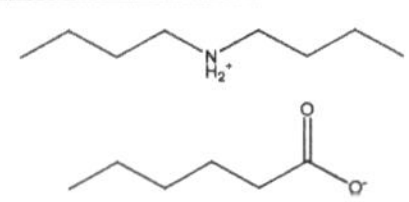 Dibutylammonium hexanoate	[DBA][C6]
	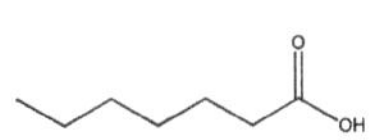 Heptanoic acid	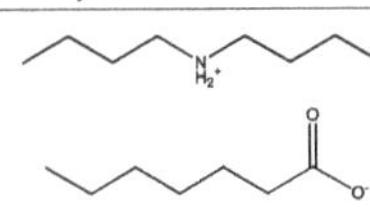 Dibutylammonium heptanoate	[DBA][C7]

3. Results and Discussion

3.1. Structural Characterization and Water Content Analysis

All six ammonium-based PILs synthesized in this work exist as liquids at room temperature. The NMR and water content results of each of these six ammonium-based PILs; diethylammonium pentanoate [DEA][C5], diethylammonium hexanoate [DEA][C6], diethylammonium heptanoate

[DEA][C7], dibutylammonium pentanoate [DBA][C5], dibutylammonium hexanoate [DEA][C6] and dibutylammonium heptanoate [DBA][C7] are presented accordingly as follows as well as in the Figures S1–S12 in the Supplementary Materials:

[DEA][C5]: ^{1}H NMR (500 MHz, $CDCl_3$): δ 0.86 [t, 3H (R-CH_3)], δ 1.29 [m, 8H (R-CH_3, R-CH_2), δ 1.51 [m, 2H (R-CH_2)], δ 2.19 (t, 2H (COOH-CH_2)], δ 2.89 [m, 4H (NH_2-CH_2)]. ^{13}C NMR (125 MHz, $CDCl_3$): δ 179.43, 41.94, 35.47, 27.60, 22.38, 13.73, 11.10. Water content: 0.93%.

[DEA][C6]: ^{1}H NMR (500 MHz, $CDCl_3$): δ 0.88 [t, 3H (R-CH_3)], 1.29 [m, 8H (R-CH_3, R-CH_2)], 1.58 [m, 4H (R-CH_2)], 2.19 [t, 2H (COOH-CH_2)], 2.89 [m, 4H (NH_2-CH_2)]. ^{13}C NMR (125 MHz, $CDCl_3$): δ 180.14, 41.76, 37.27, 31.78, 25.92, 22.49, 13.98, 11.24. Water content: 6.45%.

[DEA][C7]: ^{1}H NMR (500 MHz, $CDCl_3$): δ 0.87 [t, 3H (R-CH_3)], 1.27 [m, 8H (R-CH_3, R-CH_2)], 1.58 [m, 6H (R-CH_2)], 2.22 [t, 2H (COOH-CH_2)], 2.92 [m, 4H (NH_2-CH_2)]. ^{13}C NMR (125 MHz, $CDCl_3$): δ 180.16, 41.77, 37.38, 31.69, 29.26, 26.24, 22.56, 14.02, 11.26. Water content: 3.43%.

[DBA][C5]: ^{1}H NMR (500 MHz, $CDCl_3$): δ 0.847 [m, 9H (-CH_3)], 1.28 [m, 6H (-CH_2, -CH_3)], 1.478 [m, 2H (-CH_2)], 1.609 [m, 4H (-CH_2)], 2.106 [t, 2H (CH_2-COO-)], 2.721 [t, 4H (CH_2-NH)]. ^{13}C NMR (125 MHz, $CDCl_3$): δ 180.16, 47.36, 37.65, 28.69, 28.24, 22.70, 20.12, 13.92, 13.58. Water content: 5.70%.

[DBA][C6]: ^{1}H NMR (500 MHz, $CDCl_3$): δ 0.846 [m, 9H (-CH_3)], 1.289 [m, 8H (-CH_2, -CH_3)], 1.511 [m, 2H (-CH_2)], 1.615 [m, 4H (-CH_2)], 2.095 [t, 2H (CH_2-COO-)], 2.724 [t, 4H (CH_2-NH)].)]. ^{13}C NMR (125 MHz, $CDCl_3$): δ 180.11, 47.33, 37.92, 31.86, 28.21, 26.22, 22.52, 20.09, 13.93, 13.54. Water content: 5.56%.

[DBA][C7]: ^{1}H NMR (500 MHz, $CDCl_3$): δ 0.800 [m, 9H (-CH_3)], 1.275 [m, 10H (-CH_2, -CH_3)], 1.501 [m, 2H (-CH_2)], 1.067 [m, 4H (-CH_2)], 2.110 [t, 2H (CH_2-COO-)], 2.721 [t, 4H (CH_2-NH)].)]. ^{13}C NMR (125 MHz, $CDCl_3$): δ 180.19, 47.39, 37.99, 31.76, 29.35, 28.31, 26.52, 22.55, 20.13, 13.99, 13.58. Water content: 4.39%.

Six protic ionic liquids with the ammonium cation, [DEA] and [DBA], were synthesized through acid-base neutralization reactions with different alkyl chain length of ammonium-based cation and the corresponding organic acid. The reported water content is between 0.93% and 6.45% for the synthesized PILs. The water traces may come from the starting materials or from surrounding atmosphere during the synthesis process [35]. Besides, Chen and his team conducted investigations on water sorption in PILs at ambient environment and revealed that PILs are highly hygroscopic and have higher hydrophilicity compared to aprotic ionic liquids [37]. Meanwhile, other researchers concluded that the anion was the key factor influencing water sorption [38]. Furthermore, several researchers indicate that the presence of water molecules reduces the electrostatic attractions between the ions and consequently lowers the viscosity of ILs as the overall cohesive energy of the system is decreased [39]. Nevertheless, the thermophysical properties of our ammonium-based PILs are solely reported by using these water contents.

3.2. Thermophysical Properties Analysis

All instruments used for measuring the density, viscosity and refractive index were calibrated using standard solutions provided by the supplier. In addition, a commercial IL namely 1-hexyl-3-methylimidazolium bis(trifluoromethylsulfonyl)imide [HMIM][Tf_2N], (Merck, 382150-50-7, 98%) with readily available physical property data was tested using these instruments so as to validate the experimental data. From the measurements, the experimental values of density, viscosity and refractive index at 298.15 K displayed a very good agreement with the values from the literatures as shown in Table S1 in the Supplementary Materials.

The density values of the PILs were recorded as a function of temperature from 293.15 to 363.15 K. The plots of the experimental density of PILs are shown in Figure 1 and the experimental data is available Table S2 in the Supplementary Materials. Generally, based on the results, the densities for all six ammonium-based PILs decreased linearly as the temperature increased despite the difference in the alkyl length of the anions. At higher temperatures, the intermolecular forces between the constituted ions are weakened and therefore increased the mobility of the ions. Consequently, the unit volume for these ions increased as well [22,40,41]. On the other hand, from the experimental results, [DEA][C5] has the highest density values compared to the rest of ammonium-based PILs. This observation can be explained by the change of local packing of the PIL structure due to the smaller size of the [DEA] cation, in comparison to [DBA] cation [22,42]. Several researchers reported the similar behavior using other PIL with ethylammonium cation, which increasing trending packing efficiency with the decreasing of molecular weight [43]. Other than that, [DBA][C7] showed the lowest density value at all temperatures. As the alkyl chain increases in both cation and anion of the PIL, the structure gets bigger and bulkier which in turn promoting steric hindrance and asymmetric nature in the PIL structure and thus result in a lower density value for the PIL [41,42].

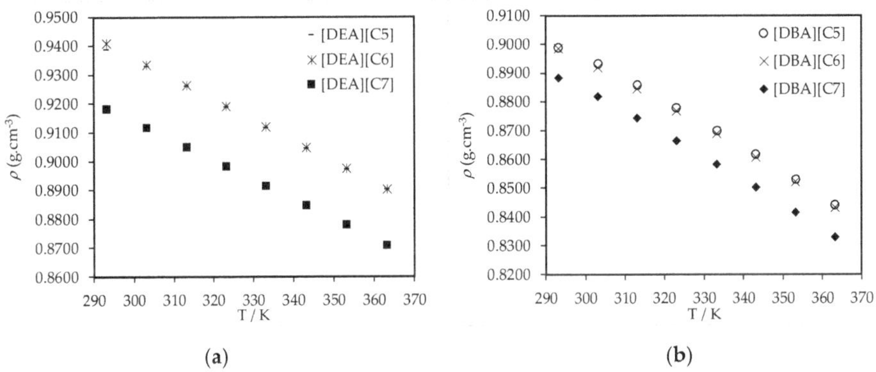

Figure 1. (**a**) Density (ρ) values of diethylammonium pentanoate [DEA][C5], diethylammonium hexanoate [DEA][C6], and diethylammonium heptanoate [DEA][C7] as a function of temperature. (**b**) Density (ρ) values of dibutylammonium pentanoate [DBA][C5], dibutylammonium hexanoate [DEA][C6] and dibutylammonium heptanoate [DBA][C7] as a function of temperature.

Viscosity is a key parameter that helps to designate the potential applications of any ILs and it is greatly influenced by intermolecular interactions namely hydrogen bonding, dispersive forces and columbic interactions [44]. Figure 2 and Table S3 in the Supplementary Materials show the dynamic viscosity data of the ammonium based PILs which have been measured in the temperature range of 293.15 to 363.15 K.

From Figure 2, analysis of the results revealed that the viscosity of all ammonium-based PILs decreased exponentially with increasing temperature in the range of temperature studied. It was found that the PILs with [DBA] cation display higher viscosity value than PILs with [DEA] cation. The increase of viscosity with increasing alkyl chain length in the ILs structure is due to the increase in van der Waals attraction between the aliphatic alkyl chain [40,45]. PILs with [DBA] cation, however show marginal increment in the viscosity values as the anion alkyl chain increases. Nevertheless, the increment of alkyl chain length on anion structure depicted the same trend for both [DEA] and [DBA] cation as the viscosity is increasing in the order of [C5] < [C6] < [C7]. On the other hand, Iglesias et al., investigated the viscosity of 2-hydroxy ethylammonium propionate (2-HEAPE), 2-hydroxy diethylammonium propionate (2-HDEAPE) and 2-hydroxy triethylammonium propionate (2-HTEAPE) and found that the viscosity value decreases with the increasing number of ethyl group at the cation [36].

This behavior might be affected by the increase of the cation structure due to the replacement of hydrogen atoms with an ethyl group.

The refractive index, n_D estimates the electronic polarizability of the molecules and shows the dielectric response to an external electric field produced by electromagnetic waves (light) [46]. Generally, it describes how fast light travels through the material. The refractive index of ammonium-based PILs was measured at temperatures of 293.15 to 333.15 K and the experimental data is available in Table S4 in the Supplementary Materials. The n_D decreased in a very narrow window with increasing of temperature as shown in Figure 3. On the other hand, the values of refractive index increased with the increase in cation and anion chain length of PILs. The increment of the refractive index in line with the increment of the alkyl chain in the PILs structures is believed to occur due to the influence of higher intermolecular interaction such as the van der Waals forces of the PILs [41].

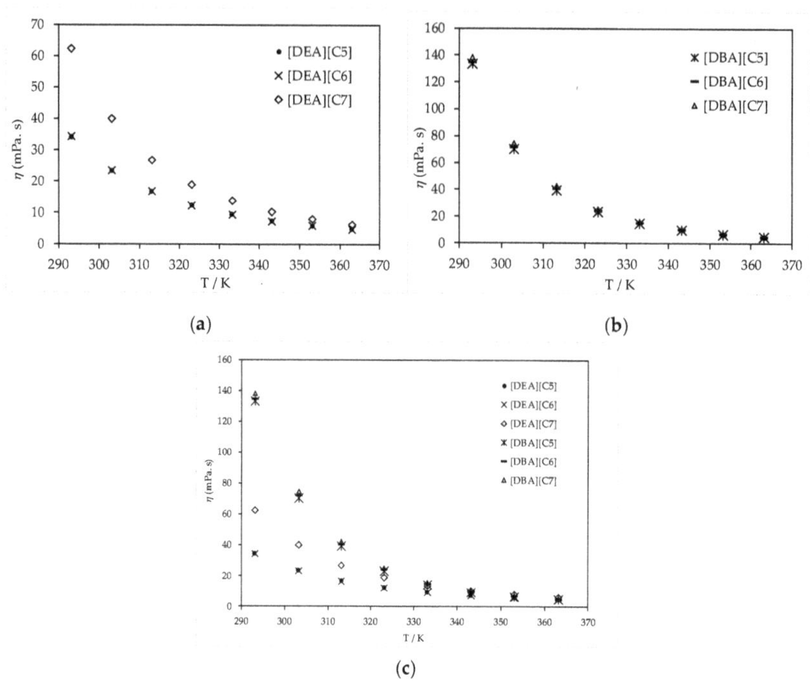

Figure 2. (**a**) Viscosity (η) of [DEA][C5], [DEA][C6] and [DEA][C7] as a function of temperature. (**b**) Viscosity (η) of [DBA][C5], [DBA][C6] and [DBA][C7] as a function of temperature. (**c**) Viscosity (η) of all six ammonium-based PILs as a function of temperature.

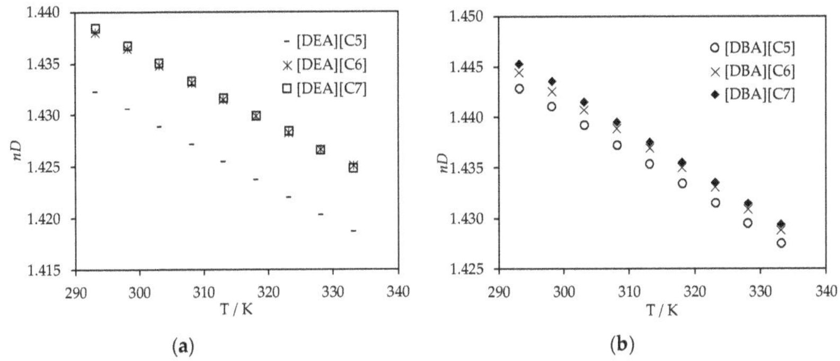

Figure 3. (**a**) Refractive index (n_D) values of [DEA][C5], [DEA][C6] and [DEA][C7] as a function of temperature. (**b**) Refractive index (n_D) values of [DBA][C5], [DBA][C6] and [DBA][C7] as a function of temperature.

3.3. Thermophysical Properties Correlations

The density, ρ dynamic viscosity, η and refractive index, n_D experimental values were fitted using the following equations [40,47]:

$$\rho = A_1 + A_2T, \tag{1}$$

$$\lg\eta = A_3 + A_4/T, \tag{2}$$

$$n_D = A_5 + A_6T, \tag{3}$$

where ρ is the density, η is dynamic viscosity and n_D is refractive index of the ammonium-based PILs, T is temperature in K, and A_1 through A_6 are correlation coefficients using the least square method. Tables 2–4 represent the estimation of values of correlation coefficients and the standard deviations, SD calculated using the equation from the literature [40].

Table 2. Fitting parameters of Equation (1) to correlate density (ρ) of synthesized PILs and calculated standard deviation (SD).

Ionic Liquids	A_1	A_2	SD
[DEA][C5]	1.1456	−0.0007	0.0010
[DEA][C6]	1.1519	−0.0007	0.0068
[DEA][C7]	1.1168	−0.0007	0.0076
[DBA][C5]	1.1331	−0.0008	0.0027
[DBA][C6]	1.1345	−0.0008	0.0027
[DBA][C7]	1.1269	−0.0008	0.0008

Table 3. Fitting parameters of Equation (2) to correlate viscosity (η) of PILs and calculated standard deviation (SD).

Ionic Liquid	A_3	A_4	SD
[DEA][C5]	−2.9587	1312.0	0.0106
[DEA][C6]	−2.9597	1312.4	0.0108
[DEA][C7]	−3.4058	1518.0	0.0133
[DBA][C5]	−5.4771	2219.3	0.0207
[DBA][C6]	−5.4156	2203.9	0.0179
[DBA][C7]	−5.3421	2186.0	0.0176

Table 4. Fitting parameters of Equation (3) to correlate refractive index (n_D) of PILs and calculated standard deviation (SD).

Ionic Liquid	A_5	A_6	SD
[DEA][C5]	1.5326	−0.0003	0.0132
[DEA][C6]	1.5335	−0.0003	0.0081
[DEA][C7]	1.5376	−0.0003	0.0120
[DBA][C5]	1.5557	−0.0004	0.0265
[DBA][C6]	1.5583	−0.0004	0.0276
[DBA [C7]	1.5623	−0.0004	0.0309

3.4. Thermal Expansion Coefficient, Standard Entropy, Lattice Potential Energy, Water Content and Thermal Study

For further understanding of the intermolecular interaction, the thermal expansion coefficient can be calculated from the experimental values of density, ρ and the data is tabulated in Table 5. Thermal expansion coefficients, α_p for the synthesized PILs can be defined as [22,40,48]:

$$\alpha_p = -1/\rho.\ (\delta\rho/\delta T) = -(A_2)/(A_1 + A_2T) \quad (4)$$

The calculated values in Table 5 show that the thermal expansion coefficients increase with the total C-numbers in the structure of PILs. This indicates that the thermal expansion coefficient is dependent on the cation symmetry and the length of alkyl substituent [49]. In view of the result obtained, thermal expansion coefficients of PILs with [DBA] cation are higher than that of PILs with [DEA] cation. Meanwhile, the behavior of thermal expansion coefficient is almost the same for all PILs with the same cation group in which the calculations show very small deviations of α_p values with temperature. Thus, thermal expansion coefficient can be considered as temperature independent as it shows similar result over the temperature range studied. Yunus et al., also reported the similar trend of variation of thermal expansion coefficient for different group of PILs [22].

Molar volume, V_m can be defined as the volume occupied by one mole of a compound at a given temperature and pressure [50]. The molar volume, V_m was obtained by the following empirical equation according to the experimental densities [45,51–53]:

$$V_m = M/(\rho.\ N_A), \quad (5)$$

where V_m is the molecular volume, M is molar mass of PILs, ρ is density of PILs at 303.15 K and N_A is the Avogadro's number.

Table 5. Thermal expansion coefficients (α_p) of the PILs calculated using Equation (4).

T/K	10^{-4} α/K^{-1}					
	[DEA][C5]	[DEA][C6]	[DEA][C7]	[DBA][C5]	[DBA][C6]	[DBA][C7]
293.15	7.4	7.4	7.7	8.9	8.9	9.0
303.15	7.5	7.4	7.7	9.0	9.0	9.1
313.15	7.6	7.5	7.8	9.1	9.1	9.2
323.15	7.6	7.6	7.9	9.1	9.2	9.2
333.15	7.7	7.6	7.9	9.2	9.2	9.3
343.15	7.7	7.7	8.0	9.3	9.3	9.4
353.15	7.8	7.7	8.0	9.4	9.4	9.5
363.15	7.9	7.8	8.1	9.5	9.5	9.6

Molar volume for all ammonium-based PILs were calculated at temperature of 303.15 K. As shown in Table 6, the molar volume, V_m is proportional to the length of the alkyl chain of the anion as well as the size of the cation. The former is caused by the addition of CH_2 group in the anion of the

PILs and similar findings have been observed in other studies, while the latter is due to the bigger size of [DBA] as compared to [DEA] [5,22].

Table 6. Molar volume, V_m; standard entropy, S°; lattice potential energy, U_{pot}; thermal decomposition, T_d; glass transition, T_g; melting point, T_m; crystallization point, T_c. V_m, S° and U_{pot} were calculated at 303.15 K.

Ionic Liquids	V_m	S°	U_{pot}	T_d	T_g	T_m	T_c
	nm^3	$J{\cdot}K^{-1}{\cdot}mol^{-1}$	$kJ{\cdot}mol^{-1}$	K	°C	°C	°C
[DEA][C5]	0.3120	418.4	449.8	412.82	−100.4	−9.7	−58.8
[DEA][C6]	0.3369	449.4	441.1	419.24	−99.5	−6.3	−62.8
[DEA][C7]	0.3704	491.2	430.6	442.00	−99.6	0.8	−60.2
[DBA][C5]	0.4302	565.8	414.7	404.78	-	−20.9	-
[DBA][C6]	0.4570	599.1	408.5	425.33	-	−12.8	-
[DBA][C7]	0.4886	638.5	401.8	435.65	-	−8.1	-

Entropy is the measurement of randomness of molecules. Generally, entropy increases with the increase in molar volume [5]. In order to study the relationship between molecular volume and standard entropy for both ionic liquids and organic acids, Glasser [54] had provided a standard equation as follows:

$$S° = 1246.5\ V_m + 29.5, \quad (6)$$

in which V_m is the molecular volume in nm^3 and S° is standard entropy.

The results in Table 6 clearly show that the standard entropy (S°) of ammonium-based PILs increased with the molar volume for all ammonium-based PILs at the studied temperature. This increment in the S° values could be related to increasing the number of carbon atoms in alkyl chain of carboxylate anion [38]. In this work, the V_m increases in the sequence of [C5] < [C6] < [C7] for both [DEA] and [DBA] cations.

Besides standard entropy, Glasser [54] also proposed a method for calculating lattice potential energies (U_{pot}) of ILs in order to predict the relative stabilities of ILs by using Equation (7) where γ and δ are fitting coefficients with values of 1981.9 $kJ{\cdot}mol^{-1}$ and 103.8 $kJ{\cdot}mol^{-1}$, respectively.

$$U_{pot} = [\gamma\ (\rho/M)^{1/3}] + \delta \quad (7)$$

There is no obvious influence of structural properties in lattice potential energy. However, electrostatic or columbic interaction is the main factor that contributes to lattice energy which is inversely related to the volume of ions [5,41,44]. Table 6 shows the calculated lattice potential energy of the studied PILs at 303.15 K. In this work, we observed that the lattice potential energy decreased with increasing carbon chain length of the carboxylate anions. This can be explained by the fact that adding a methylene group in the alkyl chain of the ammonium-based PILs will increase the entropy and consequently cause a reduction of the packing efficiency in the PILs. Thus, as a result, lattice potential energy will decrease with the increase in the alkyl chain length of the PILs.

The study of thermal analysis of a substance is crucial as it will provide information on how PILs behave as heat flows. The thermal stability of the PIL was examined by thermogravimetric analysis (TGA) and the data are tabulated in Table 6 while TGA profiles are graphically presented in Figure 4. The TGA for studied PILs was performed at scanning rate of 10 $°C{\cdot}min^{-1}$. In the TGA study, the main factor of thermal stability is dependent on the strength of the heteroatom–carbon and heteroatom–hydrogen bond [44].

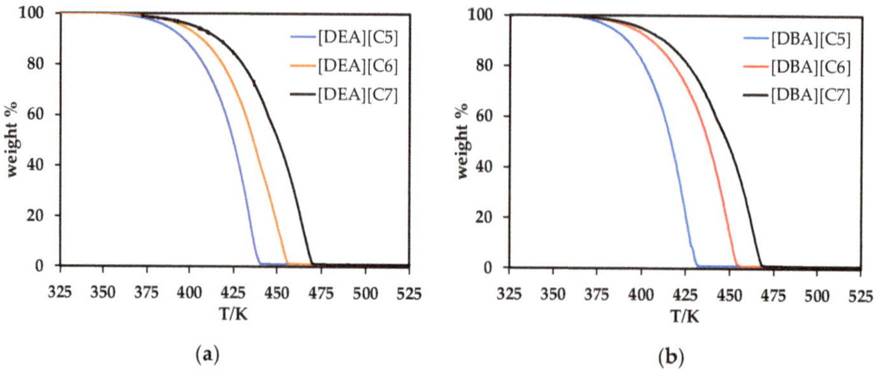

Figure 4. (**a**) Thermal decomposition curve of [DEA][C5], [DEA][C6] and [DEA][C7] at a heating rate of 10 °C·min^{-1}. (**b**) Thermal decomposition curve of [DBA][C5], [DBA][C6] and [DBA][C7] at a heating rate of 10 °C·min^{-1}.

The studied PILs showed thermal stabilities in the range of approximately 404 to 442 K. The TGA profiles of the studied PILs indicate a one-step decomposition temperature for all ammonium-based PILs. As can be seen from Figure 4 and Table 6, the thermal decomposition (T_d) increases as the alkyl chain length in anion increases within the same cation and the order of thermal stability of the ammonium-based PILs can be listed as [DEA][C5] < [DEA][C6] < [DEA][C7]. A similar order was also observed in PILs with [DBA] cation in which the decomposition steps of anion with [C5] occurred faster than anion with [C6] and [C7]. Generally, this thermal stability trend can be related to the strong intermolecular forces present in a higher alkyl chain [C7], which requires a high amount of energy to cleave the neighboring bonds. More energy is required to destruct the C–C bond as it involves a series of competitive intramolecular and intermolecular process [45,55].

DSC is a thermo-analytical technique that measures the difference in amount of heat required to increase the samples' temperature. The thermogram values of the ammonium-based PILs were recorded as a function of temperature and the extracted data were shown in Table 6. Examples of DSC curves for the ammonium-based PILs synthesized in this study are shown in Figure 5. From the DSC analysis, only PILs with [DEA] cation exhibited a glass transition temperature (T_g) under the experimental conditions. This indicates that only PILs with [DEA] cation experience heat flow on heating from amorphous glass to liquid state [5]. T_g represents the cohesive energy of the sample. Liquids with low cohesive energies possess T_g values which in turn contribute to desirable physiochemical properties such as low viscosity and high ionic conductivity [56]. There is a marginal difference in the T_g values with respect to the size of the PILs and similar trend is also observed in the literature [29,56]. Further analysis of the DSC thermograms indicates that only ammonium-based PILs with [DEA] cation possess crystallization temperature (T_c) with no obvious trend and results show that [DEA][C5] exhibits a T_c value of −58.78 °C. All ammonium-based PILs in this study show the presence of melting temperature (T_m). The T_m of PIL is largely depending on the crystal lattice strength of the PIL itself. The poor packing efficiency of the counterions destabilizes the crystal lattice of the PILs and this causes the crystal lattice energy to decrease and consequently contributes to the low T_m of the PILs [29,57]. Generally, the melting point of PILs with [DEA] cation is higher than that of PILs with [DBA] cation suggesting a better packing of the counterions in the PILs of [DEA] cations. On the other hand, all reported ILs in this paper exhibit low T_m compared to common ILs such as 3-butyl-1-methylimidazolium chloride, BMIMCl [58–60].

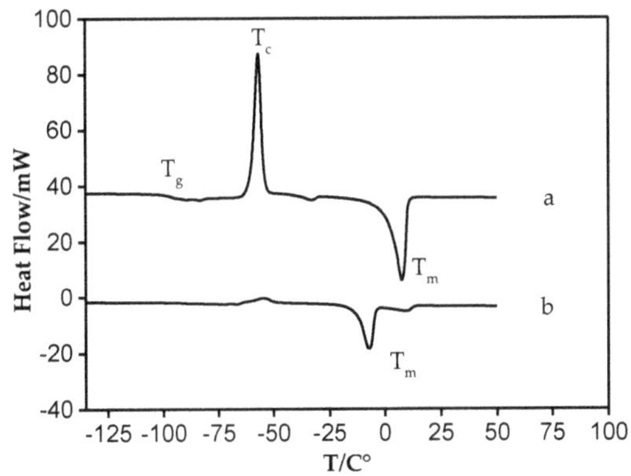

Figure 5. Differential scanning calorimetry (DSC) curves of [DEA][C7] (**a**) and [DBA][C6] (**b**) at heating rate of 10 °C.min^{-1}.

4. Conclusions

In this work, six ammonium-based protic ILs with [DEA] and [DBA] cations and combination of three alkyl length of anions (pentyl [C5], hexyl [C6], heptyl [C7]) have been successfully synthesized and characterized. The important thermophysical properties namely density and viscosity were measured from 293.15 to 363.15 °C, while the refractive index was examined from 293.15 to 333.15 K. The water contents of the PILs might have some effects on the measured thermophysical properties. The fitting parameters were proposed for the empirical correlations of density, dynamic viscosity, and refractive index. In addition, the experimental values of densities have been used to calculate the thermal expansion coefficient (α_p) and molecular volume (V_m) of the ammonium-based PILs. The standard entropy (S°) and lattice potential energy (U_{pot}) have also been calculated using equations available from the literature. Furthermore, the thermal stabilities and behaviors of the ammonium-based PILs were investigated resulting in the determination of thermal decomposition temperature (T_d), glass transition temperature (T_g), crystallization temperature (T_c) and melting point (T_m) of the ammonium-based PILs studied. All results indicate the effects of temperature, anion, alkyl chain length and size of the ammonium-based PILs towards the thermophysical properties of the newly synthesized PILs. These results can further be utilized prior to the usage of the PILs in any applications.

Supplementary Materials: The following are available online at http://www.mdpi.com/2227-9717/8/6/742/s1, Table S1: Data used for the validation of equipment for physical property measurements using [HMIM] [Tf_2N] at T = 298.15 K, Table S2: Density (ρ) values of PILs at temperatures (293.15–363.15) K, Table S3: Dynamic viscosity (η) values of PILs at temperatures (293.15–363.15) K, Table S4: Refractive index (n_D) values of PILs at temperatures (293.15–333.15 K), Figures S1–S12: NMR analysis of the PILs.

Author Contributions: Conceptualization, N.M.Y. and N.H.Z.O.Z.; methodology, N.H.Z.O.Z. and N.M.Y. validation, N.H.Z.O.Z. and A.H.A.R.; formal analysis, N.H.Z.O.Z. and N.M.Y.; resources, N.M.Y. and M.A.B.; data curation, N.M.Y.; writing—original draft preparation, N.H.Z.O.Z. and A.H.A.R.; writing—review and editing, N.M.Y.; supervision, N.M.Y.; project administration, N.M.Y. and M.A.B.; funding acquisition, N.M.Y. All authors have read and agreed to the published version of the manuscript.

Funding: This research was funded by Yayasan Universiti Teknologi PETRONAS-Fundamental Research Grant (YUTP-FRG) (cost centre 015LC0-054) under research project "Design, Synthesis and Evaluation of Protic Ionic Liquids for CO_2 Removal from Natural Gas" and the APC was funded by 015LC0-054.

Acknowledgments: Financial assistance and support from Universiti Teknologi PETRONAS and Center of Research in Ionic Liquids (CORIL), UTP are greatly acknowledged.

Conflicts of Interest: The authors declare no conflict of interest.

References

1. Endo, T.; Murata, H.; Imanari, M.; Mizushima, N.; Seki, H.; Nishikawa, K. NMR Study of cation dynamics in three crystalline states of 1-butyl-3-methylimidazolium hexafluorophosphate exhibiting crystal polymorphism. *J. Phys. Chem. B* **2012**, *116*, 3780–3788. [CrossRef]
2. Smith, J.A.; Webber, G.B.; Warr, G.G.; Atkin, R. Rheology of protic ionic liquids and their mixtures. *J. Phys. Chem. B* **2013**, *117*, 13930–13935. [CrossRef]
3. Patra, R.N.; Gardas, R.L. Effect of nitro groups on desulfurization efficiency of benzyl-substituted imidiazolium-based ionic liquids: Experimental and computational approach. *Energy Fuels* **2019**, *33*, 7659–7666. [CrossRef]
4. Singh, V.; Sharma, G.; Gardas, R.L. Thermodynamic and ultrasonic properties of ascorbic acid in aqueous protic ionic liquid solutions. *PLoS ONE* **2015**, *10*, e0126091. [CrossRef] [PubMed]
5. Sarkar, A.; Sharma, G.; Singh, D.; Gardas, R.L. Effect of anion on thermophysical properties of *N,N*-diethanolammonium based protic ionic liquids. *J. Mol. Liq.* **2017**, *242*, 249–254. [CrossRef]
6. Baicha, Z.; Salar-García, M.J.; Ortiz-Martínez, V.M.; Hernández-Fernández, F.J.; de los Ríos, A.P.; Maqueda Marín, D.P.; Collado, J.A.; Tomás-Alonso, F.; El Mahi, M. On the selective transport of nutrients through polymer inclusion membranes based on ionic liquids. *Processes* **2019**, *7*, 544. [CrossRef]
7. Abejón, R.; Rabadán, J.; Lanza, S.; Abejón, A.; Garea, A.; Irabien, A. Supported ionic liquid membranes for separation of lignin aqueous solutions. *Processes* **2018**, *6*, 143. [CrossRef]
8. Taimoor, A.A.; Al-Shahrani, S.; Muhammad, A. Ionic liquid (1-butyl-3-metylimidazolium methane sulphonate) corrosion and energy analysis for high pressure CO_2 absorption process. *Processes* **2018**, *6*, 45. [CrossRef]
9. Bogdanov, M.G.; Svinyarov, I. Distribution of N-methylimidazole in ionic liquids/organic solvents systems. *Processes* **2017**, *5*, 52. [CrossRef]
10. Keremedchieva, R.; Svinyarov, I.; Bogdanov, M.G. Ionic Liquid-Based Aqueous Biphasic Systems—A facile approach for ionic liquid regeneration from crude plant extracts. *Processes* **2015**, *3*, 769–778. [CrossRef]
11. Ahmat, X.; Ablajan, K.; Shinozaki, H. Furans-Maleimides Diels-Alder reactions in protic ionic liquid. *Chem. Res. Chin. Univ.* **2009**, *25*, 161–168.
12. Mancini, P.M.E.; Ormachea, C.M.; Rosa, C.D.D.; Kneeteman, M.N.; Domingo, L.R. Protic and nonprotic ionic liquids in polar Diels-Alder reactions using properly substituted heterocycles and carbocycles as dienophiles. A DFT study. In *Ionic Liquids: New Aspects for the Future*; IntechOpen: London, UK, 2013; Volume 9, pp. 691–695.
13. Rosa, C.D.; Ormachea, C.; Kneeteman, M.N.; Adam, C.; Mancini, P.M.E. Diels–Alder reactions of N-tosylpirroles developed in protic ionic liquids. Theoretical studies using DFT methods. *Tetrahedron Lett.* **2011**, *52*, 6754–6757. [CrossRef]
14. Greaves, T.L.; Drummond, C.J. Protic Ionic Liquids: Evolving structure–property relationships and expanding applications. *Chem. Rev.* **2015**, *115*, 11379–11448. [CrossRef] [PubMed]
15. Miran, M.S.; Yasuda, T.; Susan, M.A.B.H.; Dokko, K.; Watanabe, M. Binary protic ionic liquid mixtures as a proton conductor: High fuel cell reaction activity and facile proton transport. *J. Phys. Chem. C* **2014**, *118*, 27631–27639. [CrossRef]
16. Lee, S.-Y.; Ogawa, A.; Kanno, M.; Nakamoto, H.; Yasuda, T.; Watanabe, M. Nonhumidified intermediate temperature fuel cells using protic ionic liquids. *J. Am. Chem. Soc.* **2010**, *132*, 9764–9773. [CrossRef]
17. Greaves, T.L.; Drummond, C.J. Protic ionic liquids: Properties and applications. *Chem. Rev.* **2008**, *108*, 206–237. [CrossRef]
18. Hirao, M.; Sugimoto, H.; Ohno, H. Preparation of novel room-temperature molten salts by neutralization of amines. *J. Electrochem. Soc.* **2000**, *147*, 4168–4172. [CrossRef]
19. MacFarlane, D.R.; Forsyth, M.; Izgorodina, E.I.; Abbott, A.P.; Annat, G.; Fraser, K. On the concept of ionicity in ionic liquids. *Phys. Chem. Chem. Phys.* **2009**, *11*, 4962–4967. [CrossRef]
20. Shen, M.; Zhang, Y.; Chen, K.; Che, S.; Yao, J.; Li, H. Ionicity if protic ionic liquid: Quantitative measurement by spectroscopic methods. *J. Phys. Chem. B* **2017**, *121*, 1372–1376. [CrossRef]
21. MacFarlane, D.R.; Seddon, K.R. Ionic liquids–Progress on the fundamental issues. *Aust. J. Chem.* **2007**, *60*, 3–5. [CrossRef]

22. Yunus, N.M.; Halim, N.H.; Wilfred, C.D.; Murugesan, T.; Lim, J.W.; Show, P.L. Thermophysical properties and CO_2 absorption of ammonium-based protic ionic liquids containing acetate and butyrate anions. *Processes* **2019**, *7*, 820. [CrossRef]
23. Petrović, V.P.; Simijonović, D.; Petrović, Z.D.; Marković, S. Formation of a vanillic Mannich base—Theoretical study. *Chem. Pap.* **2015**, *69*, 1244–1252. [CrossRef]
24. Khan, A.; Lu, X.; Aldous, L.; Zhao, C. oxygen reduction reaction in room temperature protic ionic liquids. *J. Phys. Chem. C* **2013**, *117*, 18334–18342. [CrossRef]
25. Kondratenko, Y.A.; Nyanikova, G.G.; Molchanova, K.V.; Kochina, T.A. Characteristics of protic ionic liquids based on triethanolammonium salts of biologically active carboxylic acids and their impact on the growth properties of the Rhizopus oryzae fungus. *Glass Phys. Chem.* **2017**, *43*, 445–451. [CrossRef]
26. Yi, L.; Feng, J.; Li, W.-Y. Separation of phenolic compounds from coal liquefaction oil by choline chloride-glycerol deep eutectic solvents. *Energy Procedia* **2019**, *158*, 5169–5174. [CrossRef]
27. Lu, J.-G.; Li, X.; Zhao, Y.-X.; Ma, H.-L.; Wang, L.-F.; Wang, X.-Y.; Yu, Y.-F.; Shen, T.-Y.; Xu, H.; Zhang, Y.-T. CO_2 capture by ionic liquid membrane absorption for reduction of emissions of greenhouse gas. *Environ. Chem. Lett.* **2019**, *17*, 1031–1038. [CrossRef]
28. Liu, Q.; Mou, L.; Zheng, Q.; Xia, Q. Thermodynamic properties of ionic liquids. In *Progress and Developments in Ionic Liquids*; IntechOpen: London, UK, 2017. [CrossRef]
29. Shen, Y.; Kennedy, D.F.; Greaves, T.L.; Weerawardena, A.; Mulder, R.J.; Kirby, N.; Song, G.; Drummond, C.J. Protic ionic liquids with fluorous anions: Physicochemical properties and self-assembly nanostructure. *Phys. Chem. Chem. Phys.* **2012**, *14*, 7981–7992. [CrossRef]
30. Akbari, F.; Alavianmehr, M.M.; Behjatmanesh Ardakani, R.; Mohammad-Aghaie, D. Thermophysical properties of ionic liquids and their mixtures from a new equation of state. *Ionics* **2018**, *24*, 1357–1369. [CrossRef]
31. Vijayraghavan, R.; Pas, S.J.; Izgorodina, E.I.; MacFarlane, D.R. Diamino protic ionic liquids for CO_2 capture. *Phys. Chem. Chem. Phys.* **2013**, *15*, 19994–19999. [CrossRef]
32. Kurnia, K.A.; Harris, F.; Wilfred, C.D.; Abdul Mutalib, M.I.; Murugesan, T. Thermodynamic properties of CO_2 absorption in hydroxyl ammonium ionic liquids at pressures of (100–1600) kPa. *J. Chem. Thermodyn.* **2009**, *41*, 1069–1073. [CrossRef]
33. Guo, H.; Smith, T.W.; Iglesias, P. The study of hexanoate-based protic ionic liquids used as lubricants in steel-steel contact. *J. Mol. Liq.* **2020**, *299*, 112208. [CrossRef]
34. Kopczyńskia, K.; Gabryelczyka, A.; Baraniaka, M.; Łęgosza, B.; Pernaka, J.; Jankowskab, E.; Rzeszutekc, W.; Kędziorc, P.; Lota, G. Positive electrode material in lead-acid car battery modified by protic ammonium ionic liquid. *J. Energy Storage* **2019**, *26*, 100996. [CrossRef]
35. Mayrand-Provencher, L.; Lin, S.; Lazzerini, D.; Rochefort, D. Pyridinium-based protic ionic liquids as electrolytes for RuO_2 electrochemical capacitors. *J. Power Sources* **2010**, *195*, 5114–5121. [CrossRef]
36. Iglesias, M.; Gonzalez-Olmosa, R.; Cota, I.; Medina, F. Brønsted ionic liquids. Study of physico-chemical properties and catalytic activity in aldol condensations. *Chem. Eng. J.* **2010**, *162*, 802–808. [CrossRef]
37. Chen, Y.; Cao, Y.; Lu, X.; Zhao, C.; Yan, C.; Mu, T. Water sorption in protic ionic liquids: Correlation between hygroscopicity and polarity. *New J. Chem.* **2013**, *37*, 1959–1967. [CrossRef]
38. Freire, M.G.; Santos, L.M.N.B.F.; Fernandes, A.M.; Coutinho, J.A.P.; Marrucho, I.M. An overview of the mutual solubilities of water–imidazolium-based ionic liquids systems. *Fluid Phase Equilibria* **2007**, *261*, 449–454. [CrossRef]
39. Seddon, K.; Stark, A.; Torres, M.-J. Influence of Chloride, Water, and Organic Solvents on the Physical Properties of Ionic Liquids. *Pure Appl. Chem.* **2000**, *72*, 2275–2287. [CrossRef]
40. Yunus, N.M.; Abdul Mutalib, M.I.; Man, Z.; Bustam, M.A.; Murugesan, T. Thermophysical properties of 1-alkylpyridinum bis(trifluoromethylsulfonyl)imide ionic liquids. *J. Chem. Thermodyn.* **2010**, *42*, 491–495. [CrossRef]
41. Gusain, R.; Panda, S.; Bakshi, P.S.; Gardas, R.L.; Khatri, O.P. Thermophysical properties of trioctylalkylammonium bis(salicylato)borate ionic liquids: Effect of alkyl chain length. *J. Mol. Liq.* **2018**, *269*, 540–546. [CrossRef]
42. Wu, B.; Yamashita, Y.; Endo, T.; Takahashi, K.; Castner, E.W. Structure and dynamics of ionic liquids: Trimethylsilylpropyl-substituted cations and bis(sulfonyl)amide anions. *J. Chem. Phys.* **2016**, *145*, 244506. [CrossRef]

43. Pinto, R.R.; Mattedi, S.; Aznar, M. Synthesis and physical properties of three protic ionic liquids with the ethylammonium cation. *Chem. Eng. Trans.* **2015**, *43*, 1165–1170.
44. Chennuri, B.K.; Gardas, R.L. Measurement and correlation for the thermophysical properties of hydroxyethyl ammonium based protic ionic liquids: Effect of temperature and alkyl chain length on anion. *Fluid Phase Equilibria* **2016**, *427*, 282–290. [CrossRef]
45. Keshapolla, D.; Srinivasarao, K.; Gardas, R.L. Influence of temperature and alkyl chain length on physicochemical properties of trihexyl- and trioctylammonium based protic ionic liquids. *J. Chem. Thermodyn.* **2019**, *133*, 170–180. [CrossRef]
46. Zhang, X.U.; Faber, D.J.; Post, A.L.; van Leeuwen, T.G.; Sterenborg, H.J.C.M. Refractive index measurement using single fiber reflectance spectroscopy. *J. Biophotonics* **2019**, *12*, 701–713. [CrossRef] [PubMed]
47. Soave, G. Equilibrium constants from a modified Redlich-Kwong equation of state. *Chem. Eng. Sci.* **1972**, *27*, 1197–1203. [CrossRef]
48. Singh, D.; Gardas, R.L. Influence of cation size on the ionicity, fluidity, and physiochemical properties of 1,2,4-triazolium based ionic liquids. *J. Phys. Chem. B* **2016**, *120*, 4834–4842. [CrossRef]
49. Zec, N.; Vraneš, M.; Bešter-Rogač, M.; Trtić-Petrović, T.; Dimitrijević, A.; Čobanov, I.; Gadžurić, S. Influence of the alkyl chain length on densities and volumetric properties of 1,3-dialkylimidazolium bromide ionic liquids and their aqueous solutions. *J. Chem. Thermodyn.* **2018**, *121*, 72–78. [CrossRef]
50. Tariq, M.; Forte, P.A.S.; Gomes, M.F.C.; Lopes, J.N.C.; Rebelo, L.P.N. Densities and refractive indices of imidazolium- and phosphonium-based ionic liquids: Effect of temperature, alkyl chain length, and anion. *J. Chem. Thermodyn.* **2009**, *41*, 790–798. [CrossRef]
51. Liu, Q.-S.; Yang, M.; Li, P.-P.; Sun, S.-S.; Welz-Biermann, U.; Tan, Z.-C.; Zhang, Q.-G. Physicochemical properties of ionic liquids [C_3py][NTf_2] and [C_6py][NTf_2]. *J. Chem. Eng. Data* **2011**, *56*, 4094–4101. [CrossRef]
52. Liu, Q.-S.; Li, P.-P.; Welz-Biermann, U.; Chen, J.; Liu, X.-X. Density, dynamic viscosity, and electrical conductivity of pyridinium-based hydrophobic ionic liquids. *J. Chem. Thermodyn.* **2013**, *66*, 88–94. [CrossRef]
53. Liu, Q.-S.; Li, Z.; Welz-Biermann, U.; Li, C.-P.; Liu, X.-X. Thermodynamic properties of a new hydrophobic amide-based task-specific ionic liquid [EimCH_2CONHBu][NTf_2]. *J. Chem. Eng. Data* **2013**, *58*, 93–98. [CrossRef]
54. Glasser, L. Lattice and phase transition thermodynamics of ionic liquids. *Thermochim. Acta* **2004**, *421*, 87–93. [CrossRef]
55. Bandrés, I.; Royo, F.M.; Gascón, I.; Castro, M.; Lafuente, C. Anion influence on thermophysical properties of ionic liquids: 1-butylpyridinium tetrafluoroborate and 1-butylpyridinium triflate. *J. Phys. Chem. B* **2010**, *114*, 3601–3607. [CrossRef] [PubMed]
56. Greaves, T.L.; Weerawardena, A.; Fong, C.; Krodkiewska, I.; Drummond, C.J. Protic ionic liquids: solvents with tunable phase behavior and physicochemical properties. *J. Phys. Chem. B* **2006**, *110*, 22479–22487. [CrossRef]
57. Cai, G.; Yang, S.; Zhou, Q.; Liu, L.; Lu, X.; Xu, J.; Zhang, S. Physicochemical properties of various 2-hydroxyethylammonium sulfonate -based protic ionic liquids and their potential application in hydrodeoxygenation. *Front. Chem.* **2019**, *7*, 196. [CrossRef]
58. Dharaskar, S.A.; Wasewar, K.L.; Varma, M.N.; Shende, D.Z.; Yoo, C.K. Synthesis, characterization and application of 1-butyl-3-methylimidazolium tetrafluoroborate for extractive desulfurization of liquid fuel. *Arab. J. Chem.* **2016**, *9*, 578–587. [CrossRef]
59. Efimova, A.; Hubrig, G.; Schmidt, P. Thermal stability and crystallization behavior of imidazolium halide ionic liquids. *Thermochim. Acta* **2013**, *573*, 162–169. [CrossRef]
60. Dharaskar, S.A.; Wasewar, K.L.; Varma, M.N.; Shende, D.Z.; Yoo, C.K. Deep removal of sulfur from model liquid fuels using 1-butyl-3-methylimidazolium chloride. *Procedia Eng.* **2013**, *51*, 416–422. [CrossRef]

processes

MDPI

Review

Wood Ash Based Treatment of Anaerobic Digestate: State-of-the-Art and Possibilities

Alejandro Moure Abelenda and Farid Aiouache *

Department of Engineering, Lancaster University, Lancaster LA1 4YW, UK; alejandro.moure.abelenda@gmail.com
* Correspondence: f.aiouache@lancaster.ac.uk; Tel.: +44-1524-593526

Abstract: The problem of current agricultural practices is not limited to land management but also to the unsustainable consumption of essential nutrients for plants, such as phosphorus. This article focuses on the valorization of wood ash and anaerobic digestate for the preparation of a slow-release fertilizer. The underlying chemistry of the blend of these two materials is elucidated by analyzing the applications of the mixture. First, the feasibility of employing low doses (≤1 g total solids (TS) ash/g TS digestate) of wood ash is explained as a way to improve the composition of the feedstock of anaerobic digestion and enhance biogas production. Secondly, a detailed description concerning high doses of wood ash and their uses in the downstream processing of the anaerobic digestate to further enhance its stability is offered. Among all the physico-chemical phenomena involved, sorption processes are meticulously depicted, since they are responsible for nutrient recovery, dewatering, and self-hardening in preparing a granular fertilizer. Simple activation procedures (e.g., carbonization, carbonation, calcination, acidification, wash, milling, and sieving) are proposed to promote immobilization of the nutrients. Due to the limited information on the combined processing of wood ash and the anaerobic digestate, transformations of similar residues are additionally considered. Considering all the possible synergies in the anaerobic digestion and the downstream stages, a dose of ash of 5 g TS ash/g TS digestate is proposed for future experiments.

Keywords: anaerobic digestion; greenhouse gas mitigation; phosphate leaching; ammonia stripping; sorption activation; acid surfactants; dewatering; self-hardening; maturation; biofertilizer

Citation: Abelenda, A.M.; Aiouache, F. Wood Ash Based Treatment of Anaerobic Digestate: State-of-the-Art and Possibilities. *Processes* **2022**, *10*, 147. https://doi.org/10.3390/pr10010147

Academic Editor: Yingnan Yang

Received: 15 December 2021
Accepted: 6 January 2022
Published: 11 January 2022

1. Introduction

The growth of the human population and the change in their diet (e.g., more consumption of animal products) imply devoting more land to food production at an alarming rate [1]. The problem of our current agricultural practices is not limited to land management but also to the unsustainable consumption of essential nutrients for plants, such as phosphorus [2,3]. The fluctuation in prices helps preserve the long-term availability of these mineral resources [4]. In the present situation, the development of profitable strategies for utilizing the waste materials is the only way to achieve sustainable development of the society [5]. Once a material is regarded as waste, its utilization is constrained by regulations [6]. In Europe, all waste-derived products must comply with Directive 2018/851, but there are shortcuts to achieve end-of-waste (EoW) status. Generally, when a residue is produced in large amounts and has a composition suitable for a particular application, the European parliament applies EoW regulations that do not need to be transposed by the governments of each member state [7].

The production of digestate was around 180 million tons per year in the EU28, before the UK withdrawal in 2020 [8]. In the UK, 7.5 million tons of anaerobic digestate and 2.7 million tons of compost were produced in 2018 [9]. A common policy is being developed to improve the management of all nitrogenous materials employed as soil amendments [10,11]. In 2009 [12], the EU27 produced 256 million tons of municipal solid waste (MSW). Much more animal manure is applied to land as a soil amendment than anaerobic digestate because of the high CAPEX and OPEX associated with implementing anaerobic digestion (AD) in the farms [13]. The biological treatment of organic residues via AD is a promising technology for energy recovery in the form of biogas and the production of fertilizers [14]. The AD was first applied to deal with the sewage sludge (SS) produced during the primary and secondary treatment of wastewater, but its use to deal with agro-waste and the MSW with around 70% of organic material has been subsequently encouraged [15,16]. The AD technology fits with the current trends of using the biorefinery to sustainably satisfy the needs of society, which used to be covered by the petrochemical industry [17].

Vassilev et al. [18] estimated that approximately 476 million tons of biomass ash could be generated worldwide annually if the average ash content is 6.8% and the burned biomass is assumed to be 7 billion tons. This amount resembles the 780 million tonnes of coal ash generated every year [18]. According to Pitman [19], the wood ash (WA) production could be estimated by considering that 1% of the wood incinerated is left as ash. According to the UK Forestry Commission [20], 2.68 million tons of wood fuel were used in 2016. Even considering the combustion of other types of biomass, the amount of biomass ash currently produced in the UK is lower than that coming from the coal power plants. In 2016, the amount of coal ash produced in the UK was around 6 million tons [21]. However, this scenario is changing, and by 2025 the amount of biomass ash is expected to be greater than that of coal ash [22]. In 2007, Sweden produced 0.3 million tons of WA, and most of this material was disposed of in landfills [23]. The preparation of a blend of organic manure and ash to enhance the circular economy has been addressed in the literature from several points of view:

- To enhance the AD process [24].
- To improve the properties of the soil [25].
- To promote crop growth [26].
- To control the release of the nutrients [27].
- To achieve better dewatering of the digestate [28].
- To promote the self-hardening and granulation of the digestate [29].

The anaerobic digestates, animal manures, and slurries are appreciated by the farmers as organic soil amendments. Improving the nutrient profile of the digestate might be a way of increasing its value and decreasing the cost of transportation. Moreover, there are more urgent challenges that need to be addressed, such as the pollution associated with the use of these materials [10]. It is necessary to optimize the conditions of the WA-based treatment of the anaerobic digestate to improve the stability of the soil organic amendment, improve the nutrient use efficiency, and reduce the contamination of the environment. The objectives of this review are (a) to discuss the underlying chemistry of the blends of anaerobic digestates and WA, and (b) to design a process that only requires affordable and widely available resources to prepare the novel fertilizer.

2. Upstream Processing

2.1. State-of-the-Art of AD

The AD process is also known as bio gasification due to the intervention of microorganisms to convert the components of the organic matter to methane and carbon dioxide. The most accepted model, on how the reactions occur in the anaerobic digester in the absence of oxygen, describes 4 different stages: hydrolysis, acidogenesis, acetogenesis, and methanogenesis (Figure 1). The limiting step is hydrolysis, while the methanogenesis is quick, preventing the accumulation of volatile fatty acids (VFA) in the system. In this way, the concentration of VFA in the anaerobic digest ester could be used as an indicator of the correct stationary operation of the bioreactor [17].

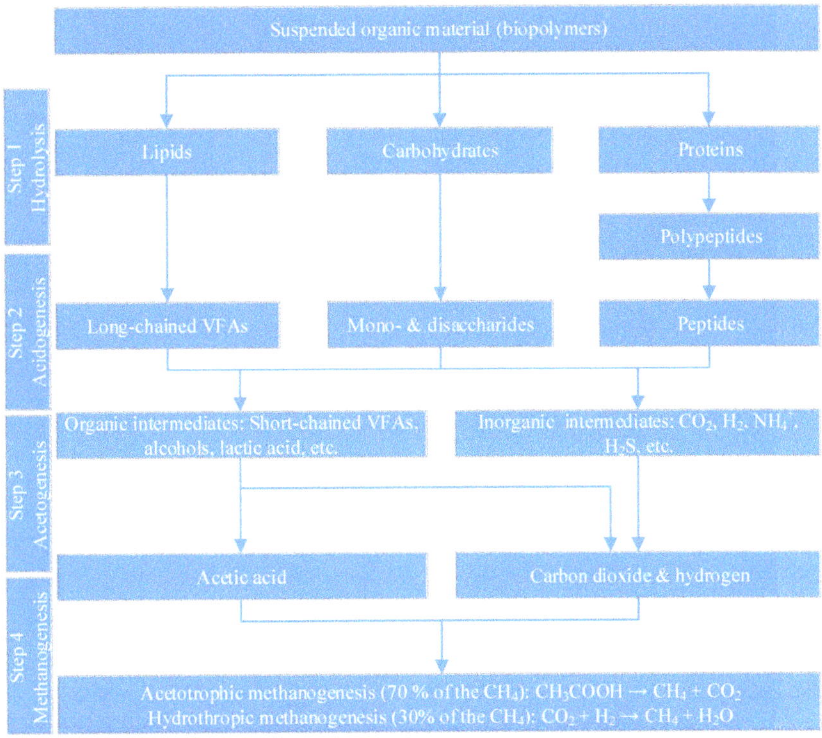

Figure 1. Breakdown of the AD process. Modified from Madsen et al. [17]. Reproduced with permission of Renewable and Sustainable Energy Reviews (Elsevier).

2.2. Use of WA as an Additive for Enhancing the AD

The addition of WA to the feedstock of AD can be understood as a way of promoting the stabilization of the anaerobic digestate because the greater the biogas production during the AD, the lower the emission of methane and carbon dioxide occurrence during storage and land application of the soil organic amendments [30]. Concerning the use of WA as an additive for AD, to the authors' knowledge, 6 experimental studies have been reported so far on this topic (Table 1). Furthermore, Alavi-Borazjani et al. [31,32] provided an insight into the feasibility of the utilization of biomass ashes in AD and biogas upgrading. According to Alavi-Borazjani et al. (2020), biomass ashes are cheap sources of alkali suitable for controlling the excessive acidification of anaerobic digesters. Biogas production is also enhanced due to the supplementation with the macro and micronutrient present in biomass ashes, which are required for the anaerobic microbes, and only a reference of WA was mentioned in relation to the biogas upgrading. Raw biogas of approximately 60% methane does not have enough quality for its introduction into the natural gas grid. Most frequently used purification technologies (water/amine scrubbing, pressure swing adsorption, cryogenic distillation, and membrane) require an excessive amount of energy and chemicals. The main targeted compounds to be removed from the biogas are CO_2 and H_2S. Given the complexity of the WA, there is a knowledge gap on the absorption capacity of bulk WA, and gathering more information to fully appreciate the valorization opportunities of this material is required [31,32].

Table 1. Summary of the experimental studies on the use of WA in the AD.

Reference	1. Type of Ash and Incinerator 2. Type of Substrate 3. Type of Anaerobic Digester 4. Ash Dose	1. Biogas Production 2. Composition of Anaerobic Digestate
[33]	1. *Bauhina monandrina* plant ash 2. Pig waste and cassava peel waste (25% *w*/*v*) and distilled water (75% *w*/*v*) 3. 45-day biochemical methane potential (BMP) experiment conducted in 2.8 L digesters 4. Ash seeding rate not specified	1. Ash-amended digester provided more than twice the biogas production during the 45-day incubation (2345 mL) 2. The pH of the digester sown with WA was 4.40, but the methanogen archaea were able to adapt to the acidic conditions of the digester media
[34]	1. Not specified type of WA 2. Brewery spent grains with rumen liquor 3. 1 L conical flasks with 100 g spent grains of 14 days of hydraulic retention time (HRT) 4. Ash to grains ratios 0:1, 1:5, 1:4, and 1:3. Several dilution ratios of the feedstock with water were tested (1:4, 1:6, 1:8, and 1:10 *w*/*v*)	1. Greatest release of biogas obtained with the lower feedstock to water ratio (1:4 *w*/*v*). Enhancement of biogas production: WA > poultry litter (PL) > urea 2. WA led to an initial increase in the carbon content, due to the accumulation of the VFA. Once the methanogenesis reached a stationary state, the carbon-to-nitrogen mass ratio (C/N) decreased, and an organic manure with better fertilizing properties than the initial feedstock of AD was obtained
[35] Cited by the 3 articles below employing WA in AD	1. MSW fly ash from industrial incinerator 850–1050 °C. The flue gases were treated with semi-dry scrubber using $Ca(OH)_2$ as a flushing agent, powdered carbon, and filter bag 2. Synthetic MSW substrate and anaerobic sludge seeding 3. 35 °C leaching/percolate (100 mL/day) bed reactor. A schematic diagram can be found in the Supplementary Material (Figure S1) 4. 0, 10 and 20 g MSW fly ash/L MSW (equivalent to 0.2 and 0.4 g/g vs. or 0.17 and 0.33 g/g TS, respectively) 5. Lo et al. [36] tested 100 g MSW bottom ash/L MSW, 2 g MSW bottom ash/g vs. or 1.67 g MSW bottom ash/g TS	1. Biogas production rate in ash-added digesters was higher than in control experiment (i.e., unamended BMP) but total yield followed the trend: 20 g/L > 0 g/L/ > 10 g/L > 100 g/L [36] 2. Drop in the pH due enhancement of the hydrolytic and acidogenic processes. After the 84 days of operation, a decrease of volatile solids (VS) and accumulation of the VFA due to a decrease of methanogenic activity
[37]	1. Wood bottom ash (WBA) 2 mm sieved and 105 °C dried 2. 2 mm sieved cattle slurry (5 weeks AD) 3. 1 L; 37 °C; 50 rpm; 20 day retention time. 4. 0 and 0.5 g ash/g total solids (TS), which is equivalent to 0.96 g WA/g vs. or 9.65 g WBA/L digested cattle slurry	1. Biogas stopped after the addition of ash, and higher production rates were found when the doe of ash decreased down to 0.25 g/g TS (0.48 g/g vs. or 4.83 g/L digested cattle slurry). The biogas increased the content of CH_4 2. The ammoniacal nitrogen (NH_4^+-N) was not affected by the ash amendment but increased. With excessive ash dose, C/N increased due to the accumulation of VFA.

Table 1. *Cont.*

Reference	1. Type of Ash and Incinerator 2. Type of Substrate 3. Type of Anaerobic Digester 4. Ash Dose	1. Biogas Production 2. Composition of Anaerobic Digestate
[38]	1. Wood fly ash from combined heat and power plant using wood residues as a fuel of a fluidized bed incinerator. 2. Thickened wastewater treatment plant (WWTP) sludge 3. 37 °C for 4 weeks. Schematic of the setup in the Supplementary Material (Figure S2) 4. 0, 0.4, 0.7 and 1 g ash/g vs. or 0, 0.28, 0.49, and 0.70 g ash/g TS or 0, 0.008, 0.015, 0.022 g ash/L, respectively	1. After 3 to 5 days, the production of biogas is comparable to the control experiment because the accumulation of VFA decreases the pH 2. The elevated pH at the beginning of the AD was responsible for the longer lag phase when adding the WA to the digester
[39]	1. Boiler ash granular (0.85–4.75 mm) and powdered (<0.075 mm): Conventional fixed grate biomass combustion system fueled by pine, spruce, and fir bark and some dry wood shavings 2. Fermented WWTP sludge 3. BMP at mesophilic conditions (38 °C). Schematic can be found in the Supplementary Material (Figure S3) 4. WA doses ranging from 0.16 to 3.7 g/g vs. fermented WWTP sludge, which is equivalent to 0.14 and 3.2 g ash/g TS or 5.72 and 133.94 g ash/L, respectively	1. WA increased the lag phase of AD and decreased the total biogas produced 2. WA accelerated the biodegradation of propionic and butyric acids. Even at a rate of 2.2 g ash/g $VS_{substrate}$ the WA poorly contributed to the total alkalinity; thus, the authors recommend using a buffer (e.g., 20 mmol/L Na_2CO_3 & $KHCO_3$ even when adding the ash). The authors found that the granular WA was able to remove NH_4^+-N from the liquid phase due to adsorption via cation exchange

2.3. *Stability and Maturity of the WA Amended Anaerobic Digestate*

In some UK and European regulations, the terms stability and maturity are used indistinctly to describe the properties of a soil organic amendment [8,40,41]. Stability is related to reactions affecting the fate of carbon, which is the most abundant element in organic materials, and the maturity focuses on all the other elements (Figure 2). Some are nutrients necessary for plant growth (e.g., N, P, K, etc.), while others are phytotoxic compounds (e.g., Cd, Hg, Pb, etc.) that limit seed germination and root development. It is important to mention that the excess of any type of nutrients has a detrimental effect on the soil biota. Because nitrogen is the most abundant of these nutrients, its mineralized form, NH_4^+-N, could measure maturity [42–44].

The aim of determining the stability is to determine the fate of the carbon present in the labile and stable organic matter, since the inert organic matter will not suffer any degradation [45]. As displayed in Figure 2, the carbon could be (a) assimilated by the microorganisms for their growth, (b) lost via respiration, also known as carbon mineralization, and to a lesser extent, (c) lost due to leaching of the low organic molecular weight compounds. The carbon use efficiency (CUE) only accounts for the C used for the microbial growth, hence this parameter could be used to directly measure the stability of an organic amendment for more efficient nutrient management. However, the most common way of determining the stability is by measuring the losses via respiration [8,41,46–48], although the test is not carried out in similar conditions to the land application but in AD conditions as per the BMP protocol.

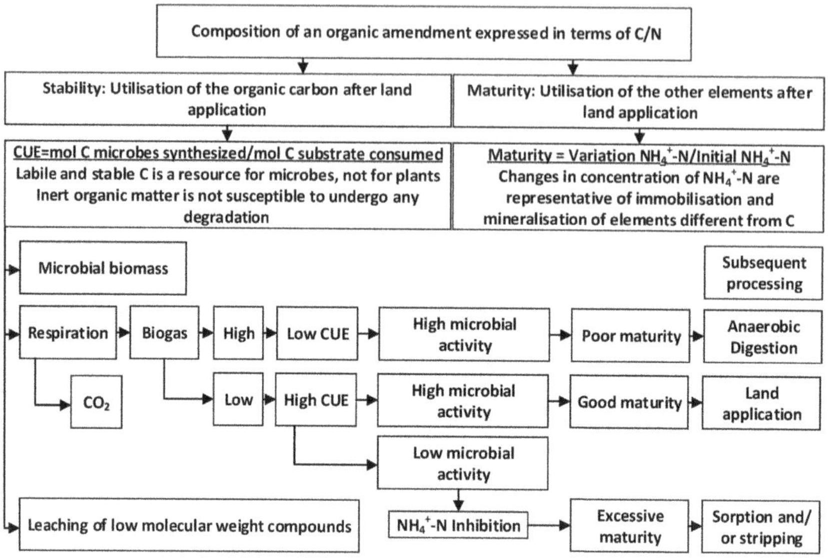

Figure 2. Evaluation of the quality of an organic amendment in terms of stability and maturity [42,49–53].

In a BMP test of an organic amendment (Figure 3), the release of biogas due to microbial activity depended on the composition expressed as the C/N and the relative amount of this material with respect to the microbial biomass (i.e., substrate-to-inoculum ratio, S/I), expressed as organic loading rate (OLR) in a continuous reactor. Together with the HRT, the OLR (Equation (1)) is one of the most important parameters in the operation of an anaerobic reactor, and both parameters are used for the design of the anaerobic digester [54,55]. The scale of the OLR for the operation of a continuous reactor, equivalent to the S/I in a batch reactor. Figure 3 included the OLR in the X axis based on the Figures 4 and 5 of Rincón et al. [56], who reported similar trends of biogas release and vs. removal. Reproduced with the permission of Biochemical Engineering Journal (Elsevier). In their original manuscript, the OLR was expressed on chemical oxygen demand (COD) basis, however, it is recommended to use the vs. instead [57].

$$\mathrm{OLR}\left(\mathrm{kg}\frac{\mathrm{VS}}{\mathrm{m^3 day}}\right) = \frac{\mathrm{Input}\left(\frac{\mathrm{kg\ substrate}}{\mathrm{day}}\right) \times \mathrm{Labile\ matter}\left(\frac{\mathrm{kg\ VS}}{\mathrm{kg\ substrate}}\right)}{\mathrm{Net\ digester\ volume\ (m^3)}} \tag{1}$$

The highest CUE, and therefore the highest stability of the organic amendment, were obtained at the lowest C/N (Figure 3) because the microbes have all the nutrients that they need to build their cell structures and do not need to get rid of excess carbon. On the other hand, when a low amount of substrate is available (OLR < 6), the microbes are in starvation mode, and they assimilate the carbon for their growth more efficiently, compared to higher OLR conditions under which more carbon is lost in microbial respiration due to excessive microbial activity (Figure 2). Since less carbon ends up in the microbial biomass, which is measured as part of the VS, the CUE decreases. Despite the different physiology of the terrestrial microorganisms from the consortium of the AD, the trend of CUE measured in an aerobic environment is useful to explain the fate of the carbon during batch and continuous AD operations (Figure 3). There are similar processes involved in the nutrient turnover in the AD and once the soil organic amendment is applied to land. The maturity profile is offered in the Supplementary Material (Figure S4).

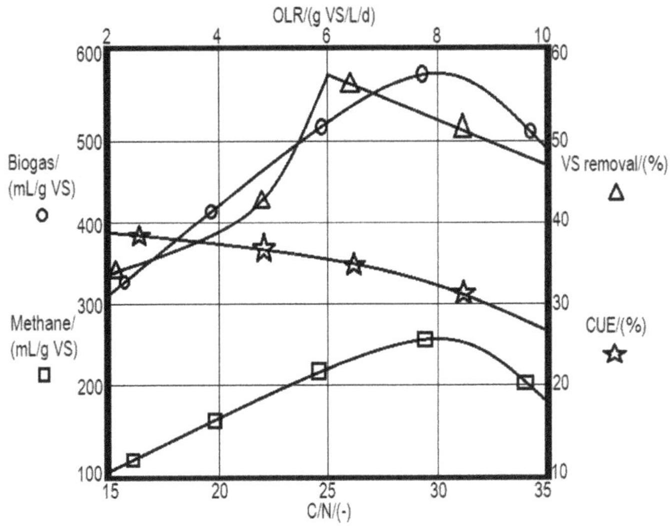

Figure 3. Application of the concept of CUE from soil science to the AD process. The relation between the CUE and the C/N for terrestrial decomposers was taken from the Figure 1a of Sinsabaugh et al. [53]. Reproduced with the permission of Ecology Letters (John Wiley & Sons). The relation between the biogas yield, the methane yield, and the vs. removal with the C/N for the operation of a discontinuous reactor was taken from Figure 3 of Wang et al. [58], who carried out 30-day AD experiments with a S/I of 0.5 (expressed in terms of VS). Reproduced with the permission of Bioresource Technology (Elsevier).

Theorem 1. *Microbes get rid of any nutrient in excess available in the medium (e.g., anaerobic digester, soil solution, etc.), hence the composition of the organic amendment should be as similar as possible to the composition of the microbes.*

If carbon is in excess (C/N $\geq$ 35), the microbes cast aside this element via respiration or extracellular polymeric substances (EPS) segregation [59]. The composition of the AD feedstock has been traditionally measured using the C/N, and this parameter has been employed for other types of fermentations, such as composting and land application of the organic soil amendment [60]. Möller & Müller [61,62] emphasized that a more representative ratio to express the composition of the organic soil amendment would be organic carbon (C_{org}) to organic nitrogen. Some researchers reported carbon as the main component of the wood ashes [62,63]. According to Forbes et al. [64], the continuum of combustion products such as char, ash, and charcoal are referred to as black carbon. This feature makes the properties of the wood ashes more suitable to be employed as a land amendment, to restore the soil as a carbon sink, maintain an adequate microbial activity in the soil, appropriate management of the nutrients to boost crop growth, and prevent pollution via leaching and gaseous emissions [65]:

Proof of Theorem 1. According to the British Standards Institution's Publicly Available Specification (BSI PAS 110:2014) for anaerobic digestate [46], the upper limit for a stable organic amendment in the 28-day test is 450 mL biogas/g VS, which is in agreement with Figure 3. It should be noted that this threshold value was established in the conditions of no inhibition of the test, which could be related to an inappropriate S/I (i.e., OLR > 10 g VS/L/d) and/or low nutrient content (i.e., C/N > 35). Profiles of biogas production indicating inhibitory conditions of the BMP test are displayed in the Supplementary Materials (Figures S5–S7). It is important to highlight that nitrogen plays a role in enhancing biogas production but supplementing the organic amendment with other nutrients is advised [66], which is the reason for seeding the AD feedstock with ashes [67]. Equation (2) represents how the heterotrophic microbial biomass ($C_5H_7NO_2$) is chemically built from glucose ($C_6H_{12}O_6$), as a source of carbon, and NH_3, as a source of nitrogen:

$$3C_6H_{12}O_{6(s)} + 8O_{2(g)} + 2NH_{3(g)} \rightarrow 2C_5H_7NO_{2(s)} + 8CO_{2(g)} + 14H_2O_{(l)} \quad (2)$$

The maximum theoretical CUE is approximately 56% (2 × 5/3/6 × 100) in aerobic fermentation [68], in agreement with Figure 3. The reason for measuring the stability as a release of biogas could be the fact that there is no accurate quantification of the microbial biomass available in the biodigester, and this is often measured as suspended vs. [69]. The polymerase chain reaction to quantify only the active cells in the anaerobic digester is in the development stage and requires equipment advances [70]. The CUE measured in soil science is a more straightforward technique that relies on the extraction of the microbial carbon with a solution of 0.5 M K_2SO_4 after fumigation with chloroform [71]. □

3. Downstream Processing

As in the previous section about upstream processing, this section of downstream processing explains the possible synergistic effects found by adding the WA to the anaerobic digestate. The WA can be added to the anaerobic digestate after the biodigester to ease the management of the organic material by developing a process complying with green chemistry principles [72], such as minimum input of energy and resources, due to all the synergies involved. The role of the WA as sorbent is essential to produce a controlled-release fertilizer derived from the anaerobic digestate. It should be noted that there are other materials (e.g., ion exchange resins) that could also be employed for the same purpose [73]. The sorption (adsorption and/or absorption; [62]) will take place from the moment in which the wood ash touches the anaerobic digestate; thereby this phenomenon is implicitly involved in each step of the process of Figure 4.

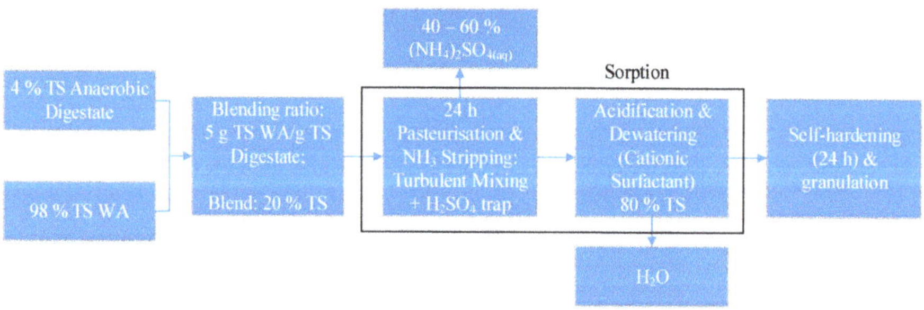

Figure 4. Schematic block flow diagram of the proposed downstream processing of the anaerobic digestate with WA to produce a granular organic fertilizer (%TS = percentage of total solids). Elaborated considering the different treatments, synergies, and technologies that can be implemented with a blend of WA and anaerobic digestate [28,29,35,37–39,74,75].

The dose of WA depends on the intended properties of the fertilizer to be manufactured. The following are some of the considerations that were taken into account to establish the blending ratio of 5 g TS WA/g TS anaerobic digestate (Figure 4):

1. Enhancement of AD could be achieved by preparing the feedstock with coal fly ash (CFA) using a dose as high as 2% (*w*/*w*) [76].
2. Precipitation of struvite and adsorption of phosphate achieved by preparing a suspension with up to 3% (*v*/*w*) ash in swine wastewater [77].
3. Application to land of both raw materials following a blending of up to 5% (*w*/*w*) of ash [78].

4. Mitigation of CH_4, CO_2, NH_3, N_2O associated with the storage of cattle slurry with 4.6% TS content by adding charcoal or biochar at a rate of 4.5% (*w*/*v*) [79] and around 11% (*w*/*w*) [80].
5. Adjustment of the pH to 5.5 of untreated pig slurry and co-digested pig slurry by adding 2% and 3.5% (*w*/*w*) of powdered aluminium sulfate ($Al_2(SO_4)_3$), respectively [81,82]
6. Supplementation of the anaerobic digestate by means of a WA dose up to 9.99% (*w*/*w*) or 3.09 g TS ash/g TS digestate to improve the nutrient ratio (C:N:P), the availability of phosphorus, and the microbial activity in the soil [60].
7. Agreement with the regulations regarding the maximum content of heavy metals present in the anaerobic digestate [46,83]. The share of WA should not be greater than 15.51% (*w*/*w*) relative to the anaerobic digestate or 1.47 g TS WA/g TS AD. These results were obtained by considering the maximum content of heavy metals in the WA values established in the UK Quality Protocol of PL ash [84]. The content of Zn was found to be the limiting factor. The assumption of these calculations (Tables S1 and S2) are described in the Supplementary Material.
8. Prevention of a large volume of dewatered digestate obtained via filtration by using as much CFA as the dry matter of the digestate (i.e., 1 g TS CFA/g TS digestate) to assist the dewatering process [28].
9. Moure Abelenda et al. [85–88] tested alkaline and acid conditions to minimize the volatilization of NH_3, and carbon and PO_4^{3-} solubilization. They obtained better results (i.e., lower availability of nitrogen, carbon, and phosphorus) under acid conditions (4.39 g TS WA/g TS digestate) than under alkaline conditions (5.51 g TS WA/g TS digestate).
10. Alkaline stabilization of sewage sludge via liming with a dose of CaO as high as 40% (*w*/*w*) or 8 g TS CaO/g TS sewage sludge to decrease the pathogens (Méndez et al., 2002). A dose of 3.82 g TS CaO/g TS digestate or 224.5 g CaO/L digestate (5.88% TS) was required for reaching a pH 12 and removing 51.2% of the NH_4^+-N due to NH_3 volatilization [74]. Limoli et al. [74] reported that a low dose of 45 g/L increased the TS content of the manure digestate by 42.7%. When the organic material had higher dry matter (25.4% TS content), a dose of 50 g CaO/kg SS represented an increase in the TS content of approximately 30% and just 2 units of pH. This liming effect reduced the availability of heavy metals in the SS [89].
11. Reducing phosphate availability by adding 5.6 kg of CFA to each kilogram of dairy slurry [90] could present a dose of greater than 110 g TS CFA/g TS slurry if the moisture content of the organic manure is 95%.
12. Preparation of granules prepared with 100% (*w*/*w*) biomass ash showed the best mechanical properties. Decreasing the content to 80% bio ash and 20% dewatered SS (45% moisture) significantly affected the compressive strength of the pellets [29]. The lowest dose of bio ash and $Ca(OH)_2$ that Pesonen et al. [29] tested corresponded to a 5.19 g TS bio ash + $Ca(OH)_2$/g TS hygienized SS.

3.1. Pasteurization and Sterilization

According to the animal by-product regulations for biogas plants in the UK [91], unless the AD is done in the conditions of Table 2, pasteurization (70 °C for 1 h) of the digestate is required before using the digestate for any application [46]. A considerable reduction in the number of pathogens can be achieved during fermentation, especially if it is done in thermophilic conditions [17]. Also, the best control of the digester was obtained in the one-stage non-mixing thermophilic reactor [92].

Table 2. Conditions to produce a sanitized digestate when using animal by-products (e.g., manure and slurry) as feedstock in AD [91].

Minimum Temperature	Minimum Time	Maximum
57 °C	5 h	50 mm

Greater doses of WA than that used to improve the performance of the AD can be employed with pasteurization and sterilization purposes. While sterilization implies the destruction of all life forms in the anaerobic digestate, the pasteurized material might contain beneficial or harmless microorganisms [46]. The aim is to keep the pH of the organic manure above 12 for at least 2 h [93,94]. This is often achieved using a rate of application of WA to organic material of 0.1 g $Ca(OH)_2$ per g TS [93,95]. It should be noted that calcium represents approximately 18% of the total weight of the WA [96], thus this element is one of the main components of this material [97]. This alkaline treatment, similar to the use of lime, reduces the storage time of organic manures by 3 months while still preventing the contamination of the crops by pathogens [98,99]. The shares of the calcium as oxide, hydroxide, and carbonate in the wood ashes are mainly determined by the temperature of incineration and storage conditions. Under 500 °C, carbonates and bicarbonates predominate, while oxides require temperatures around 1000 °C [96]. During storage, the reaction with the moisture and the CO_2 in the atmosphere leads to the formation of hydroxides and carbonates, although converting to CaO is possible via calcination of the ashes at temperatures over 500 °C [100,101].

3.2. Nitrogen Recovery Technologies

In this subsection, only the processes of NH_3 stripping, manufacturing of $(NH_4)_2CO_3$, and struvite ($MgNH_4PO_4 \cdot 6\ H_2O$) precipitation and sorption of the nitrogen in the soil organic amendments are described for the exploitation of this element as fertilizer because these technologies are the most convenient for the implementation with wood ashes [102]:

3.2.1. NH_3 Stripping Processes from the WA Anaerobic Digestate

Limoli et al. [74] tested the addition of the CaO to the anaerobic digestate to increase the pH and promote the volatilization of NH_3. Less than 1% of the NH_4^+ & NH_3 is volatilized as part of the biogas released during the AD [61]. Limoli et al. [74] described a stepwise mechanism involving the ammonium dissociation ($NH_4^+ \rightarrow NH_3 + H^+$) and the mass transfer in the water-air interface. Since the WA primarily consists of calcium and other alkaline elements, this material can be used to increase the pH of the anaerobic digestate and promote the volatilization of NH_3. Limoli et al. [74] did not consider how much of this nitrogen ends up in a trap containing a sulfuric acid (hereinafter H_2SO_4 trap). The resulting 40–60% ammonium sulfate ($(NH_4)_2SO_4$) solution used can be as commercial-grade fertilizer (Equation (3)), although this will depend on the organic contamination of the liquid fertilizer [103]. If the gas-liquid contact system employs a high volume of air to strip the NH_3 from the anaerobic digestate, the cost of the process increases since the absorption in the H_2SO_4 solutions is hindered. It should be noted that the recovery of the NH_3 from the gaseous stream has complications, and often the H_2SO_4 traps require calibration to determine the efficiency of this technology [104].

$$2NH_{3(g)} + H_2SO_{4(aq)} \rightarrow 2NH^+_{4(aq)} + SO^{2-}_{4(aq)} \rightarrow (NH_4)_2SO_{4(s)} \quad (3)$$

The reason for the costly amount of alkali required to basify the anaerobic digestate is the fact that there are 3 buffer equilibria (Equations (4)–(6)) responsible for the pH in the anaerobic digestate [61]:

$$NH^+_{4\ (aq)} \leftrightarrow H^+_{\ (aq)} + NH_{3(aq)} \quad (4)$$

$$CO_{2(aq)} + H_2O_{(l)} \leftrightarrow H_2CO_{3(aq)} \leftrightarrow H^+_{\ (aq)} + HCO^-_{3\ (aq)} \leftrightarrow 2H^+_{\ (aq)} + CO^{2-}_{3\ (s)} \quad (5)$$

$$CH_3COOH_{(aq)} \leftrightarrow 2H^+{}_{(aq)} + CH_3COO^-{}_{(aq)} \quad (6)$$

Since oxides, hydroxides, and carbonates are the most common forms of metals in wood ashes [96], the reactions of the WA in an aqueous solution, involving the alkali and alkaline metals, should be (Equations (7) and (8)):

$$K_2O_{(s)} + H_2O_{(l)} \rightarrow 2K^+_{(aq)} + O^{2-}_{(aq)} + H_2O_{(l)} \rightarrow 2K^+_{(aq)} + 2OH^-_{(aq)} \quad (7)$$

$$CaO_{(s)} + H_2O_{(l)} \rightarrow Ca^{2+}_{(aq)} + O^{2-}_{(aq)} + H_2O_{(l)} \rightarrow Ca^{2+}_{(aq)} + 2OH^-_{(aq)} \quad (8)$$

After the dissociation of the alkaline oxide, the oxide anion (O^{2-}) rapidly reacts with water [105,106], leading to the formation of a hydroxide anion (OH^-). Also, basic cations (e.g., Ca^{2+} and K^+) increase the pH because the electric charge of the liquid digestate needs to be neutral, thus the concentration of H^+ is lower [61].

$$K_2CO_{3(s)} + H_2O_{(l)} \rightarrow 2K^+_{(aq)} + CO^{2-}_{3\,(aq)} + H_2O_{(l)} \rightarrow 2K^+_{(aq)} + CO_{2\,(g)} + 2OH^-_{(aq)} \quad (9)$$

$$CaCO_{3(s)} + H_2O_{(l)} \rightarrow Ca^{2+}_{(aq)} + CO^{2-}_{3\,(aq)} + H_2O_{(l)} \rightarrow Ca^{2+}_{(aq)} + CO_{2\,(g)} + 2OH^-_{(aq)} \quad (10)$$

3.2.2. Manufacturing of $(NH_4)_2CO_3$

It is worth mentioning that Equations (4) and (5) are connected by the precipitation of $(NH_4)_2CO_3$, which naturally increases the pH of the anaerobic digestate. This is, in fact, another route to recover the NH_4^+ contained in the anaerobic digestate. The NH_4HCO_3 is a marketable fertilizer that contains 18% of nitrogen and minimizes the losses of nitrogen during the processing of the anaerobic digestate. Unlike in the stripping of NH_3 by increasing the pH of the anaerobic digestate and subsequent reaction with the H_2SO_4, raising the pH of the digestate by adding WA is not useful because the high pH prevents the release of the CO_2. It should be noted, however, that the WA could be, though as a source of $CO_3{}^{2-}$-C to increase the share of inorganic carbon in the anaerobic digestate. In the process designed by Drapanauskaite et al. [107], the simulated liquid fraction of the anaerobic digestate contained 20 times more $HCO_3{}^-$ than NH_4^+. The $CO_3{}^{2-}$-C content is usually not reported in AD studies [55] but the total alkalinity [49] (expressed as a concentration of $CaCO_3$), although this parameter includes the buffer effect of the VFA (e.g., acetic acid represented in Equation (6)) and the NH_4^+-N as well [108]. Astals et al. [49] reported that the mass ratio $CO_3{}^{2-}$-C/NH_4^+-N was lower than 10 in the anaerobic digestate of pig manure and glycerol. In the case of studies focused on using the anaerobic digestate as an organic soil amendment, most studies only report the amount of C_{org}, and they consider the amount of $CO_3{}^{2-}$-C negligible (Theorem 1) for the calculation of the C/N [109–111]. Otherwise, they directly calculate the C_{org}/N without recognizing any role for the $CO_3{}^{2-}$-C [42,112]. Considering that the biogas is approximately 40% v/v and the alkaline pH of the anaerobic digestate [61], a considerable amount of CO_2 needs to remain dissociated in the form of $CO_3{}^{2-}$-C. The quantification of $CO_3{}^{2-}$-C is interesting from the point of view of controlling the removal of NH_4^+-N, because to achieve the formation of $NH_{3(aq)}$ in a stripping process, a significant amount of alkaline reagent needs to be spent on shifting the equilibria described in Equations (4)–(6) [113]. However, experimental studies developing process for the recovery of the NH_4^+-N via stripping do not clearly stablish the content of $CO_3{}^{2-}$-C in the anaerobic digestate [74,114,115].

In the novel process simulation performed by Drapanauskaite et al. [107], the anaerobic digestate was subjected to distillation at 3.3 bar with the condenser operating at 49 °C to produce a liquid stream rich in $HCO_3{}^-$ and NH_4^+. The solid NH_4HCO_3 was obtained in a crystallizer at 12 °C and was finally recovered via drying [107]. According to Drapanauskaite et al. [107], approximately 100% of the NH_4^+-N was used to manufacture the NH_4HCO_3. Drapanauskaite et al. [107] compared the open-loop process of $(NH_4)_2SO_3$, which implies using H_2SO_4 as an external chemical, with the NH_4HCO_3 process where both reagents ($HCO_3{}^-$ and NH_4^+) are initially present in the anaerobic digestate. In addition to the lower capital cost of the manufacturing of NH_4HCO_3, the operating cost was also lower due to the lower consumption of utilities.

3.2.3. Struvite Isolation Using WA as a Source of Magnesium

Another strategy to remove the excess of NH_4^+-N from the anaerobic digestate using the WA is via isolation of struvite [116]. However, the production of this slow-release fertilizer is expensive due to the cost of the source of magnesium and the alkaline agent needed to reach the target pH for the crystallization of this material [77,100,101]. Huang et al. [77] reported that 0.31 US Dollar/kg PO_4^{3-}-P can be saved if straw ash is employed instead of NaOH. Both Sakthivel et al. [101] and Huang et al. [77] recovered more than 96% of the phosphate in the ureolysis urine and swine wastewater, respectively, using different doses of WA. While Sakthivel et al. [101] employed 2.7 mol Mg/mol P; Huang et al. [77] used 1.2 mol Mg/mol P. This might be related to the fact that only 50% of the magnesium in the WA of Sakthivel et al. [101] was water-soluble (WS), thus only this amount was available for the formation of struvite. It would be possible to increase the availability of the magnesium via calcination of the WA at temperatures higher than 600 °C [101]. Another option would be to perform aeration when adding the WA to the anaerobic digestate to remove the CO_3^{2-}-C [77]. Drosg et al. [116] explained that the magnesium should be added in excess, according to the molar ratio 1.3:1:0.9 for Mg:N:P. This also agrees with the dose of magnesium proposed by Miles & Ellis [117] (1.25:1:1), and both nutrient ratios recommended for the precipitation of struvite are slightly different from the stoichiometry of the chemical reaction (Equation (11)). It should be noted that Equation (11) implies the release of H^+ [61,118], but other authors prefer to represent the drop of the pH as consumption of OH^- [116].

$$NH_4^+{}_{(aq)} + Mg^{2+}{}_{(aq)} + H_nPO_4^{n-3}{}_{(aq)} + 6H_2O_{(l)} \rightarrow MgNH_4PO_4{\cdot}6H_2O_{(s)} + nH^+{}_{(aq)} \quad (11)$$

According to Drosg et al. [116], materials such as anaerobic digestate contain more NH_4^+ & NH_3, hence the addition of orthophosphoric acid (H_3PO_4) is necessary to reach the target nutrient ratio. Escudero et al. [119] studied how the removal of NH_4^+-N from anaerobically treated effluents was affected by the source of orthophosphate PO_4^{3-}-P and Mg^{2+} and the nutrient ratio. They found an NH_4^+ removal of 95% in 30 s when using the best sources of magnesium (e.g., $MgCl_2{\cdot}6H_2O$ and $MgSO_4{\cdot}7H_2O$) and a molar ration Mg:N:P of 1:1:1 [119]. Sakthivel et al. [101] reported that the solid precipitate was not pure struvite due to the high content of $CaCO_3$ in the WA. They found a phosphorus content of 3% in the precipitate that is lower than the struvite (13%) or the diammonium phosphate (46%). Although the phosphorus content in the precipitate was greater than that in the initial WA, the estimated value of the precipitate was lower than the WA because 60% of the potassium initially present in the WA remained undissolved in the phosphate-depleted ureolysed urine [101]. Huang et al. [77] reported the competition reaction between struvite (Figure 4) and K-struvite ($MgKPO_4{\cdot}6H_2O$). These authors reported a greater share of K-struvite at pH 10, which was higher than the optimum pH range (7.5 to 9) for the precipitation of struvite [118]. According to Huang et al. [77], at pH 9.5, and with a dose of plant ash of 6 mol K/mol NH_4^+-N, the amounts of struvite and K-struvite are the same. In the structure of the struvite (Figure 5), the Mg^{2+} can coordinate with the PO_4^{3-} and the NH_4^+ despite being surrounded by six water molecules arranged according to an octahedral geometry.

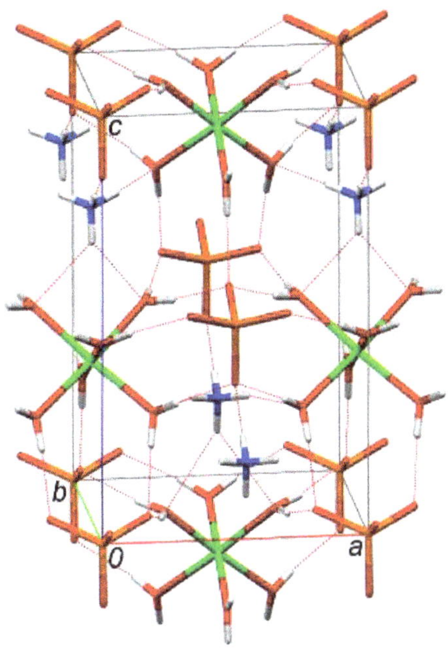

Figure 5. Structure of the struvite represented by Prywer et al. [120]: the magnesium (green) is surrounded by 6 molecules of water (oxygen in red and hydrogen in white) and the nitrogen (blue) and the phosphorus (yellow). Using the coordinate system adopted by Prywer et al. [120] and the hydrogen-bond network marked with dotted lines. Reproduced with the permission of Crystals (MDPI).

Theorem 2. *The optimum pH for the precipitation of struvite is the pH of zero point charge (pH_{zpc}), at which the surface charge of the source of magnesium (e.g., WA) is neutralized, thus the magnesium is able to sorb WS NH_4^+ and WS PO_4^{3-} simultaneously.*

The pH_{zpc} could be identified by plotting Q (Equation (12)) versus the pH, a second-order fitting calculated (generic form: $Q = a(pH)^2 + b(pH) + c$), and the minimum value of the quadratic regression determined (Figure 6). Unlike the measurement of the zeta potential, the pH_{zpc} is a simpler technique that does not require specialized equipment. Nevertheless, both parameters can be used interchangeably since the zeta potential is the electrical voltage at the interface that separates mobile fluid from the fluid that remains attached to the particle surface [121].

$$Q = \frac{1}{W}\left(C_a - \left[H^+\right] + \left[OH^-\right]\right) \tag{12}$$

Q (mol/L/g dry adsorbent), surface charge.
W (g/L), dry mass of WBA-based adsorbent in the aqueous system (i.e., analyte).
C_a (mol/L), concentration of the acid titrant in the aqueous system.
$[H^+]$ & $[OH^-]$, concentration of H^+ and OH^- resulting from the direct measurement of the pH in the aqueous system ($pH = -\log([H^+])$; $[H^+]\cdot[OH^-] = 10^{-14}$).

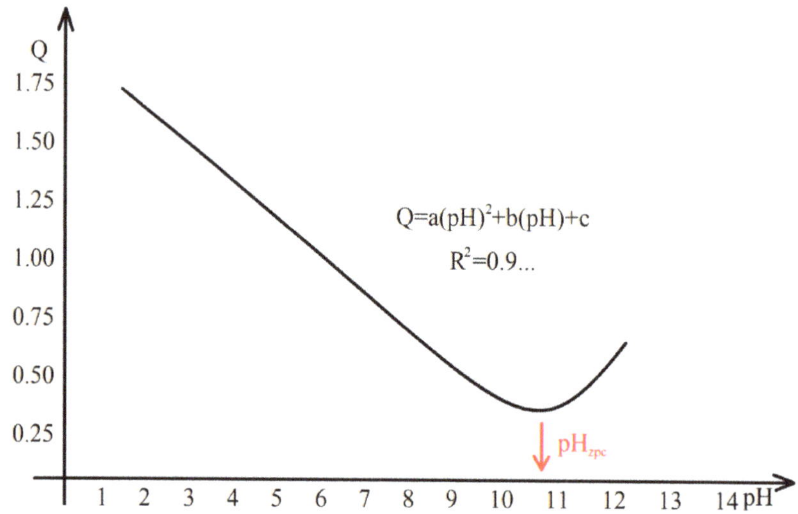

Figure 6. Determination of the pH_{zpc} of the WBA representing the surface charge of the WBA adsorbent, Q, as a function of the titrate aqueous solution. Elaborated from the description provided by Leechart et al. [63] and Shah et al. [122].

Proof of Theorem 2. While the systems had a lower pH than the pH_{zpc}, the surface of the WA-based adsorbents became positively charged (Equation (13)), favoring the adsorption of anionic species.

$$\text{Surface}^{+}_{(n-1)} \underset{+H^{+}}{\overset{+OH^{-}}{\rightleftarrows}} \underset{pH_{zpc}}{\underbrace{\text{Surface}^{0}_{n}(s)}} \underset{+H^{+}}{\overset{+OH^{-}}{\rightleftarrows}} \text{Surface}^{-}_{(n+1)} \tag{13}$$

source: [63].

Furthermore, since the pH of minimum solubility of CaO and MgO are 11.0 and 12.4, respectively, the surface of CaO and MgO could support the adsorption of the dye when the pH of the system was greater than pH_{zpc} (Equations (14) and (15)).

$$5Ca^{2+}{}_{(aq)} + 4OH^{-}{}_{(aq)} + 3HPO_4^{2-}{}_{(aq)} \rightarrow Ca_5(OH)(PO_4)_{3(s)} + 3H_2O_{(l)} \tag{14}$$

$$5Mg^{2+}{}_{(aq)} + 4OH^{-}{}_{(aq)} + 3HPO_4^{2-}{}_{(aq)} \rightarrow Mg_5(OH)(PO_4)_{3(s)} + 3H_2O_{(l)} \tag{15}$$

□

The sorption and precipitation processes are investigated as a way of reducing the cost of the purification of the water and preventing the eutrophication of underground water. Yagi & Fukushi [123] studied the mechanisms of adsorption and precipitation of calcium phosphate onto monohydrocalcite ($CaCO_3 \cdot H_2O$). According to Yagi & Fukushi [123], the sorption behavior was determined mainly by concentrating phosphate in the wastewater. The $CaCO_3 \cdot H_2O$ is known to be more reactive than the conventional $CaCO_3$ [123]. The clean effluent obtained by Yagi & Fukushi [123] could be coupled with air or CO_2 bubbling to remove any trace of Ca^{2+} ion via precipitation of $CaCO_3$ [124]. Similarly, Brennan et al. [80] tested whether the addition of lime to the cattle slurry minimized the emission of NH_3, N_2O, CH_4, and CO_2 and the leaching of WS $PO_4{}^{3-}$.

3.3. Acidification of the Blend of WA and Anaerobic Digestate to Improve the Nutrient Management

The commercial acids, such as H_2SO_4, HCl, HNO_3, and H_3PO_4, or easily fermentable compounds which ultimately lead to the formation of organic acids due to the microbial metabolism, such as the acetic acid ($C_6H_{12}O_6 \rightarrow 3\,CH_3COOH$), represent the most obvious choices to reduce the pH of the organic manures [125–127]. However, better stabilizations have been reported with the use of salts such as $Al_2(SO_4)_3$ [82], $FeCl_2$ [80] and $FeCl_3$ [128], and $CaCl_2$ [129,130]. The $CaCl_2$ does not necessarily need to be produced with quicklime (CaO; Equation (16)) but with slaked lime (consisting mainly in $Ca(OH)_2$), limestone ($CaCO_3$), or a mixture of these compounds that might be present in the WA. Similarly, the $Al_2(SO_4)_3$ (Equation (17)) or the $Fe_2(SO_4)_3$ could be produced from CFA [131]. It is noteworthy to mention that the $Al_2(SO_4)_3$, $FeCl_2$, and $FeCl_3$ (Equation (18)) are among the best acidifying additives for the stabilization of animal slurry anaerobic digestate [115,131,132]. These salts are widely employed in wastewater treatment for coagulation and precipitation of compounds [82,116]. This use is in line with the properties of WA as sorbent [133–136]. The activation of raw materials (e.g., ores and ashes) for the manufacturing of the salts might be why they are able to decrease the pH of the organic manures.

$$CaO_{(s)} + 2HCl_{(aq)} \rightarrow CaCl_{2(aq)} + H_2O_{(l)} \quad (16)$$

$$2Al(OH)_{3(s)} + 3H_2SO_{4(aq)} \rightarrow Al_2(SO_4)_{3(s)} + 6H_2O_{(l)} \quad (17)$$

$$Fe_3O_{4(s)} + 8HCl_{(aq)} \rightarrow FeCl_{2(s)} + 2FeCl_{3(s)} + 4H_2O_{(l)} \quad (18)$$

The $CaCl_2$ is one of the key components of the specially formulated products to stabilize the organic manures [137], and this compound is also widely used as an extracting agent for soil analysis. A solution of 0.01 M of $CaCl_2$ provides the same ionic strength [138] as the average salt concentration in many soil solutions (i.e., the medium in which surface and aqueous solution reactions occur in the soil). The rationale of solubilizing the elements to minimize the gaseous emissions agrees with the higher stability of the organic manures with higher moisture content [132]. On the other hand, when manures are diluted with water, most of the H^+ ions tend to remain attracted to the solid particles and are not released to the aqueous fraction. The simple dilution of manure with water reduces the emissions [132] but does not acidify the medium. Small amounts of $CaCl_2$ provide Ca^{2+} ions that replace some of the H^+ ions adsorbed in the solid particles, thus decreasing the pH of the medium. The stabilization of the organic slurry via the addition of $CaCl_2$ is explained by the solubilization of nutrients in the moisture of the organic slurry, ion exchange capacity of Ca^{2+}, H^+ release from the surface of the solid particles, and pH drop. The acid effect of these salts once dissolved in the aqueous phase of the organic slurries can be explained by a number of mechanisms. For example, the addition of $FeCl_3$ acidifies the aqueous solution by removing the sulfide and the phosphate, according to Equations (19)–(21) [61].

$$FeCl_{3(s)} \rightarrow Fe^{3+}{}_{(aq)} + 3Cl^{-}{}_{(aq)} \quad (19)$$

$$H_2S_{(aq)} + 2Fe^{3+}{}_{(aq)} \rightarrow S_{(s)} + 2Fe^{2+}{}_{(aq)} + 2H^{+}{}_{(aq)} \quad (20)$$

$$H_nPO_4^{n-3}{}_{(aq)} + Fe^{3+}{}_{(aq)} \rightarrow FePO_{4(s)} + nH^{+}{}_{(aq)} \quad (21)$$

3.3.1. Activation of the WA as Sorbent to Improve the Properties of the Blend with the Anaerobic Digestate as Slow-Released Fertilizer

Commercial acids are also employed to activate the ash and to prepare a cheap and sustainable sorbent. For example, Mor et al. [139], who used the calcination at 500 °C and HCl to activate the rice husk ash, reported an increase in the adsorption of the WS $PO_4{}^{3-}$ with a maximum removal (91.7%) of WS $PO_4{}^{3-}$ at pH 2 and the lowest removal at pH 10. According to Mor et al. [139], because at high pH, there are more OH^- interacting with the surface, thus decreasing the WS $PO_4{}^{3-}$ adsorption. It is necessary to bear in mind that the removal of PO_4^{3-} [140] and the NH_4^+ [119] from the WS phase via struvite crystallization is not promoted at acidic pH. The dose of acid needs to be as much as it is required to reach the pH_{zpc} without a further decrease (Figure 5). Leechart et al. [63] found that the pH_{zpc} for the WBA employed was 10.8. When a pretreatment was applied to the WBA with distilled water (WBA-H_2O) or sulfuric acid (WBA-H_2SO_4), the pH_{zpc} was 10.9 in both cases [63].

Unlike acidification, carbonation uses CO_2 rather than commercial acids to decrease the pH of the adsorbent. In addition to the reagents, the suitability of each activation treatment depends on the origin of the adsorbent and the primary mechanism for capturing the nutrients (Table 3). Janiszewska et al. [141] employed the carbonation at 800 °C by putting it in touch with the biochar, which had been prepared at 400 °C in an N_2 atmosphere, with a constant flow of CO_2 (150 mL/min) for 180 min. According to Janiszewska et al. [141], the CO_2 activation employed aimed to increase the porosity of the carbonized wastes. The mechanism relies on the volatilization of a significant part of the carbon content to enable the formation of a microporous structure in the biochar [141]. Leechart et al. [63] used the Brunauer, Emmett, and Teller (BET) isotherm to determine the sorptive surface area. It should be noted that the BET isotherm considers the possibility that the monolayer in the Langmuir adsorption isotherm could act as support for further adsorption. The nitrogen adsorption-desorption isotherm, employed for the BET study of these adsorbents, was characterized by clear hysteresis loops due to the capillary condensation occurring in the mesopore range [63]. The H_2O treatment removed some tars and led to a higher development of porosity and surface area than the treatment with 0.1 N H_2SO_4. The WBA-H_2O still had a lower surface area than AC. The average pore diameter of WBA, WBA-H_2O, and WBA-H_2SO_4 was of mesoporous size (i.e., 2–50 nm) while the activated carbon had micropores (<2 nm).

Table 3. Implications of the activation treatments that affect the suitability for a particular application (e.g., preparation of a slow-release fertilizer, liming agent, recovery of nutrients from wastewater, etc.). Elaborated by considering the information of [62,63,100,122,140].

Activation Procedure	Carbon Content	pH	Reactivity
Carbonization	Increased, due to loss of volatile compounds containing H, O, S, and N	Slightly increased, due to the accumulation of ash with alkali and alkaline elements	Decreased. Mainly charred materials and recalcitrant compounds
Carbonation	Increase, in the form of inorganic carbon	Decrease, due to the neutralization of the alkaline element with carbonic acid	Decreased because carbonates are less soluble than oxides.
Calcination	Decrease, since all volatile compounds have already been lost in the carbonization	Increase, due to the release of CO_2	Increase, due to the formation of oxides
Acidification	Decrease, due to the dissociation of the CO_3 -C and subsequent CO_2 emissions	Decrease, due to the dissociation of the commercial acids	Slightly increase [1]
Wash	Slightly decrease, due to the removal of impurities	Slightly decreased due to the solubilization of alkalis	Slightly decrease, due to extraction of oxides
Milling and sieving	Slightly increase. Might enhance carbonation reactions during storage	Slightly increase. Might enhance the reaction for acidic and alkaline salts	Increase, due to greater availability of the elements

[1] The oxides of the alkaline elements are more soluble and reactive than the carbonates. It is considered that the acidification of the alkaline oxides (e.g., CaO) with commercial acids, rather than with CO_2, surpasses the preparation of adsorbents (e.g., $CaCO_3$) and lead to the formation of coagulants and flocculants (e.g., $CaCl_2$; Equation (16)).

Ma et al. [142] described a more complicated activation procedure for producing an adsorbent based on wheat straw. They reported how the pH and the level of swelling (i.e., hydration) affect the adsorption of WS NH_4^+ and WS PO_4^{3-} [142]. According to the results of Ma et al. [142], the optimum pH for the adsorption of these ions is between 4 and 8. The explanation that Ma et al. [142] provided for the trend of the WS NH_4^+ is that at acidic pH, the H^+ ions will compete with the WS NH_4^+ for the anionic groups in the wheat straw derived adsorbent. On the other hand, at alkaline pH, more WS NH_4^+ will be in the form of WS NH_3 that will subsequently escape from the aqueous solution. According to Ma et al. [142], the adsorption of WS PO_4^{3-} showed less dependence on the pH. Ma et al. [142] described the impact of the pH and the swelling capacity (i.e., hydration) of the performance of the wheat straw-derived adsorbent. The high swelling of the wheat straw-derived adsorbent obtained at pH 3 and above was associated with the high diffusion of water molecules and ions into the pores and through the internal skeleton of the adsorbent. The highest swelling was obtained at a pH 7 because acidic and alkaline solutions have H^+ and OH^- ions, respectively, that reduce the difference in osmotic pressure between inside and outside the porous structure of the wheat straw derived adsorbent [142]. As described by Awad et al. [143], the swelling phenomenon is often associated with an increase in the adsorption of the organic materials in the WS phase. This can also be seen in the structure of the struvite (Figure 5), in which the nutrients NH_4^+ and PO_4^{3-} remain attached to the magnesium hexaaqua ($Mg^{2+}(H_2O)_6$).

3.3.2. Dewatering of the Blend of WA and Anaerobic Digestate

The studies of Ma et al. [142] and Mor et al. [139] provided contrary results regarding how the pH and the level of swelling/hydration affect the adsorption. The dehydration without acidification was tested by Zheng et al. [28]. They used a cationic surfactant (cetyltrimethylammonium bromide; CTAB) together with CFA to ease the filtration of the anaerobic digestate (Figure 7). The CTAB neutralized the negative charges on the surface of the flocs of the anaerobic digestates, reduced the electrostatic repulsion between the flocs, and promoted coagulation-flocculation. Moreover, the neutralization of the charges allowed the release of the extracellular polymeric EPS of the cells and fiber of the digestates and bound water molecules. The proteins and polysaccharides are components of the EPS that largely contributed to enhance the binding ability of water to the flocs of the digestate. Award et al. [143] divided the EPS into subgroups of main sources of carbon, nitrogen, and phosphorus. An example of EPS-phosphorus moieties would be nucleic acids [144]. After the release of the hydrated EPS from the flocs surface, the bound water that was coordinated with the EPS became free water [28]. On top of that, the CFA acted as a rigid lattice to reduce the filter cake compressibility and to provide drainage pathways for water to exit the filter cake [28].

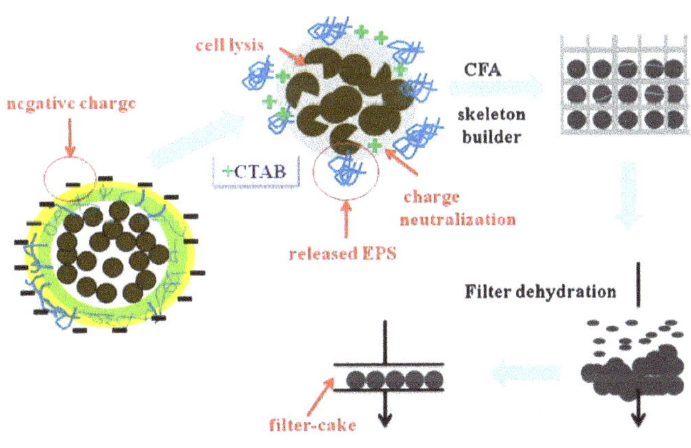

Figure 7. Mechanism proposed by Zheng et al. [28] for the flocculation of the anaerobic digestate by adding a cationic surfactant (CTAB) and subsequent filtration using CFA to improve the drainage of the filter cake to achieve the dewatering. The CTAB neutralized the surface charges of the flocs of the anaerobic digestate and enabled the release of the hydrated extracellular polymeric substances (EPS). The CFA acted as a skeleton builder of the filter cake and enhanced the physical solid-liquid separation by decreasing the compressibility of the filter cake. Reprinted with the permission of Separation and Purification Technology (Elsevier).

The dewatering of the anaerobic digestate is required to reduce the cost of storage, transportation, and land application [9,10]. Although the AD plants are located in strategic sites, usually with several farms around and with plenty of nearby areas to apply the anaerobic digestate to land, the transportation of this material to remote locations is still required once the nutrient quotas are reached in the fields around the plants [107,144]. The most common approach for the solid and the liquid fractions obtained after the separation is their use as fertilizer. There is particular concern about the emission from the resulting solid fraction [145–150] due to the higher area for gas exchange compared to the liquid fraction (Figure 8). While the gaseous emitting area of the slurry is the horizontal cross-section of the storage tank, the shape of the pile of dried organic manure was described by Dinuccio et al. [146,147] as a truncated cone heap with a greater area to volume ratio.

(a)

(b)

Figure 8. (**a**) Surface area of the liquid pig slurry (3.9% TS) in an open storage pond [151]. Reprinted with the permission of Biosystems Engineering (Elsevier). (**b**) Surface area of dewatered pig slurry (65.4% TS after composting without a bulking agent) [152]. Reprinted with the permission of Powder Technology (Elsevier).

Regueiro et al. [82] concluded that the $Al_2(SO_4)_3$ was able to reduce the gaseous emissions (NH_3, CO_2, CH_4, and N_2O) even after the solid-liquid separation (i.e., dewatering of the solid fraction) of raw and co-digested pig slurries. The $Al_2(SO_4)_3$ increased the TS in the liquid fraction but had the opposite effect on the solid fraction obtained after the solid-liquid separation. Regueiro et al. [82] attributed this fact to the role of the sulfate [150], and explained that the increase of the TS in the liquid fraction via the formation of low molecular weight carbohydrates derived from the acid hydrolysis of cellulose or hemicellulose. The acid pH provided by the $Al_2(SO_4)_3$ enhanced the retention of nutrients in the liquid fraction, minimizing the gaseous emissions from the solid fraction. It should be noted that the nutrients are more stable in the liquid fraction than in the solid fraction, due to the high moisture content (i.e., lower concentration) and lower surface area. Kavanagh et al. [132] explained that a slurry with 4% TS is more stable than the undiluted slurry with 7% TS. Thereby, the lower amount of TS in the solid fraction, the lesser the gaseous emissions. Regueiro et al. [82] found that the gaseous emissions were also minimized in the liquid fraction, due to its lower pH, compared to the liquid fraction obtained from the solid-liquid separation without acidification. It should be noted that the content of nitrogen, expressed in grams of nitrogen per gram of TS, decreased in the solid and the liquid fractions due to the increase of the TS, which resulted from the addition of $Al_2(SO_4)_3$ [82].

3.4. Commercial Processes for Manufacturing of a Granular Fertilizer based on Blends of Ash and Organic Manures

The features of the process of Limoli et al. [74] employing CaO and H_2SO_4 and the process of Zheng et al. [28] using CFA and CTAB were compared against processes implementing the granulation, as this step could improve the aesthetic properties of the blended fertilizer of WA and anaerobic digestate to be retailed (Table 4). There is a lack of holistic approach in the processes described below as none of them entirely complies with the following objectives:

1. Use of the most cost-efficient way of achieving solid-liquid separation.
2. Find the optimum carbon and nutrient profile of the main stream coming out of the process, intended to be used as an organic amendment.
3. Self-hardening to provide this material with the best mechanical properties before and after the granulation.
4. Minimize any waste streams with valuable nutrients or other pollutants that need to be removed before disposal.

Table 4. Identification of the synergies and possibilities in the process involving the mixture of anaerobic digestate or sewage sludge with WA or alike materials.

Process	Raw Materials	Blending Ratio	Advantages	Disadvantages
RecoPhos [153].	Sewage sludge and H_3PO_4.	Incineration to produce SS ash. Reaction of the SS ash with the H_3PO_4 using phosphorus molar ratio 1:12 of sewage sludge ash: H_3PO_4	Marketable granular product with similar properties to the triple superphosphate	The NH_3 volatilized during the drying is not recovered. Emission of nitrogen oxides during incineration.
ADFerTech [154].	Anaerobic digestate, dolomite ($CaMg(CO_3)_2$), organic binders and coating.	Dolomite was added to the liquid fraction (>91% moisture) in a dose ranging from 10 to 200 g/L.	Improve the aesthetic properties of the liquid fraction of anaerobic digestate. Decrease the cost of transportation and storage.	Additives of the liquid fraction of the anaerobic digestate are suitable for land application.
Limoli et al. [74].	Anaerobic digestate, CaO, and H_2SO_4.	>95% moisture of anaerobic digestate. Dose of CaO to operate the stripping at pH 10.	Recovery of the NH_3 volatilized. Enable the self-hardening and granulation of the NH_3-depleted organic amendment.	Low fluency to be employed in a traditional stripping column (i.e., packed tower). High COD content of the filtrate.
Zhengh et al. [28].	Anaerobic digestate, CFA, and CTAB.	>93% moisture of anaerobic digestate Mass of CTAB up to half of the TS of the digestate Mass of CFA up to the TS of the digestate	Reduce the energy consumption of the filtration of anaerobic digestate. Possible to enhance the mechanical separation with adsorption.	Presence of CTAB and heavy metals of the CFA in the filtrate. High COD content in the filtrate.
Pesonen et al. [29]	Sewage sludge, wood-peat ash, and $Ca(OH)_2$.	Up to 40% of SS (45% moisture) and up to 30% $Ca(OH)_2$. The dose of ash can go up to 100%.	No need to include $Ca(OH)_2$ to have high compressive strength. Low presence of heavy metals	Dewatering and sanitation The NH_3 released is not captured
Jewiarz et al. [155].	Anaerobic digestate and woody biomass.	18–20% moisture of the anaerobic digestate. Up to 75% WA.	Save in energy for drying the anaerobic digestate by thermal drying in fueling the drum drier with the biomass to produce the ash.	The NH_3 released is not captured. Not possible to include the biofertilizer together with the ash due to the high pH.

The RecoPhos P 38 fertilizer is produced from SS following the same manufacturing procedure of the triple superphosphate ($Ca(H_2PO_4)_2$) from the phosphate rock [153]. This process aligns with the management strategy proposed by Chojnacka et al. [156] for organic wastes. Firstly, the valorization is carried out via complete dehydration and subsequent incineration to reduce the organic pollutants and to obtain ash with similar characteristics to the phosphate ores. Secondly, the solubility of the phosphorus contained in the ash is enhanced via reaction with H_3PO_4 (Equation (22)), prior to the preparation of the granular fertilizer. This technology could be improved since carbon and nitrogen are lost during the fluidized bed incineration. Furthermore, the isolation of the phosphorus contained in the ash only can be done via solubilization with phosphoric acid and subsequent drying. The RecoPhos might not be an energy-intensive process if, for instance, the drying of the SS prior to the incineration is done with the heat released in the combustion.

$$Ca_4Mg_5(PO_4)_{6(s)} + 12H_3PO_{4(ac)} \rightarrow 4Ca(H_2PO_4)_{2(s)} + 5Mg(H_2PO_4)_{2(s)} \quad (22)$$

The ADFerTech process of Fivelman [154] is an energy-intensive process because, in addition to the preliminary solid-liquid separation, the subsequent processing of the liquid fraction involves drying (a) after the adsorption to enable the self-hardening and (b) after the agglomeration to have a 3% moisture content in the final granules. In this way, Fivelman [154] employed the adsorption technology to enhance the solid-liquid separation of the liquid fraction of the digestate. The granular fertilizer was produced with the TS remaining in the liquid fraction after adding organic polymer binders (e.g., carboxymethyl cellulose) to promote the aggregation of the nutrient-laden particles of the adsorbent. Fivelman [154] claimed that the amount of nutrients that remain in the liquid stream after the adsorption is low enough to be discharged directly to the local water bodies without further treatment. The main powder adsorbent that they used was dolomite in doses ranging from 10 to 200 g/L (mixed for 5 min at 20 °C), which was able to retain around 250 mg nitrogen and 300 mg phosphorus per gram of dolomite.

Steenari and Lindqvist [157] explained that the self-hardening or solidification of a gram of WA, when adding between 0.3 and 0.5 g of water, depends on a number of ongoing reactions which can last from days to months. It should be noted the importance of the (moisture) adsorption capacity of the WA to promote self-hardening [158]. The hydration of CaO to produce $Ca(OH)_2$ is faster than the subsequent formation of $CaCO_3$ due to the mass transfer resistance, limiting the diffusion of CO_2 from the atmosphere. However, the $CaCO_3$ is more stable due to its lower solubility and alkalinity, providing the ash with better properties for cementation and for being used as a long-lasting liming agent. Pellets predominantly made with organic matter such as cattle manure rely less on these reactions to obtain appropriate durability (i.e., percentage of the mass of the pellet that remains after tumbling [159]). Due to the fact that organic matter can retain more water [157], Zafari et al. [159] obtained the highest durability (~97%) in the cattle manure pellets when these were prepared with the open-end die method at 50% (w/w) moisture content, 40 °C, and 6 MPa. According to Alemi et al. [160], both the density and porosity increased when the pellets of cattle manure and urea were prepared with higher moisture content (from 11 to 24% w/w) and a higher compressive force (from 2000 to 5000 N). Even after the preparation, Alemi et al. [160] reported an increase of up to 8% (w/w) in the moisture content due to the absorption of the water in the surrounding environment with a relative humidity of up to 80%.

Pesonen et al. [29] studied the production of fertilizer based on biomass ash which was obtained from the combustion of a mixture of 65% wood and 35% peat. As a nitrogen source, they used SS, which had been previously sanitized and dewatered (until a 45% moisture) in a resource-intensive process. Pesonen et al. [29] included $Ca(OH)_2$ in the fertilizer as well, with the main purpose of promoting self-hardening and increasing the compressible load that the granules can hold. Nevertheless, this material did not increase the compressible strength to prevent the deformation of the granules. They carried out the granulation in a rotary drum, after which the granules were left in a fume hood for 28 days at 21 °C. This allowed to reach 3% of moisture due to the loss of free water, since the compounds responsible for the self-hardening require higher temperatures to decompose. Pesonen et al. [29] concluded that the solidification was not enhanced because the main reason for this phenomenon was the reaction of CaO with the water supplying $Ca(OH)_2$, and, to a lesser extent, the subsequent formation of $CaCO_3$ due to the absorption of the CO_2 in the atmosphere, which is a much slower reaction.

According to Steenari and Lindqvist [157], ashes with a content of combustible matter greater than 10% are not suitable for self-hardening, which could explain why the organic waste was not a good binder for the curing process and the solidification of the ash was hindered in the study of Pesonen et al. [29]. The acidification could be useful to promote dehydration [74] and mechanical properties (e.g., durability and compressive strength) of the solid obtained after the self-hardening. Based on the study of Steenari and Lindqvist [157], from the point of view of the self-hardening of the blend, a greater amount of ash would be desirable since this would minimize the organic binder. This agrees with the results of Rao et al. [161], who investigated the preparation of organo-mineral fertilizers with the compost of pig waste solids (20% *w*/*w*), PL (26% *w*/*w*), spent mushroom compost (26% *w*/*w*), cocoa husks (18% *w*/*w*), and moistened shredded paper (10% *w*/*w*). The additives that they tested were feather meal (≤29% *w*/*w*), dried blood (≤29% *w*/*w*), $(NH_4)_2SO_4$ (≤27% *w*/*w*), nitro-chalk or $5Ca(NO_3)_2 \cdot NH_4NO_3 \cdot 10H_2O$ (≤27% *w*/*w*), phosphate rock (≤15% *w*/*w*), $Ca(H_2PO_4)_2$ (≤21% *w*/*w*), and K_2SO_4 (≤21% *w*/*w*). Only when the share of compost in the mixture was greater than 50% (*w*/*w*), pelletization was not possible. On the other hand, Pampuro et al. [152] reported that none of the woody bulking agents they tested (woody biochar and wood chips) improved the compression resistance of the pellets made with composted pig solid fraction. However, they found that the pellets made with woody biochar and composted pig waste had higher durability than those made only with composted pig waste. According to Pampuro et al. [152], the greater durability of the woody biochar amended pellets was due to the low particle size of this material (<0.5 mm). The greater surface area of smaller particle promote sorption processes during the compaction for densification. Pampuro et al. [152] concluded that smaller particles enable the manufacturing of pellets of higher quality. Based on the results of Mudryk et al. [162], Jewiarz et al. [155] proposed the entire process for the production of this granular fertilizer. Although initial dewatering of the digestate was required, the greatest reduction of the moisture content, from 70% to 25%, was achieved in a direct contact drum dryer which employed the flue gases produced in the combustion of the biomass as a drying agent [155]. The process took advantage of the ash produced in the incineration but also it could employ the drying step to sanitize the digestate and to produce a pathogen-free material (i.e., biological stabilization). If the digestate contains a lot of fibrous material, the fragmentation is performed with a hammer mill after the drying. The process could be improved by blending the digestate with the biomass ash before the drying of the digestate. In this way, the absorption of the CO_2 and other components of the flue gases in contact with the liquid blend with high pH could be possible, given the high alkalinity of the wood ashes and that this material represented up to 75% of the final product. According to the description of the process provided by Jewiarz et al. [155], the blend should solidify for 3 h before the granulation. Steenari & Lindqvist [157] explained that curing the WA is necessary because the reactive oxides and soluble salts might negatively affect plants (e.g., pH shock and burning tissues) if the untreated WA is applied directly to land. Additionally, Jewiarz et al. [155] suggested inoculating the fungi Trichoderm to enhance nitrogen assimilation by the plants, although this biofertilizer cannot be in the same blend containing the ash because it would not survive at a high pH.

3.5. Inoculation of Biofertilizers in the Blended Fertilizer Prepared with Anaerobic Digestate and WA

After the removal of the pathogens by processing the anaerobic digestate with WA [163], it is possible to inoculate microbial species [164], which enhances the fertilizing effect of the blended fertilizer [155]. These microbes regarded as biofertilizers are divided into four categories [165]. The following are some examples of microbial species that can improve the fertilization in the rhizosphere:

1. Vascular Arbuscular Mycorrhiza fungus:
    - *Glomus mosseae* [165,166].
    - *Glomus intraradices* [165].
    - *Glomus fasciculatum* [166].
2. N-fixer bacterias can be symbiotic and non-symbiotic:

- ○ *Azotobacter chroococcum* [165,167].
- ○ *Azospirillum lipoferum* [167].
- ○ *Azospirillum brasilense* [166,167].

3. Phosphate-solubilizing bacteria:
   - ○ *Bacillus megatherium* [165,167].
4. Potassium-solubilizing bacteria:
   - ○ *Bacillus mucilaginous* [165].

Mahfouz & Shamf-Eldin [167] concluded that the growth of the plant Fennel (Foeniculum vulgare Mill.) was greater when using half the recommended dose of the NPK mineral fertilizer (i.e., synthetic material containing nitrogen, phosphorus, and potassium) and a mixture of *Azotobacter chroococcum*, *Azospirillum lipoferum*, and *Bacillus megatherium*, compared to the full NPK dose. The total carbohydrates in the dry plant material were impacted by the use of the mixture of biofertilizers. Similarly, the biofertilizers also increase the yield of essential oil in the plant [167]. According to Brar et al. [168], the best physical properties of the soil (e.g., water infiltration) and highest crop yield in maize-wheat rotation were obtained using a balanced application of mineral fertilizers and organic farmyard manure [168]. In addition to greater crop growth, better fertilization is also associated with less greenhouse gas emissions since the nutrients are employed more efficiently once the organic material is applied to the soil [169].

4. Conclusions

There are many possibilities for employing the WA as an anaerobic digestate additive. The treatments described in this manuscript ranged from enhancing anaerobic fermentation in order to reach the correct levels of stability and maturity to improving the properties of organic amendment as a controlled-release fertilizer, with additional processing steps after the biodigester. It remains to completely link the WA dose employed for the AD (up to 1 g TS ash/g TS digestate) with the purpose of the subsequent processing stage. Considering all the possible synergies that could surge from treating the anaerobic digestate with the WA, an approximate dose of 5 g TS WA/g TS anaerobic digestate is proposed for future experiments on the manufacturing of a granular fertilizer with the best mechanical properties (e.g., compressibility strength, durability, etc.).

Supplementary Materials: The following are available online at https://www.mdpi.com/article/10.3390/pr10010147/s1, Figure S1: Setup employed by Lo et al. [35]; Figure S2: Process flow diagram of the study of Bauer at el. [38]; Figure S3: Process flow diagram of the study of Cimon et al. [39]; Figure S4: Profile of maturity of anaerobic digestate as per the results of Wang et al. [58]; Figure S5: Inhibited BMP test due to high COD described by Banks et al. [47,48]; Figure S6: Inhibited BMP test due to low S/I described by Banks et al. [47,48]; Figure S7: Summary of the BMP results with biochar and WA reported by of Cimon et al. [39]; Table S1: Upper limit of heavy metals for anaerobic digestate and PLA established in UK regulations [46,84]; Table S2: Blending ration of WA and anaerobic digestate as per the UK regulations.

Author Contributions: Conceptualization, A.M.A. and F.A.; resources, F.A.; data curation, A.M.A.; writing—original draft preparation, A.M.A.; writing—review and editing, F.A.; funding acquisition, F.A. All authors have read and agreed to the published version of the manuscript.

Funding: This research was funded by the Engineering and Physical Sciences Research Council (EPSRC), grant number EP/N509504/1, and the Natural Environment Research Council (NERC), grant number NE/L014122/1, of the United Kingdom.

Institutional Review Board Statement: Not appilable.

Informed Consent Statement: Not appilable.

Data Availability Statement: Not appilable.

Conflicts of Interest: The authors declare no conflict of interest. The funders had no role in designing the study; with the collection, analyses, or interpretation of data; with the writing of the manuscript, or in the decision to publish the results.

Abbreviations

NH_4^+-N	ammoniacal nitrogen
AD	anaerobic digestion
BET	Brunauer, Emmett, and Teller
BMP	biochemical methane potential
CUE	carbon use efficiency
C/N	carbon-to-nitrogen mas ratio
CTAB	cetyltrimethylammonium bromide
COD	chemical oxygen demand
EoW	end-of-waste
CFA	coal fly ash
EPS	extracellular polymeric substances
HRT	hydraulic retention time
MSW	municipal solid waste
OLR	organic loading rate
C_{org}	organic carbon
SS	sewage sludge
S/I	substrate to inoculum ration
TS	total solids
VFA	volatile fatty acids
VS	volatile solids
WS	water-soluble
WA	wood ash
WBA	wood bottom ash
WBA-H_2O	wood bottom ash treated with de-ionized water
WBA-H_2SO_4	wood bottom ash treated with sulfuric acid
WWTP	wastewater treatment plant

References

1. Banwart, S. Save our soils. *Nature* **2011**, *474*, 151–152. [CrossRef]
2. Elser, J.; Bennett, E. A broken biogeochemical cycle. *Nature* **2011**, *478*, 29–31. [CrossRef]
3. Vaccari, D.A. Phosphorus: A Looming Crisis. *Sci. Am.* **2009**, *300*, 54–59. [CrossRef] [PubMed]
4. Cordell, D.; Drangert, J.O.; White, S. The story of phosphorus: Global food security and food for thought. *Glob. Environ. Chang.* **2009**, *19*, 292–305. [CrossRef]
5. Hsu, E. Cost-benefit analysis for recycling of agricultural wastes in Taiwan. *Waste Manag.* **2021**, *120*, 424–432. [CrossRef]
6. European Parliament. Directive 2018/851 Amending Directive 2008/98/EC on Waste Framework. Available online: https://eur-lex.europa.eu/legal-content/EN/TXT/?uri=uriserv:OJ.L.2018.150.01.0109.01.ENG (accessed on 14 December 2021).
7. Corden, C.; Bougas, K.; Cunningham, E.; Tyrer, D.; Kreißig, J.; Zetti, E.; Gamero, E.; Wildey, R.; Crookes, M. Digestate and Compost as Fertilisers: Risk Assessment and Risk Management Options. Available online: https://ec.europa.eu/environment/chemicals/reach/pdf/40039%20Digestate%20and%20Compost%20RMOA%20-%20Final%20report%20i2_20190208.pdf (accessed on 14 December 2021).
8. Saveyn, H.; Eder, P. End-of-Waste Criteria for Biodegradable Waste Subjected to Biological Treatment (Compost & Digestate): Technical Proposals. Available online: https://publications.jrc.ec.europa.eu/repository/handle/JRC87124 (accessed on 14 December 2021).
9. Victor, L. AD and Composting Industry Market Survey Report 2020. Available online: https://wrap.org.uk/sites/default/files/2021-01/AD%20%26%20Composting%20Market%20Survey%20Report.pdf (accessed on 14 December 2021).
10. Wood, E.; James, K.; Barker, E. Comparison of the Environmental Impacts of Nitrogenous Materials. Available online: https://wrap.org.uk/sites/default/files/2021-01/Nitrogenous%20Materials%20Report%202020.pdf (accessed on 14 December 2021).
11. Klages, S.; Heidecke, C.; Osterburg, B.; Bailey, J.; Calciu, I.; Casey, C.; Dalgaard, T.; Frick, H.; Glavan, M.; D'Haene, K.; et al. Nitrogen Surplus—A Unified Indicator for Water Pollution in Europe? *Water* **2020**, *12*, 1197. [CrossRef]
12. Blumenthal, K. Generation and Treatment of Municipal Waste. Available online: https://op.europa.eu/en/publication-detail/-/publication/bd8a43dc-8076-4134-987d-c3081c8311e8 (accessed on 14 December 2021).

13. Hou, Y.; Velthof, G.L.; Case, S.D.C.; Oelofse, M.; Grignani, C.; Balsari, P.; Zavattaro, L.; Gioelli, F.; Bernal, M.P.; Fangueiro, D.; et al. Stakeholder perceptions of manure treatment technologies in Denmark, Italy, the Netherlands and Spain. *J. Clean. Prod.* **2018**, *172*, 1620–1630. [CrossRef]
14. Holm-Nielsen, J.B.; Al Seadi, T.; Oleskowicz-Popiel, P. The future of anaerobic digestion and biogas utilization. *Bioresour. Technol.* **2009**, *100*, 5478–5484. [CrossRef] [PubMed]
15. Bond, T.; Templeton, M.R. History and future of domestic biogas plants in the developing world. *Energy Sustain. Dev.* **2011**, *15*, 347–354. [CrossRef]
16. Dagnall, S. UK strategy for centralised anaerobic digestion. *Bioresour. Technol.* **1995**, *52*, 275–280. [CrossRef]
17. Madsen, M.; Holm-Nielsen, J.B.; Esbensen, K.H. Monitoring of anaerobic digestion processes: A review perspective. *Renew. Sustain. Energy Rev.* **2011**, *15*, 3141–3155. [CrossRef]
18. Vassilev, S.V.; Baxter, D.; Vassileva, C.G. An overview of the behaviour of biomass during combustion: Part I. Phase-mineral transformations of organic and inorganic matter. *Fuel* **2013**, *112*, 391–449. [CrossRef]
19. Pitman, R.M. Wood ash use in forestry—A review of the environmental impacts. *Forestry* **2006**, *79*, 563–588. [CrossRef]
20. UK Forestry Commission. Forestry Statistics. Available online: https://www.forestresearch.gov.uk/tools-and-resources/statistics/forestry-statistics/forestry-statistics-2016-introduction/ (accessed on 14 December 2021).
21. UK Quality Ash Association. UKQAA Ash Availability Report. Available online: http://www.ukqaa.org.uk/information/statistics/ (accessed on 14 December 2021).
22. UK Quality Ash Association. About Us. Available online: http://www.ukqaa.org.uk/about/ (accessed on 14 December 2021).
23. Andersson, L. Regular Recycling of Wood Ash to Prevent Waste Production. RecAsh. Available online: https://www.osti.gov/etdeweb/servlets/purl/20886561 (accessed on 14 December 2021).
24. Guerrero, L.; Da Silva, C.; Barahona, A.; Montalvo, S.; Huiliñir, C.; Borja, R.; Peirano, C.; Toledo, M.; Carvajal, A. Fly ash as stimulant for anaerobic digestion: Effect over hydrolytic stage and methane generation rate. *Water Sci. Technol.* **2019**, *80*, 1384–1391. [CrossRef]
25. Fernández-Delgado Juárez, M.; Waldhuber, S.; Knapp, A.; Partl, C.; Gómez-Brandón, M.; Insam, H. Wood ash effects on chemical and microbiological properties of digestate- and manure-amended soils. *Biol. Fertil. Soils* **2013**, *49*, 575–585. [CrossRef]
26. Ibeto, C.N.; Lag-Brotons, A.J.; Marshall, R.; Semple, K.T. The Nutritional Effects of Digested and Undigested Organic Wastes Combined with Wood Ash Amendments on Carrot Plants. *J. Soil Sci. Plant Nutr.* **2020**, *20*, 460–472. [CrossRef]
27. Fenton, O. Chemical Amendment of Slurry to Control Phosphorus Losses in Runoff. Available online: https://www.teagasc.ie/media/website/publications/2011/5669-Chemical-Amendment-of-Slurry.pdf (accessed on 14 December 2021).
28. Zheng, Y.; Ke, L.; Xia, D.; Zheng, Y.; Wang, Y.; Li, H.; Li, Q. Enhancement of digestates dewaterability by CTAB combined with CFA pretreatment. *Sep. Purif. Technol.* **2016**, *163*, 282–289. [CrossRef]
29. Pesonen, J.; Kuokkanen, V.; Kuokkanen, T.; Illikainen, M. Co-granulation of bio-ash with sewage sludge and lime for fertilizer use. *J. Environ. Chem. Eng.* **2016**, *4*, 4817–4821. [CrossRef]
30. Appels, L.; Lauwers, J.; Degrve, J.; Helsen, L.; Lievens, B.; Willems, K.; Van Impe, J.; Dewil, R. Anaerobic digestion in global bio-energy production: Potential and research challenges. *Renew. Sustain. Energy Rev.* **2011**, *15*, 4295–4301. [CrossRef]
31. Alavi-Borazjani, S.A.; Capela, I.; Tarelho, L.A.C. Valorization of biomass ash in biogas technology: Opportunities and challenges. *Energy Rep.* **2020**, *6*, 472–476. [CrossRef]
32. Alavi-Borazjani, S.A.; Tarelho, L.A.C.; Capela, I. A Brief Overview on the Utilization of Biomass Ash in Biogas Production and Purification. *Waste Biomass Valorization* **2021**, *12*, 6375–6388. [CrossRef]
33. Adeyanju, A.A. Effect of seeding of wood-ash on biogas production using pig waste and cassava peels. *J. Eng. Appl. Sci.* **2008**, *3*, 242–245.
34. Onwosi, C.O.; Okereke, G.U. Effect of water dilution and nutrient supplements (wood ash, urea and poultry droppings) on biogas production from brewers spent grain. In *Current Research Topics in Applied Microbiology and Microbial Biotechnology*; World Scientific: Seville, Spain, 2009; pp. 232–235. [CrossRef]
35. Lo, H.M.; Liu, M.H.; Pai, T.Y.; Liu, W.F.; Lin, C.Y.; Wang, S.C.; Banks, C.J.; Hung, C.H.; Chiang, C.F.; Lin, K.C.; et al. Biostabilization assessment of MSW co-disposed with MSWI fly ash in anaerobic bioreactors. *J. Hazard. Mater.* **2009**, *162*, 1233–1242. [CrossRef] [PubMed]
36. Lo, H.M.; Kurniawan, T.A.; Sillanpää, M.E.T.; Pai, T.Y.; Chiang, C.F.; Chao, K.P.; Liu, M.H.; Chuang, S.H.; Banks, C.J.; Wang, S.C.; et al. Modeling biogas production from organic fraction of MSW co-digested with MSWI ashes in anaerobic bioreactors. *Bioresour. Technol.* **2010**, *101*, 6329–6335. [CrossRef] [PubMed]
37. Podmirseg, S.M.; Seewald, M.S.A.; Knapp, B.A.; Bouzid, O.; Biderre-Petit, C.; Peyret, P.; Insam, H. Wood ash amendment to biogas reactors as an alternative to landfilling? A preliminary study on changes in process chemistry and biology. *Waste Manag. Res.* **2013**, *31*, 829–842. [CrossRef] [PubMed]
38. Bauer, T.; Pelkonen, M.; Lagerkvist, A. Co-digestion of sewage sludge and wood fly ash. *Environ. Technol.* **2020**, *9*, 106564. [CrossRef] [PubMed]
39. Cimon, C.; Kadota, P.; Eskicioglu, C. Effect of biochar and wood ash amendment on biochemical methane production of wastewater sludge from a temperature phase anaerobic digestion process. *Bioresour. Technol.* **2020**, *297*, 122440. [CrossRef]
40. Bachmann, N. Design and engineering of biogas plants. In *The Biogas Handbook: Science, Production, and Applications*; Elsevier: Amsterdam, The Netherlands, 2013; pp. 191–211. [CrossRef]

41. Linke, B. Kinetic study of thermophilic anaerobic digestion of solid wastes from potato processing. *Biomass Bioenergy* **2006**, *30*, 892–896. [CrossRef]
42. Alburquerque, J.A.; de la Fuente, C.; Bernal, M.P. Chemical properties of anaerobic digestates affecting C and N dynamics in amended soils. *Agric. Ecosyst. Environ.* **2012**, *160*, 15–22. [CrossRef]
43. Astals, S.; Nolla-Ardèvol, V.; Mata-Alvarez, J. Thermophilic co-digestion of pig manure and crude glycerol: Process performance and digestate stability. *J. Biotechnol.* **2013**, *166*, 97–104. [CrossRef]
44. Bernal, M.P.; Alburquerque, J.A.; Moral, R. Composting of animal manures and chemical criteria for compost maturity assessment. A review. *Bioresour. Technol.* **2009**, *100*, 5444–5453. [CrossRef]
45. Strosser, E. Methods for determination of labile soil organic matter: An overview. *J. Agrobiol.* **2011**, *27*, 49–60. [CrossRef]
46. WRAP. BSI PAS 110:2014 Specification for Whole Digestate, Separated Liquor and Separated Fibre Derived from the Anaerobic Digestion of Source-Segregated Biodegradable Materials. 2014. Available online: https://wrap.org.uk/resources/guide/bsi-pas-110-producing-quality-anaerobic-digestate (accessed on 5 January 2022).
47. Banks, C.J.; Haeven, S.; Zhang, Y.; Sapp, M. Review of the Application of the Residual Biogas Potential Test. Available online: http://www.organics-recycling.org.uk/uploads/article2652/PAS110%20digestate%20stability%20review.pdf (accessed on 14 December 2021).
48. Walker, M.; Banks, C.; Heaven, S.; Frederickson, J. Residual Biogas Potential Test for Digestates. Available online: https://www.wrap.org.uk/sites/files/wrap/Residual%20Biogas%20Potential.pdf (accessed on 28 February 2017).
49. Astals, S.; Nolla-Ardèvol, V.; Mata-Alvarez, J. Anaerobic co-digestion of pig manure and crude glycerol at mesophilic conditions: Biogas and digestate. *Bioresour. Technol.* **2012**, *110*, 63–70. [CrossRef]
50. Dijkstra, P.; Salpas, E.; Fairbanks, D.; Miller, E.B.; Hagerty, S.B.; Jan, K.; Groenigen, V.; Hungate, B.A.; Marks, J.C.; Koch, G.W.; et al. High carbon use efficiency in soil microbial communities is related to balanced growth, not storage compound synthesis. *Soil Biol. Biochem.* **2015**, *89*, 35–43. [CrossRef]
51. Geyer, K.M.; Kyker-Snowman, E.; Grandy, A.S.; Frey, S.D. Microbial carbon use efficiency: Accounting for population, community, and ecosystem-scale controls over the fate of metabolized organic matter. *Biogeochemistry* **2016**, *127*, 173–188. [CrossRef]
52. Manzoni, S.; Taylor, P.; Richter, A.; Porporato, A. Environmental and stoichiometric controls on microbial carbon-use efficiency in soils. *New Phytol.* **2012**, *196*, 79–91. [CrossRef]
53. Sinsabaugh, R.L.; Manzoni, S.; Moorhead, D.L.; Richter, A. Carbon use efficiency of microbial communities: Stoichiometry, methodology and modelling. *Ecol. Lett.* **2013**, *16*, 930–939. [CrossRef]
54. UK Government. Quality Protocol: Compost—End of Waste Criteria for the Production and Use of Quality Compost from Source-Segregated Biodegradable Waste. Available online: https://www.gov.uk/government/publications/quality-protocol-for-the-production-and-use-of-compost-from-waste (accessed on 14 December 2021).
55. WRAP. PAS 100:2011 Specification for Composted Materials. Available online: http://www.organics-recycling.org.uk/page.php?article=3483#:~{}:text=PAS%20100%3A2018%20requires%20producers%20to%20set%20up%20a,also%20relates%20to%20the%20new%20%C3%82%C2%91compost%20quality%C3%82%C2%92%20clause (accessed on 15 December 2021).
56. Wang, X.; Yang, G.; Feng, Y.; Ren, G.; Han, X. Optimizing feeding composition and carbon—nitrogen ratios for improved methane yield during anaerobic co-digestion of dairy, chicken manure and wheat straw. *Bioresour. Technol.* **2012**, *120*, 78–83. [CrossRef]
57. Rincón, B.; Borja, R.; González, J.M.; Portillo, M.C.; Sáiz-Jiménez, C. Influence of organic loading rate and hydraulic retention time on the performance, stability and microbial communities of one-stage anaerobic digestion of two-phase olive mill solid residue. *Biochem. Eng. J.* **2008**, *40*, 253–261. [CrossRef]
58. Angelidaki, I.; Alves, M.; Bolzonella, D.; Borzacconi, L.; Campos, J.L.; Guwy, A.J.; Kalyuzhnyi, S.; Jenicek, P.; Van Lier, J.B. Defining the biomethane potential (BMP) of solid organic wastes and energy crops: A proposed protocol for batch assays. *Water Sci. Technol.* **2009**, *59*, 927–934. [CrossRef] [PubMed]
59. Mannina, G.; Presti, D.; Montiel-Jarillo, G.; Carrera, J.; Suárez-Ojeda, M.E. Recovery of polyhydroxyalkanoates (PHAs) from wastewater: A review. *Bioresour. Technol.* **2020**, *297*, 122478. [CrossRef] [PubMed]
60. Richards, S.; Marshall, R.; Lag-Brotons, A.J.; Semple, K.T.; Stutter, M. Phosphorus solubility changes following additions of bioenergy wastes to an agricultural soil: Implications for crop availability and environmental mobility. *Geoderma* **2021**, *401*, 115150. [CrossRef]
61. Möller, K.; Müller, T. Effects of anaerobic digestion on digestate nutrient availability and crop growth: A review. *Eng. Life Sci.* **2012**, *12*, 242–257. [CrossRef]
62. Laohaprapanon, S.; Marques, M.; Hogland, W. Removal of organic pollutants from wastewater using wood fly ash as a low-cost sorbent. *Clean—Soil Air Water* **2010**, *38*, 1055–1061. [CrossRef]
63. Leechart, P.; Nakbanpote, W.; Thiravetyan, P. Application of 'waste' wood-shaving bottom ash for adsorption of azo reactive dye. *J. Environ. Manag.* **2009**, *90*, 912–920. [CrossRef]
64. Forbes, M.S.; Raison, R.J.; Skjemstad, J.O. Formation, transformation and transport of black carbon (charcoal) in terrestrial and aquatic ecosystems. *Sci. Total Environ.* **2006**, *370*, 190–206. [CrossRef] [PubMed]
65. Zimmerman, A.R.; Gao, B.; Ahn, M.Y. Positive and negative carbon mineralization priming effects among a variety of biochar-amended soils. *Soil Biol. Biochem.* **2011**, *43*, 1169–1179. [CrossRef]
66. Menon, A.; Wang, J.Y.; Giannis, A. Optimization of micronutrient supplement for enhancing biogas production from food waste in two-phase thermophilic anaerobic digestion. *Waste Manag.* **2017**, *59*, 465–475. [CrossRef]

67. Romero-Güiza, M.S.; Vila, J.; Mata-Alvarez, J.; Chimenos, J.M.; Astals, S. The role of additives on anaerobic digestion: A review. *Renew. Sustain. Energy Rev.* **2016**, *58*, 1486–1499. [CrossRef]
68. AECOM Inc.; Metcal & Eddy Inc. Fundamentals of biological treatment. In *Wastewater Engineering: Treatment and Resource Recovery*, 5th ed.; Tchobanoglous, G., Stensel, H.D., Tsuchihashi, R., Burton, F., Eds.; Mc Graw Hill: New York, NY, USA, 2014; p. 576.
69. Shanmugam, P.; Horan, N.J. Simple and rapid methods to evaluate methane potential and biomass yield for a range of mixed solid wastes. *Bioresour. Technol.* **2009**, *100*, 471–474. [CrossRef]
70. Traversi, D.; Villa, S.; Lorenzi, E.; Degan, R.; Gilli, G. Application of a real-time qPCR method to measure the methanogen concentration during anaerobic digestion as an indicator of biogas production capacity. *J. Environ. Manag.* **2012**, *111*, 173–177. [CrossRef] [PubMed]
71. Vance, E.D.; Brookes, P.C.; Jenkinson, D.S. An extraction method for measuring soil microbial biomass C. *Soil Biol. Biochem.* **1987**, *19*, 703–707. [CrossRef]
72. Seadon, J.K. Sustainable waste management systems. *J. Clean. Prod.* **2010**, *18*, 1639–1651. [CrossRef]
73. Zhang, Q.; Liu, Z.; Petracchini, F.; Lu, C.; Li, Y.; Zhang, Z.; Paolini, V.; Zhang, H. Preparation of Slow-Release Insecticides from Biogas Slurry: Effectiveness of Ion Exchange Resin in the Adsorption and Release of Ammonia Nitrogen. *Processes* **2021**, *9*, 1461. [CrossRef]
74. Limoli, A.; Langone, M.; Andreottola, G. Ammonia removal from raw manure digestate by means of a turbulent mixing stripping process. *J. Environ. Manag.* **2016**, *176*, 1–10. [CrossRef] [PubMed]
75. Vaneeckhaute, C.; Lebuf, V.; Michels, E.; Belia, E.; Vanrolleghem, P.A.; Tack, F.M.G.; Meers, E. Nutrient Recovery from Digestate: Systematic Technology Review and Product Classification. *Waste Biomass Valorization* **2017**, *8*, 21–40. [CrossRef]
76. Abbas, Y.; Yun, S.; Wang, K.; Ali Shah, F.; Xing, T.; Li, B. Static-magnetic-field coupled with fly-ash accelerant: A powerful strategy to significantly enhance the mesophilic anaerobic-co-digestion. *Bioresour. Technol.* **2021**, *327*, 124793. [CrossRef] [PubMed]
77. Huang, H.; Zhang, D.D.; Li, J.; Guo, G.; Tang, S. Phosphate recovery from swine wastewater using plant ash in chemical crystallization. *J. Clean. Prod.* **2017**, *168*, 338–345. [CrossRef]
78. Bougnom, B.P.; Niederkofler, C.; Knapp, B.A.; Stimpfl, E.; Insam, H. Residues from renewable energy production: Their value for fertilizing pastures. *Biomass Bioenergy* **2012**, *39*, 290–295. [CrossRef]
79. Miranda, C.; Soares, A.S.; Coelho, A.C.; Trindade, H.; Teixeira, C.A. Environmental implications of stored cattle slurry treatment with sulphuric acid and biochar: A life cycle assessment approach. *Environ. Res.* **2021**, *194*, 110640. [CrossRef]
80. Brennan, R.B.; Healy, M.G.; Fenton, O.; Lanigan, G.J. The effect of chemical amendments used for phosphorus abatement on greenhouse gas and ammonia emissions from dairy cattle slurry: Synergies and pollution swapping. *PLoS ONE* **2015**, *10*, e0111965. [CrossRef]
81. Regueiro, I.; Coutinho, J.; Fangueiro, D. Alternatives to sulfuric acid for slurry acidification: Impact on slurry composition and ammonia emissions during storage. *J. Clean. Prod.* **2016**, *131*, 296–307. [CrossRef]
82. Regueiro, I.; Coutinho, J.; Gioelli, F.; Balsari, P.; Dinuccio, E.; Fangueiro, D. Acidification of raw and co-digested pig slurries with alum before mechanical separation reduces gaseous emission during storage of solid and liquid fractions. *Agric. Ecosyst. Environ.* **2016**, *227*, 42–51. [CrossRef]
83. UK Government. Quality Protocol of Anaerobic Digestate. End of Waste Criteria for the Production and Use of Quality Outputs from Anaerobic Digestion of Source-Segregated Biodegradable Waste. Available online: https://www.gov.uk/government/publications/quality-protocol-anaerobic-digestate (accessed on 14 December 2021).
84. UK Government. Quality Protocol Poultry Litter Ash. End of Waste Criteria for the Production and Use of Treated Ash from the Incineration of Poultry Litter, Feathers and Straw. Available online: https://www.gov.uk/government/publications/quality-protocol-poultry-litter-ash (accessed on 14 December 2021).
85. Moure Abelenda, A.; Semple, K.T.; Lag-Brotons, A.J.; Herbert, B.M.J.; Aggidis, G.; Aiouache, F. Effects of Wood Ash-Based Alkaline Treatment on Nitrogen, Carbon, and Phosphorus Availability in Food Waste and Agro-Industrial Waste Digestates. *Waste Biomass Valorization* **2020**, *12*, 3355–3370. [CrossRef]
86. Moure Abelenda, A.; Semple, K.T.; Lag-Brotons, A.J.; Herbert, B.M.J.; Aggidis, G.; Aiouache, F. Kinetic study of the stabilization of an agro-industrial digestate by adding wood fly ash. *Chem. Eng. J. Adv.* **2021**, *7*, 100127. [CrossRef]
87. Moure Abelenda, A.; Semple, K.T.; Lag-Brotons, A.J.; Herbert, B.M.J.; Aggidis, G.; Aiouache, F. Impact of sulphuric, hydrochloric, nitric, and lactic acids in the preparation of a blend of agro-industrial digestate and wood ash to produce a novel fertiliser. *J. Environ. Chem. Eng.* **2021**, *9*, 105021. [CrossRef]
88. Moure Abelenda, A.; Semple, K.T.; Lag-Brotons, A.J.; Herbert, B.M.; Aggidis, G.; Aiouache, F. Alkaline Wood Ash, Turbulence, and Traps with Excess of Sulfuric Acid Do Not Strip Completely the Ammonia off an Agro-waste Digestate. *Edelweiss Chem. Sci. J.* **2021**, *4*, 19–24. [CrossRef]
89. Jamali, M.K.; Kazi, T.G.; Arain, M.B.; Afridi, H.I.; Memon, A.R.; Jalbani, N.; SHAH, A. Use of Sewage Sludge After Liming as Fertilizer for Maize Growth. *Pedosphere* **2008**, *18*, 203–213. [CrossRef]
90. Brennan, R.B.; Fenton, O.; Rodgers, M.; Healy, M.G. Evaluation of chemical amendments to control phosphorus losses from dairy slurry. *Soil Use Manag.* **2011**, *27*, 238–246. [CrossRef]
91. UK Government. Using Animal by-Products at Compost and Biogas Sites. Available online: https://www.gov.uk/guidance/using-animal-by-products-at-compost-and-biogas-sites (accessed on 14 December 2021).

92. Kim, M.; Ahn, Y.H.; Speece, R.E. Comparative process stability and efficiency of anaerobic digestion; mesophilic vs. thermophilic. *Water Res.* **2002**, *36*, 4369–4385. [CrossRef]
93. AECOM Inc.; Metcalf & Eddy Inc. Advanced alkaline stabilisation technologies. In *Wastewater Engineering: Treatment and Resource Recovery*, 5th ed.; Tchobanoglous, G., Stensel, H.D., Tsuchihashi, R., Burton, F., Eds.; Mc Graw Hill: New York, NY, USA, 2014; pp. 1501–1502.
94. UK Government. Sewage Sludge in Agriculture: Code of Practice for England, Wales and Northern Ireland. Available online: https://www.gov.uk/government/publications/sewage-sludge-in-agriculture-code-of-practice/sewage-sludge-in-agriculture-code-of-practice-for-england-wales-and-northern-ireland (accessed on 14 December 2021).
95. Kim, J.S.; Lee, Y.Y.; Kim, T.H. A review on alkaline pretreatment technology for bioconversion of lignocellulosic biomass. *Bioresour. Technol.* **2016**, *199*, 42–48. [CrossRef]
96. Demeyer, A.; Voundi Nkana, J.C.; Verloo, M.G. Characteristics of wood ash and influence on soil properties and nutrient uptake: An overview. *Bioresour. Technol.* **2001**, *77*, 287–295. [CrossRef]
97. Ribbing, C. Environmentally friendly use of non-coal ashes in Sweden. *Waste Manag.* **2007**, *27*, 1428–1435. [CrossRef] [PubMed]
98. Nag, R.; Auer, A.; Nolan, S.; Russell, L.; Markey, B.K.; Whyte, P.; O'Flaherty, V.; Bolton, D.; Fenton, O.; Richards, K.G.; et al. Evaluation of pathogen concentration in anaerobic digestate using a predictive modelling approach (ADRISK). *Sci. Total Environ.* **2021**, *800*, 149574. [CrossRef] [PubMed]
99. UK Government. Handling of Manure and Slurry to Reduce Antibiotic Resistance. Available online: https://www.gov.uk/guidance/handling-of-manure-and-slurry-to-reduce-antibiotic-resistance (accessed on 14 December 2021).
100. Al-Mallahi, J.; Sürmeli, R.Ö.; Çalli, B. Recovery of phosphorus from liquid digestate using waste magnesite dust. *J. Clean. Prod.* **2020**, *272*, 122616. [CrossRef]
101. Sakthivel, S.R.; Tilley, E.; Udert, K.M. Wood ash as a magnesium source for phosphorus recovery from source-separated urine. *Sci. Total Environ.* **2012**, *419*, 68–75. [CrossRef] [PubMed]
102. Numviyimana, C.; Warchoł, J.; Ligas, B.; Chojnacka, K. Nutrients Recovery from Dairy Wastewater by Struvite Precipitation Combined with Ammonium Sorption on Clinoptilolite. *Materials* **2021**, *14*, 5822. [CrossRef]
103. Laureni, M.; Palatsi, J.; Llovera, M.; Bonmatí, A. Influence of pig slurry characteristics on ammonia stripping efficiencies and quality of the recovered ammonium-sulfate solution. *J. Chem. Technol. Biotechnol.* **2013**, *88*, 1654–1662. [CrossRef]
104. Ndegwa, P.M.; Vaddella, V.K.; Hristov, A.N.; Joo, H.S. Measuring Concentrations of Ammonia in Ambient Air or Exhaust Air Stream using Acid Traps. *J. Environ. Qual.* **2009**, *38*, 647–653. [CrossRef]
105. Barrett, J. Periodicity of aqueous chemistry I: S- and p-block chemistry. In *Inorganic Chemistry in Aqueous Solution*; Barrett, J., Ed.; Royal Society of Chemistry: Cambridge, UK, 2003; pp. 98–123.
106. Zumdahl, S.S. Peroxide Ion. Available online: https://www.britannica.com/science/peroxide-ion (accessed on 5 January 2022).
107. Drapanauskaite, D.; Handler, R.M.; Fox, N.; Baltrusaitis, J. Transformation of Liquid Digestate from the Solid-Separated Biogas Digestion Reactor Effluent into a Solid NH4HCO3 Fertilizer: Sustainable Process Engineering and Life Cycle Assessment. *ACS Sustain. Chem. Eng.* **2021**, *9*, 580–588. [CrossRef]
108. Jantsch, T.G.; Mattiasson, B. A simple spectrophotometric method based on pH-indicators for monitoring partial and total alkalinity in anaerobic processes. *Environ. Technol.* **2003**, *24*, 1061–1067. [CrossRef]
109. Alburquerque, J.A.; de la Fuente, C.; Campoy, M.; Carrasco, L.; Nájera, I.; Baixauli, C.; Caravaca, F.; Roldán, A.; Cegarra, J.; Bernal, M.P. Agricultural use of digestate for horticultural crop production and improvement of soil properties. *Eur. J. Agron.* **2012**, *43*, 119–128. [CrossRef]
110. Owamah, H.I.; Dahunsi, S.O.; Oranusi, U.S.; Alfa, M.I. Fertilizer and sanitary quality of digestate biofertilizer from the co-digestion of food waste and human excreta. *Waste Manag.* **2014**, *34*, 747–752. [CrossRef] [PubMed]
111. Tambone, F.; Scaglia, B.; D'Imporzano, G.; Schievano, A.; Orzi, V.; Salati, S.; Adani, F. Assessing amendment and fertilizing properties of digestates from anaerobic digestion through a comparative study with digested sludge and compost. *Chemosphere* **2010**, *81*, 577–583. [CrossRef]
112. Alburquerque, J.A.; de la Fuente, C.; Ferrer-Costa, A.; Carrasco, L.; Cegarra, J.; Abad, M.; Bernal, M.P. Assessment of the fertiliser potential of digestates from farm and agroindustrial residues. *Biomass Bioenergy* **2012**, *40*, 181–189. [CrossRef]
113. Mu, Z.X.; He, C.S.; Jiang, J.K.; Zhang, J.; Yang, H.Y.; Mu, Y. A modified two-point titration method for the determination of volatile fatty acids in anaerobic systems. *Chemosphere* **2018**, *204*, 251–256. [CrossRef]
114. Guštin, S.; Marinšek-Logar, R. Effect of pH, temperature and air flow rate on the continuous ammonia stripping of the anaerobic digestion effluent. *Process Saf. Environ. Prot.* **2011**, *89*, 61–66. [CrossRef]
115. Liu, L.; Pang, C.; Wu, S.; Dong, R. Optimization and evaluation of an air-recirculated stripping for ammonia removal from the anaerobic digestate of pig manure. *Process Saf. Environ. Prot.* **2015**, *94*, 350–357. [CrossRef]
116. Drosg, B.; Fuchs, W.; Al Seadi, T.; Madsen, M.; Linke, B. Nutrient Recovery by Biogas Digestate Processing. Available online: http://task37.ieabioenergy.com/technical-brochures.html#form (accessed on 14 December 2021).
117. Miles, A.; Ellis, T.G. Struvite precipitation potential for nutrient recovery from anaerobically treated wastes. *Water Sci. Technol.* **2001**, *43*, 259–266. [CrossRef]
118. Campos, J.L.; Crutchik, D.; Franchi, Ó.; Pavissich, J.P.; Belmonte, M.; Pedrouso, A.; Mosquera-Corral, A.; Val del Río, Á. Nitrogen and Phosphorus Recovery from Anaerobically Pretreated Agro-Food Wastes: A Review. *Front. Sustain. Food Syst.* **2019**, *2*, 91. [CrossRef]

119. Escudero, A.; Blanco, F.; Lacalle, A.; Pinto, M. Struvite precipitation for ammonium removal from anaerobically treated effluents. *J. Environ. Chem. Eng.* **2015**, *3*, 413–419. [CrossRef]
120. Prywer, J.; Sieroń, L.; Czylkowska, A. Struvite Grown in Gel, Its Crystal Structure at 90 K and Thermoanalytical Study. *Crystals* **2019**, *9*, 89. [CrossRef]
121. Schwarz, S.; Lunkwitz, K.; Keßler, B.; Spiegler, U.; Killmann, E.; Jaeger, W. Adsorption and stability of colloidal silica. *Colloids Surf. A Physicochem. Eng. Asp.* **2000**, *163*, 17–27. [CrossRef]
122. Shah, I.; Adnan, R.; Ngah, W.S.W.; Mohamed, N. Iron impregnated activated carbon as an efficient adsorbent for the removal of methylene blue: Regeneration and kinetics studies. *PLoS ONE* **2015**, *10*, e0122603. [CrossRef] [PubMed]
123. Yagi, S.; Fukushi, K. Removal of phosphate from solution by adsorption and precipitation of calcium phosphate onto monohydrocalcite. *J. Colloid Interface Sci.* **2012**, *384*, 128–136. [CrossRef] [PubMed]
124. Courtney, C.; Brison, A.; Randall, D.G. Calcium removal from stabilized human urine by air and CO_2 bubbling. *Water Res.* **2021**, *202*, 117467. [CrossRef]
125. Anukam, A.; Mohammadi, A.; Naqvi, M.; Granström, K. Methods of accelerating and optimizing process efficiency. *Processes* **2019**, *7*, 504. [CrossRef]
126. Fangueiro, D.; Hjorth, M.; Gioelli, F. Acidification of animal slurry—A review. *J. Environ. Manag.* **2015**, *149*, 46–56. [CrossRef]
127. Mao, C.; Feng, Y.; Wang, X.; Ren, G. Review on research achievements of biogas from anaerobic digestion. *Renew. Sustain. Energy Rev.* **2015**, *45*, 540–555. [CrossRef]
128. Hjorth, M.; Christensen, K.V.; Christensen, M.L.; Sommer, S.G. Solid-liquid separation of animal slurry in theory and practice. *Sustain. Agric.* **2009**, *2*, 953–986. [CrossRef]
129. Shi, Y.; Parker, D.B.; Cole, N.A.; Auvermann, B.W.; Mehlhorn, J.E. Surface amendments to minimize ammonia emissions from beef cattle feedlots. *Trans. Am. Soc. Agric. Eng.* **2001**, *44*, 677–682. [CrossRef]
130. Vandré, R.; Clemens, J. Studies on the relationship between slurry pH, volatilization processes and the influence of acidifying additives. *Nutr. Cycl. Agroecosyst.* **1996**, *47*, 157–165. [CrossRef]
131. Fan, M.; Brown, R.C.; Wheelock, T.D.; Cooper, A.T.; Nomura, M.; Zhuang, Y. Production of a complex coagulant from fly ash. *Chem. Eng. J.* **2005**, *106*, 269–277. [CrossRef]
132. Kavanagh, I.; Burchill, W.; Healy, M.G.; Fenton, O.; Krol, D.J.; Lanigan, G.J. Mitigation of ammonia and greenhouse gas emissions from stored cattle slurry using acidifiers and chemical amendments. *J. Clean. Prod.* **2019**, *237*, 117822. [CrossRef]
133. Fomenko, A.I.; Sokolov, L.I. A study of sorption of phosphate ions from aqueous solutions by wood ash. *Russ. J. Appl. Chem.* **2015**, *88*, 652–656. [CrossRef]
134. Hamidi, N.H.; Ahmed, O.H.; Omar, L.; Ch'ng, H.Y. Soil Nitrogen Sorption Using Charcoal and Wood Ash. *Agronomy* **2021**, *11*, 1801. [CrossRef]
135. Mosoarca, G.; Vancea, C.; Popa, S.; Boran, S.; Tanasie, C. A green approach for treatment of wastewater with manganese using wood ash. *J. Chem. Technol. Biotechnol.* **2020**, *95*, 1781–1789. [CrossRef]
136. Rosenfeld, P.E.; Henry, C.L. Activated Carbon and Wood Ash Sorption of Wastewater, Compost, and Biosolids Odorants. *Water Environ. Res.* **2001**, *73*, 388–393. [CrossRef]
137. Kavanagh, I.; Fenton, O.; Healy, M.G.; Burchill, W.; Lanigan, G.J.; Krol, D.J. Mitigating ammonia and greenhouse gas emissions from stored cattle slurry using agricultural waste, commercially available products and a chemical acidifier. *J. Clean. Prod.* **2021**, *294*, 126251. [CrossRef]
138. Houba, V.J.G.; Temminghoff, E.J.M.; Gaikhorst, G.A.; van Vark, W. Soil analysis procedures using 0.01 M calcium chloride as extraction reagent. *Commun. Soil Sci. Plant Anal.* **2000**, *31*, 1299–1396. [CrossRef]
139. Mor, S.; Chhoden, K.; Ravindra, K. Application of agro-waste rice husk ash for the removal of phosphate from the wastewater. *J. Clean. Prod.* **2016**, *129*, 673–680. [CrossRef]
140. Park, N.D.; Michael Rutherford, P.; Thring, R.W.; Helle, S.S. Wood pellet fly ash and bottom ash as an effective liming agent and nutrient source for rye grass (*Lolium perenne* L.) and oats (*Avena sativa*). *Chemosphere* **2012**, *86*, 427–432. [CrossRef]
141. Janiszewska, D.; Olchowski, R.; Nowicka, A.; Zborowska, M.; Marszałkiewicz, K.; Shams, M.; Giannakoudakis, D.A.; Anastopoulos, I.; Barczak, M. Activated biochars derived from wood biomass liquefaction residues for effective removal of hazardous hexavalent chromium from aquatic environments. *GCB Bioenergy* **2021**, *13*, 1247–1259. [CrossRef]
142. Ma, Z.; Li, Q.; Yue, Q.; Gao, B.; Li, W.; Xu, X.; Zhong, Q. Adsorption removal of ammonium and phosphate from water by fertilizer controlled release agent prepared from wheat straw. *Chem. Eng. J.* **2011**, *171*, 1209–1217. [CrossRef]
143. Awad, A.M.; Shaikh, S.M.R.; Jalab, R.; Gulied, M.H.; Nasser, M.S.; Benamor, A.; Adham, S. Adsorption of organic pollutants by natural and modified clays: A comprehensive review. *Sep. Purif. Technol.* **2019**, *228*, 115719. [CrossRef]
144. Ronga, D.; Mantovi, P.; Pacchioli, M.T.; Pulvirenti, A.; Bigi, F.; Allesina, G.; Pedrazzi, S.; Tava, A.; Dal Prà, A. Combined effects of dewatering, composting and pelleting to valorize and delocalize livestock manure, improving agricultural sustainability. *Agronomy* **2020**, *10*, 661. [CrossRef]
145. Dinuccio, E.; Balsari, P.; Berg, W. GHG emissions during the storage of rough pig slurry and the fractions obtained by mechanical separation. *Aust. J. Exp. Agric.* **2008**, *48*, 93–95. [CrossRef]
146. Dinuccio, E.; Gioelli, F.; Balsari, P.; Dorno, N. Ammonia losses from the storage and application of raw and chemo-mechanically separated slurry. *Agric. Ecosyst. Environ.* **2012**, *153*, 16–23. [CrossRef]

147. Dinuccio, E.; Berg, W.; Balsari, P. Gaseous emissions from the storage of untreated slurries and the fractions obtained after mechanical separation. *Atmos. Environ.* **2008**, *42*, 2448–2459. [CrossRef]
148. Dinuccio, E.; Biagini, D.; Rosato, R.; Balsari, P.; Lazzaroni, C. Organic matter and nitrogen balance in rabbit fattening and gaseous emissions during manure storage and simulated land application. *Agric. Ecosyst. Environ.* **2019**, *269*, 30–38. [CrossRef]
149. Gioelli, F.; Dinuccio, E.; Balsari, P. Residual biogas potential from the storage tanks of non-separated digestate and digested liquid fraction. *Bioresour. Technol.* **2011**, *102*, 10248–10251. [CrossRef] [PubMed]
150. Gioelli, F.; Dinuccio, E.; Cuk, D.; Rollè, L.; Balsari, P. Acidification with sulfur of the separated solid fraction of raw and co-digested pig slurry: Effect on greenhouse gas and ammonia emissions during storage. *Anim. Prod. Sci.* **2016**, *56*, 343–349. [CrossRef]
151. Balsari, P.; Airoldi, G.; Dinuccio, E.; Gioelli, F. Ammonia emissions from farmyard manure heaps and slurry stores-Effect of environmental conditions and measuring methods. *Biosyst. Eng.* **2007**, *97*, 456–463. [CrossRef]
152. Pampuro, N.; Bagagiolo, G.; Priarone, P.C.; Cavallo, E. Effects of pelletizing pressure and the addition of woody bulking agents on the physical and mechanical properties of pellets made from composted pig solid fraction. *Powder Technol.* **2017**, *311*, 112–119. [CrossRef]
153. Weigand, H.; Bertau, M.; Hübner, W.; Bohndick, F.; Bruckert, A. RecoPhos: Full-scale fertilizer production from sewage sludge ash. *Waste Manag.* **2013**, *33*, 540–544. [CrossRef]
154. Fivelman, Q. ADFerTech: Granular Fertiliser from Anaerobic Digestate Liquor. Available online: http://www.adfertech.com/ (accessed on 14 December 2021).
155. Jewiarz, M.; Wróbel, M.; Fraczek, J.; Mudryk, K.; Dziedzic, K. Digestate, ash and Trichoderm based fertilizer-production line concept design. *MATEC Web Conf.* **2018**, *168*, 04004. [CrossRef]
156. Chojnacka, K.; Moustakas, K.; Witek-Krowiak, A. Bio-based fertilizers: A practical approach towards circular economy. *Bioresour. Technol.* **2020**, *295*, 122223. [CrossRef] [PubMed]
157. Steenari, B.M.; Lindqvist, O. Stabilisation of biofuel ashes for recycling to forest soil. *Biomass Bioenergy* **1997**, *13*, 39–50. [CrossRef]
158. Illikainen, M.; Tanskanen, P.; Kinnunen, P.; Körkkö, M.; Peltosaari, O.; Wigren, V.; Österbacka, J.; Talling, B.; Niinimäki, J. Reactivity and self-hardening of fly ash from the fluidized bed combustion of wood and peat. *Fuel* **2014**, *135*, 69–75. [CrossRef]
159. Zafari, A.; Hosein Kianmehr, M. Effect of Temperature, Pressure and Moisture Content on Durability of Cattle Manure Pellet in Open-end Die Method. *J. Agric. Sci.* **2012**, *4*, 203–208. [CrossRef]
160. Alemi, H.; Kianmehr, M.H.; Borghaee, A.M. Effect of Pellet Processing of Fertilizer on Slow-Release Nitrogen in Soil. *Asian J. Plant Sci.* **2010**, *9*, 74–80. [CrossRef]
161. Rao, J.R.; Watabe, M.; Stewart, T.A.; Millar, B.C.; Moore, J.E. Pelleted organo-mineral fertilisers from composted pig slurry solids, animal wastes and spent mushroom compost for amenity grasslands. *Waste Manag.* **2007**, *27*, 1117–1128. [CrossRef] [PubMed]
162. Mudryk, K.; Frączek, J.; Wróbel, M.; Jewiarz, M.; Dziedzic, K. Agglomeration of Ash-Based Fertilizer Mixtures from Biomass Combustion and Digestate. In *Renewable Energy Sources: Engineering, Technology, Innovation*; Krzysztof, M., Sebastian, W., Eds.; Springer: Cham, Switzerland, 2018; pp. 823–834.
163. US EPA. Alkaline Stabilization of Biosolids. Available online: https://www.epa.gov/biosolids/fact-sheet-alkaline-stabilization-biosolids (accessed on 14 December 2021).
164. Rajendran, K.; Devaraj, P. Biomass and nutrient distribution and their return of Casuarina equisetifolia inoculated with biofertilizers in farm land. *Biomass Bioenergy* **2004**, *26*, 235–249. [CrossRef]
165. Wu, S.C.; Cao, Z.H.; Li, Z.G.; Cheung, K.C.; Wong, M.H. Effects of biofertilizer containing N-fixer, P and K solubilizers and AM fungi on maize growth: A greenhouse trial. *Geoderma* **2005**, *125*, 155–166. [CrossRef]
166. Aseri, G.K.; Jain, N.; Panwar, J.; Rao, A.V.; Meghwal, P.R. Biofertilizers improve plant growth, fruit yield, nutrition, metabolism and rhizosphere enzyme activities of Pomegranate (*Punica granatum* L.) in Indian Thar Desert. *Sci. Hortic.* **2008**, *117*, 130–135. [CrossRef]
167. Mahfouz, S.A.; Shamf-Eldin, M.A. Effect of mineral vs. biofertilizer on growth, yield, and essential oil content of fennel (Foeniculum vulgare Mill.). *Int. Agrophysics* **2007**, *21*, 361–366. [CrossRef]
168. Singh Brar, B.; Singh, J.; Singh, G.; Kaur, G. Effects of Long Term Application of Inorganic and Organic Fertilizers on Soil Organic Carbon and Physical Properties in Maize–Wheat Rotation. *Agronomy* **2015**, *5*, 220–238. [CrossRef]
169. Johansen, A.; Carter, M.S.; Jensen, E.S.; Hauggard-Nielsen, H.; Ambus, P. Effects of digestate from anaerobically digested cattle slurry and plant materials on soil microbial community and emission of CO_2 and N_2O. *Appl. Soil Ecol.* **2013**, *63*, 36–44. [CrossRef]

processes

MDPI

Review

Process Intensification in Bio-Ethanol Production–Recent Developments in Membrane Separation

Izumi Kumakiri [1,*], Morihisa Yokota [1], Ryotaro Tanaka [1], Yu Shimada [1], Worapon Kiatkittipong [2], Jun Wei Lim [3], Masayuki Murata [4,5] and Mamoru Yamada [4,5]

1 Graduate School of Sciences and Technology for Innovation, Faculty of Engineering, Yamaguchi University, 2-16-1 Tokiwadai Ube, Yamaguchi 755-8611, Japan; morichan@yamaguchi-u.ac.jp (M.Y.); b037vf@yamaguchi-u.ac.jp (R.T.); b061vfv@yamaguchi-u.ac.jp (Y.S.)
2 Department of Chemical Engineering, Faculty of Engineering and Industrial Technology, Silpakorn University, Nakhon Pathom 73000, Thailand; kiatkittipong_w@su.ac.th
3 Department of Fundamental and Applied Sciences, HICoE-Centre for Biofuel and Biochemical Research, Institute of Self-Sustainable Building, Universiti Teknologi PETRONAS, Seri Iskandar 32610, Perak, Malaysia; junwei.lim@utp.edu.my
4 Research Center for Thermotolerant Microbial Resources, Yamaguchi University, Yamaguchi 753-8315, Japan; muratam@yamaguchi-u.ac.jp (M.M.); m-yamada@yamaguchi-u.ac.jp (M.Y.)
5 Graduate School of Sciences and Technology for Innovation, Faculty of Agriculture, Yamaguchi University, Yamaguchi 753-8515, Japan
* Correspondence: izumi.k@yamaguchi-u.ac.jp

Citation: Kumakiri, I.; Yokota, M.; Tanaka, R.; Shimada, Y.; Kiatkittipong, W.; Lim, J.W.; Murata, M.; Yamada, M. Process Intensification in Bio-Ethanol Production–Recent Developments in Membrane Separation. *Processes* **2021**, 9, 1028. https://doi.org/10.3390/pr9061028

Academic Editor: Giampaolo Manzolini

Received: 25 May 2021
Accepted: 8 June 2021
Published: 10 June 2021

Abstract: Ethanol is considered as a renewable transport fuels and demand is expected to grow. In this work, trends related to bio-ethanol production are described using Thailand as an example. Developments on high-temperature fermentation and membrane technologies are also explained. This study focuses on the application of membranes in ethanol recovery after fermentation. A preliminary simulation was performed to compare different process configurations to concentrate 10 wt% ethanol to 99.5 wt% using membranes. In addition to the significant energy reduction achieved by replacing azeotropic distillation with membrane dehydration, employing ethanol-selective membranes can further reduce energy demand. Silicalite membrane is a type of membrane showing one of the highest ethanol-selective permeation performances reported today. A silicalite membrane was applied to separate a bio-ethanol solution produced via high-temperature fermentation followed by a single distillation. The influence of contaminants in the bio-ethanol on the membrane properties and required further developments are also discussed.

Keywords: bio-ethanol; thermotolerant yeast; membrane separation; ethanol-selective membrane; energy demand

1. Introduction

The transport sector is one of the largest contributors of greenhouse gas. Fossil-fuel is still the major energy source today and a shift to more sustainable fuels is indispensable. Huge efforts have been made to convert renewable biomass, in particular, lignocellulosic feedstocks, to biofuel. Bio-ethanol has been drawing attention among various types of bio-fuel, because ethanol can be mixed with petrol or used as is as a transport fuel.

Bio-ethanol production via fermentation can be divided into several stages [1], as shown in Figure 1. Pretreatment is required when using cellulosic biomass, and enzymatic saccharification is required for starchy or cellulosic biomass. Various technologies have been proposed and developed for each stage to make the conversion process more efficient and cost-effective. The usage of residue is another key to making the bio-ethanol conversion process more economically and environmentally friendly [2–4].

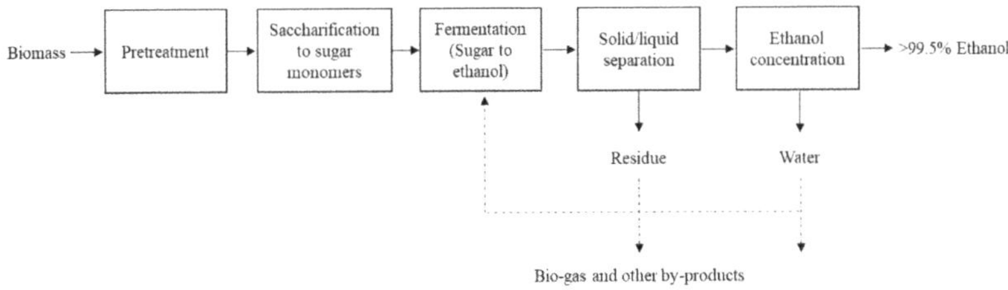

Figure 1. Overview of a bio-ethanol production process.

Energy consumption and distribution per each stage in Figure 1 can vary depending on the type of raw biomass, the process configuration, and other various conditions. The ethanol concentration after fermentation is low and is often less than 10%. In the case of producing a fuel grade ethanol, which requires purity over 99.5 mol% (99.8 wt%) [5,6], the fermented broth needs to be concentrated. Distillation is a well-established and widely used technology; however, when applied to anhydrous bio-ethanol production it can occupy more than one-third of the total energy consumption [7,8].

The energy demand at the conversion process should be limited as much as possible to enhance the benefit of using bio-ethanol. Developing environmentally friendly and easier operation/maintenance technologies will make the conversion process more attractive. Because bio-refineries require various technologies, collaboration efforts among different research areas are essential.

In the following, the trend in bio-ethanol demand is summarized using Thailand as an example. Then, advances in fermentation technology, comparisons of membrane integrated downstream processes with conventional distillation, and the current status of membrane properties in separating bio-ethanol are discussed. Among various technologies, this article focused on thermotolerant yeast and membrane-based technologies. The number of publications on high-temperature fermentation and bio-ethanol production using dehydration membranes has increased over the last decade, as shown in Figure 2. The literature survey was performed by Google Scholar [9] using the keywords indicated in the figure caption. Reports on ethanol-selective membranes, on the contrary, is rather small compared to the number of publications on dehydration membranes. However, a process simulation performed here showed their potential contribution to realizing increased energy-efficient bio-ethanol production. The status of ethanol-selective membranes is explained by comparing different membrane materials in the following.

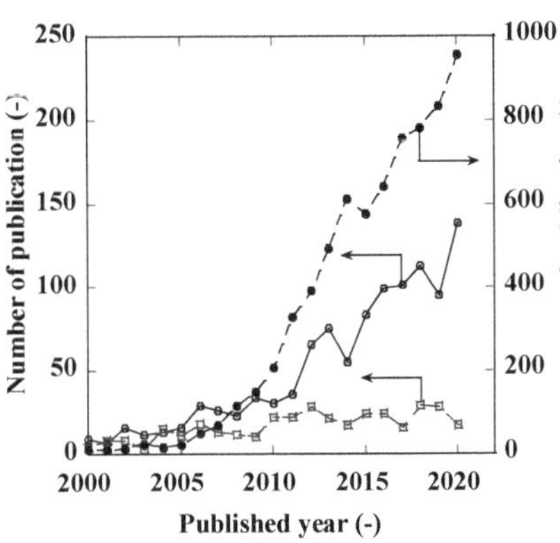

Figure 2. Number of publications surveyed by Google Scholar on the 3 June 2021. The keywords used for the survey are as follows: •, thermotolerant and bioethanol; ○, dehydration, membrane, separation, bioethanol, and hybrid process; □, ethanol selective, membrane, and separation.

2. Bioethanol Production in Thailand

Thailand is the seventh-largest ethanol producer, following the USA, Brazil, European Union (EU) countries, China, India, and Canada, as shown in Table 1. However, interestingly, Thailand's ethanol blend rate is the second highest ranked in the world, with an average of 13.7% in 2019, following only by Brazil, which currently mandated as high as 27% (E27), additionally with a high portion of hydrous ethanol [10].

Table 1. The World's top eight ethanol producers in 2019 and their blend rate in gasoline (Data taken from [11–15]).

Country	Production (Million Liters per Day)	Blend Rate in Gasoline (vol%)
United State	163.86	10.55
Brazil	89.40	27
EU	14.93	6.16
China	9.33	2.4
India	5.50	4.5
Canada	5.19	6.6
Thailand	4.36	13.7
Argentina	3.01	11.7

In Thailand, ethanol is initially employed to blend with gasoline as an octane enhancer for replacement of methyl *tert*-butyl ether (MTBE). It should be noted that ethanol and other alcohols produced from, e.g., in situ conversion of refinery cuts [16], were also used successfully as substitutes of gasoline ether oxygenates.

Presently, the common ethanol to gasoline blending proportion is 10%, 20%, and 85% and are referred to as E10, E20, and E85, respectively; around 66% of petrol vehicles in Thailand are compatible with E20 fuel [17]. The primary raw materials for ethanol production are molasses, cassava, and sugarcane juice [18].

Thailand's ethanol production and domestic utilization have continuously increased, as shown in Figure 3 [18,19]. The increase in gasohol consumption has resulted from government policy and subsidization from the state oil fund. However, the Thai government has adjusted the ethanol consumption target. Previously, according to the Alternative Energy Development Plan (AEDP), 2015, the target is 4.1 billion liters by 2036. In the current plan (AEDP, 2018), the ethanol consumption target was reduced to 2.7 billion liters because the ethanol production raw materials may not be sufficient [20,21].

Promoting ethanol production from other raw materials is necessary to overcome feedstock insufficiency problem. The recent trend for ethanol production has focused on non-edible feedstock, especially lignocellulosic materials. The Thailand Institute of Scientific and Technological Research (TISTR) has found that bagasse, rice straw, and corncobs are efficient and feasible raw materials for ethanol production at the industrial level [22]. However, the pretreatment process is the major challenge of second-generation ethanol production technology. Besides second-generation ethanol production technology, Thailand is also in search of R&D on microalgae as a third-generation ethanol production technology [23]. The support of second and third-generation ethanol production from the Thai government corresponds to the Thailand Integrated Energy Blueprint (TIEB), which sets ethanol production target from these two generations biofuels equal to 10 kilotons of oil equivalent (ktoe) by 2036 [23].

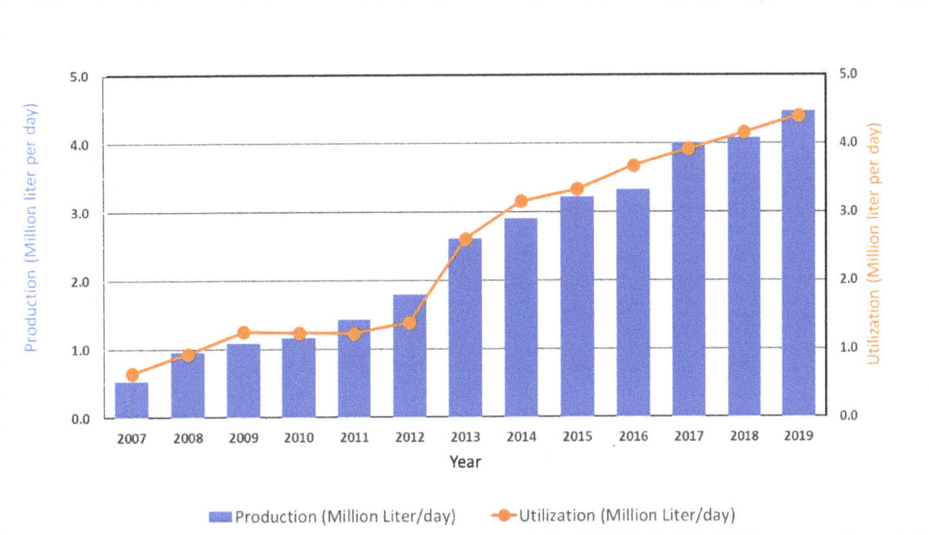

Figure 3. Ethanol production and consumption in Thailand (2007–2019).

3. New Approaches on the Microbial Biomass Conversion

Saccharomyces cerevisiae is used worldwide for the fermentation of ethanol. In the conventional fermentation-distillation-dehydration process with first-generation biomass, it is not profitable unless the production scale of ethanol is 15,000 kL/year or more in Japan, based on two different demonstration projects performed in Hokkaido from 2007 to 2012. In areas where it is difficult to secure a huge amount of feedstock and the cost of transportation is high, as it is in Japan, the local production for local consumption by small-scale ethanol production is desirable. However, in order to be profitable in a small ethanol production facility, it is essential to develop an innovative technology for each step of the ethanol production process. Here, we introduce a high-temperature fermentation (HTF) technology that has several benefits, such as reduced cooling costs, reduced microbial contamination, and reduced amounts of hydrolytic enzymes used in simultaneous saccharification and fermentation (SSF) [24,25]. Fermentation is an exothermic reaction that requires cooling of the reactor, as the temperature inside the reactor rises to around 313 K. Otherwise, such high temperatures cause prevention of the fermentation ability of ethanol-producing microorganisms or cell death. A stable HTF at about 313 K that does not require temperature control is thus highly desirable. In addition, SSF that can be performed by a relatively simple facility with easy operations is suitable for small-scale production.

On the other hand, thermotolerant microorganisms capable of efficiently fermenting and producing ethanol at high temperatures are indispensable for HTF technology. We have isolated and characterized many thermotolerant yeasts from joint research from southeast Asian countries, centering on Thailand [25]. Among them, *Kluyveromyces marxianus* DMKU3-1042, isolated in Thailand, can grow at 321 K and showed high ethanol productivity up to about 316 K when glucose was used as a carbon source. Additionally, unlike *S. cerevisiae*, *K. marxianus* utilizes a wide range of carbon sources [26] and can convert polysaccharide inulin to ethanol [27]. In addition, a *K. marxianus* strain that has a high ethanol production capacity from xylose contained in second generation biomass [28], a *Pichia kudriavzevii* strain [29] that is suitable for HTF using cassava starch as a raw material, and a *Spathaspora passalidarum* strain [30] that has no glucose suppression with high ethanol production capacity from xylose have been isolated. Other research groups have isolated thermotolerant, ethanol-fermenting yeasts from India, Bangladesh, Vietnam, and Brazil [31–34].

As mentioned above, the temperature in the reactor increases by metabolic and mechanical heat sources, being close to a non-permissible level for non-thermotolerant yeast, with *S. cerevisiae* being the most frequently applied yeast. However, thermotolerant yeasts can grow and ferment under such high temperatures, which allows us to perform temperature-non-controlled fermentation. For example, the ethanol production rate drastically decreases with non-thermotolerant yeasts when the temperature is increased from 303 to 313 K. On the contrary, the ethanol production rate is almost independent of this temperature range [26]. When a bench-scale fermentation without temperature controls using 2 L of 9% glucose medium was tested, thermotolerant *K. marxianus* DMKU 3-1042 [35,36] produced ethanol equivalent to that under the temperature-controlled condition at 303 K. Moreover, a fermenter-scale fermentation with 4000 L of 18% sugarcane was tested to achieve 7% ethanol production [37]. This fermentation is favorable because the cooling cost tends to be higher in tropical countries or increases in summer in many other countries. In addition, temperature-non-controlled fermentation can keep producing ethanol, even if any power outage occurs. As HTF reduces the risk of undesirable micro-organisms growth, the air-lock of the fermenter does not need to be perfect, which makes the fermenter design simpler.

Another interesting advantage of HTF is the possibility to combine fermentation and vacuum distillation in one unit. Such a process can reduce the manufacturing time and the cost of equipment. The higher fermentation temperature increases the saturated vapor pressure of ethanol. The higher saturated vapor pressure facilitates the vapor distillation that requires less pressure than the saturated vapor pressure. A system consisting of a fermentation and a distillation tank, the primary and secondary ethanol recovery units, a vacuum pump, and a drain unit was constructed [37]. Ethanol is concentrated as the process proceeds from the primary to the secondary ethanol recovery unit, and the air in the tank is discharged outside during the vacuum distillation; some ethanol is trapped in the drain unit. When fermentation with *K. marxianus* DMKU 3-1042 and distillation at 70 mbar and 314 K were applied, about 35% and 60% were recovered in the primary and secondary bottles, respectively [37]. The process of the simultaneous fermentation and distillation under low pressure was continuously repeated three times with 12% rice-hydrolysate. There are some additional benefits in this system: (a) microbes avoid exposure to high concentrations of ethanol or acetic acid, or strong oxidative stress and (b) fermentation can be continued during distillation, increasing ethanol yields. The fermentation and vacuum distillation combined system removes solid materials in the fermented broth from the liquid fraction. The ethanol can be concentrated further by applying membranes or other separation technologies.

4. Various Membrane Separation Processes

Employing distillation is a conventional way to increase the ethanol concentration after fermentation. Azeotropic distillation is required to obtain over 96 wt% (89 mol%) ethanol. Adsorption columns can be used to produce anhydrous ethanol [38]. Recently, a hybrid process consisting of distillation and membrane dehydration was proposed as an energy-saving alternative. In the hybrid process, azeotropic distillation was replaced with membrane separation [39–41]. In the late 1990s, a successful industrial application of A-type zeolite membranes, a type of inorganic membrane, to dry solvents was reported [42]. Since then, the number of industrial applications of membrane integrated processes has been growing, and more than two hundred units are under operation today.

Energy consumption in bio-ethanol production may be reduced further by employing membranes in different ways. Various membrane processes, including pervaporation (PV) [42], membrane distillation (MD) [43,44], vapor permeation (VP) [41,45], nanofiltration (NF) [46], reverse osmosis (RO) [47], and forward osmosis (FO) [48,49] have been intensively studied to concentrate liquid mixtures, together with the developments of new types of membranes. Membranes can be integrated with saccharification or fermentation to concentrate sugars and remove fermentation inhibitors [48,49]. Membranes can also be used to concentrate ethanol after fermentation. In the latter case, dehydration PV membranes have been studied the most. While a phase change occurs during the permeation through PV and MD membranes, NF and RO processes are a pressure-driven separation with no phase change. Considering the large latent heat of water, NF and RO can be more energy-efficient than PV.

Permeation equations, for example a solution-diffusion model, relate these different membrane processes [43,50]. Nakao compared the PV and RO processes using the transport model and proposed a combination of water-selective RO membranes and ethanol-selective RO membrane [51], as shown in Figure 4. The energy requirement to concentrate 10% ethanol to 96% with ethanol recovery over 95% was studied. The study showed that the RO process is much more energy efficient than PV and requires 1/1000 of the energy required by distillation.

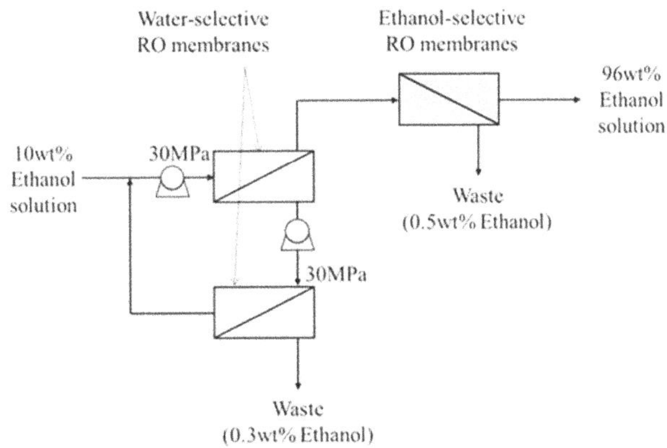

Figure 4. Example of combining two types of RO membranes (modified from [51]).

The RO-based separation process is with a simple design as illustrated in Figure 4. The major components are pumps to raise the feed side pressure, membrane modules and valves to regulate pressure and flow rate., A heating or cooling system is dispensable. One of the drawbacks of RO is the high pressure required at the feed side to overcome the osmotic pressure across the membrane. Conventional polymeric membranes are difficult to apply under the high pressure of 30 MPa assumed in the above process calculations [51]. Therefore, inorganic membranes have been investigated. One of the first attempts at an inorganic RO membrane is the application of an A-type zeolite membrane to water/ethanol separation [47]. The membrane showed pressure stability at least up to 5 MPa and water-selective permeation, but the flux was too small. Recently, various types of inorganic RO membranes have been developed for desalination purpose [52,53]. However, further developments are required to realize a RO-membrane based downstream process for bio-ethanol production.

Besides dehydration VP/PV membranes [42,54], there are membranes permeating more ethanol than water. Figure 5 shows the liquid-gas equilibrium of an ethanol/water mixture. Examples of permeate concentration through membranes as a function of ethanol concentration in the feed solution are added to the figure. A-type zeolite [54] and Silicalite [55] membranes were used as examples of water- and ethanol-selective membranes, respectively. Dashed lines in the figure show an example of the first tray composition in distillation. Several trays are required to concentrate, e.g., 10% ethanol to over 80%, by distillation. On the contrary, over 80% ethanol can be obtained after a single permeation through a hydrophobic membrane [55–59]. Membrane separation is not limited by the azeotrope. Water-selective membranes permeates over 99.9% water for a large ethanol concentration range in the feed. In the following, a few configurations using water-selective and ethanol-selective VP/PV membranes are compared with conventional distillation.

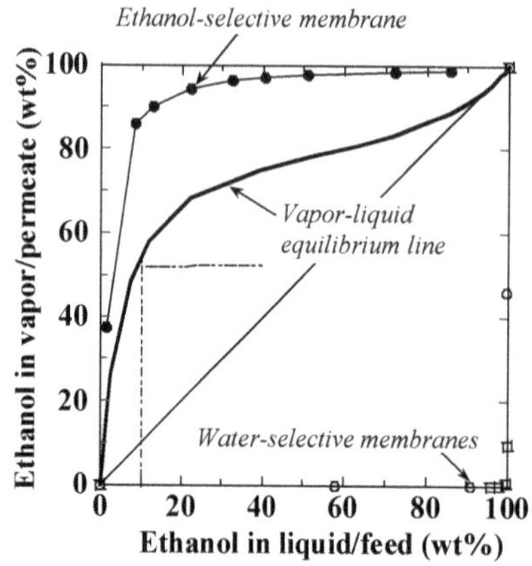

Figure 5. Liquid-vapor equilibrium of ethanol-water mixture with example separation performances of water- and ethanol-selective membranes: solid line, liquid-gas equilibrium; dashed lines, examples of stages in the distillation column; •, ethanol concentration in the permeate of an ethanol-selective silicalite membrane [55], □; ○, ethanol concentration in the permeate of water-selective A-type zeolite membranes (□, [54]; ○, this study).

5. Combination of Distillation and Water-Selective VP/PV Membranes

A hybrid process combining distillation and membrane dehydration is getting increasingly accepted as an energy-saving alternative to concentrate close-boiling mixtures, such as ethanol solution and iso-propyl alcohol solution [41,60]. In this hybrid process, the feed to the membranes can be a liquid mixture (PV) [61] or a vapor mixture (VP) [41]. In some of the industrially applied hybrid process using A-type zeolite membranes, a fraction of vapor from the top of a distillation tower is fed to the membrane. If no heat is added to the membrane unit, the fluid may be liquefied while flowing over the membrane. While A-type zeolite membranes show quite high water-selective separation properties in both PV and VP separations, the application of this membrane is limited to dry solvents with less than 15% water due to the insufficient stability of the membrane in water-rich conditions. Recently, various other types of zeolite membranes, such as zeolite T, CHA, MOR and ZSM-5, having higher stability in water-rich and in acidic media have been reported [62–64]. The application of dehydration membranes in the hybrid process can be extended to fluids containing higher amounts of water with these new types of membranes. However, the impact of extension on the total energy consumption, process size, and other factors are not clear.

The energy consumption and ethanol recovery rate were compared by changing the inlet ethanol concentration to a dehydration membrane. Concentrating 10% ethanol to 99.5% was assumed in the calculation using a cape-open to cape-open (COCO) program [59]. Figure 6 shows some of the process schemes compared in this study. Distillation followed by azeotropic distillation was used as a standard case (Figure 6a). The downstream process can be combined with continuous fermentation, as illustrated in the figure. Figure 6b shows a schematic view of the hybrid process, where azeotropic distillation is replaced with membrane dehydration. The membrane operational temperature was assumed to be the same as the distillation tower top temperature. A vapor mixture was fed to the membrane in this assumption. Neither membrane flux nor required membrane area were considered. The membrane separation factor was assumed to be 2000, which means that 0.05% ethanol in the feed mixture permeates through the membrane. A sweep gas was applied to the permeate side of the membrane instead of the vacuum lines often used in the VP/PV tests [50,55,62]. The sweep gas flow was assumed to be nine times higher than the total flux through the membrane. The outlet of the sweep gas was emitted to the atmosphere in the calculation, which reduced the ethanol recovery.

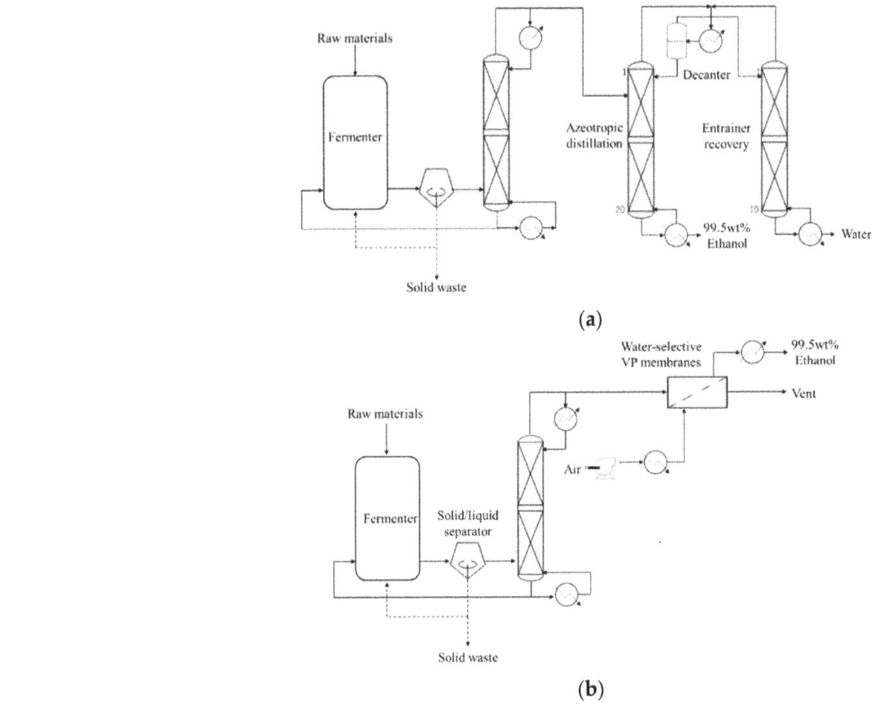

Figure 6. Distillation and dehydration membrane combined process: (**a**), diagram of distillation followed by azeotropic distillation; (**b**), diagram of distillation and membrane dehydration with sweep system combined.

With the above assumptions, azeotropic distillation required 2339 W/kg-ethanol to concentrate 80% ethanol to 99.5%. On the contrary, the energy consumption by membrane dehydration was 287 W/kg-ethanol, which is one order smaller than that of distillation. A significant reduction in the separation energy by replacing azeotropic distillation with membrane dehydration was also reported with different process configurations and assumptions [60]. The calculation showed that the dehydration membrane has a significant advantage, even with applying sweep gas at the permeate side.

Lowering the ethanol concentration at the inlet of a dehydration membrane reduces the distillation load. For example, changing the inlet ethanol concentration from 85 to 75 wt%, the energy demand at the distillation became 1820 to 1790 W/kg-EtOH. On the contrary, lowering the ethanol concentration in the feed to a membrane enhances the membrane duty. As the outlet ethanol concentration is fixed to 99.5%, the total amount of water permeating through a membrane increases when the feed ethanol concentration is smaller. Accordingly, a higher sweep flow rate is required to maintain the pressure difference of water across the membrane, which increased the compressor energy and the energy to heat the sweep gas to the membrane operation temperature. As a result, the energy demand by changing the inlet concentration to the membrane became minor. For example, reducing the ethanol concentration at the inlet of the membrane unit from 85 to 75% changed the total energy demand of the hybrid process to concentrate 10% ethanol to 99.5% from 1840 to 1830 Wh/kg-EtOH. The impact of extending the application of the dehydration membrane is not significant as compared with the replacement of azeotropic distillation with membrane dehydration under the process configuration and the assumptions used in this study.

6. Combination of Ethanol-Selective and Water-Selective VP/PV Membranes

Hydrophobic membranes selectively permeate ethanol over water. Table 2 shows some examples of ethanol-selective membranes, which can be divided into polymeric membranes, polymer and filler composite membranes, which are called mixed matrix membranes (MMMs), and inorganic membranes. Separation factor, α, is defined as:

$$\alpha = (y_{Etahnol}/y_{Water})/(x_{Ethanol}/x_{Water}) \quad (1)$$

where x and y are the mass fraction of each component in the feed and in the permeate, respectively. Permeate concentrations in the table were calculated from the separation factor and feed composition given in the references. Polymeric membranes, including commercially available membranes, showed slightly higher ethanol-selectivity than the liquid-vapor equilibrium [65–67]. The ethanol-selectivity can be enhanced by mixing hydrophobic fillers to the polymers (MMMs) [65,67,68]. Inorganic membranes, consisting of pure filler materials, in the table showed a higher ethanol-separation ability with higher flux [55–57]. Both silicalite and beta-zeolite membranes are types of zeolite membranes. These membranes are made of pure silica and have ordered pores of sub-nanometers. The ethanol-selective permeation is based on the strong adsorption of ethanol to the zeolitic pores, which inhibits the water permeation. Beta-zeolite has larger pores than silicalite. The pores may be too large to be plugged by adsorbed ethanol and let some water permeating through, which results in a lower ethanol-selectivity in beta zeolite membranes than in silicalite membranes. The reported flux values of silicalite membranes have some variations as the membrane micro-morphologies depend on the preparation conditions, such as the support types used and hydrothermal synthesis conditions. Nevertheless, several groups reported that when silicalite membranes were applied to ca. 10% ethanol solution, the ethanol concentration in the permeate was over 80% [55,56,58]. Based on these results, the separation factor of ethanol-selective membrane was assumed to be 36 in the following simulations.

Table 2. Examples of ethanol-selective membranes.

Membrane Type	Membrane Material	Feed Ethanol Conc. (wt%)	Temperature (K)	Permeate Ethanol Conc. (wt%) *	Separation Factor, α (-)	Flux (Kg m^{-2} h^{-1})	Ref.
Polymeric membranes	PDMS † (Pervatech, Netherland)	5	323	26	6.7 ± 1.0	2.6 ± 0.4	[66]
	PERVAP 4060 (Sulzer Chemtech, Switzerland)	5	323	27	7.1 ± 1.3	1.3 ± 0.3	[66]
	PERVAP 4060 (Sulzer Chemtech, Switzerland)	5	303	32	9	0.6	[67]
Mixed matrix membranes (MMMs)	PDMS-silicalite hollow spheres (30 wt% **)	6	313	49	15.3	0.07	[68]
	PDMS-hydrophobized Al_2O_3 + (1 wt% **)	5	303	37	11	0.06	[67]
	PTMSP # -silica (1.5 wt% **)	10	323	63	15.3	0.40	[65]
Inorganic membranes	Silicalite	12	333	89	58	0.76	[55]
	Silicalite	10	323	91	92	3.00	[56]
	All silica beta-zeolite	10	323	58	12.3	6.29	[57]

* calculated from the feed concentration and the separation factor given in the paper, ** amount of filler mixed to the polymeric matrix, + surface modification using triethoxy(octy)silane, † PDMS: polydimethylsiloxane, Silicalite: all-silica MFI type zeolite. # PTMSP: poly(1-trimethylsilyl-1-propyne).

Combining ethanol-selective and water-selective VP/PV membranes, it is possible to eliminate the distillation column completely. Figure 6 shows schemes of a downstream process employing these two types of membranes. An ethanol-selective membrane was applied to concentrate 10% ethanol to 80%, then a water-selective membrane was applied to dehydrate ethanol to 99.5%. Differently from water-permeating membranes described in the above section, the ethanol vapor permeating through the first membrane unit needs to be collected. As industrial water with a temperature of 300 K was not sufficient to condense ethanol vapor, a chiller of 253 K with 50% efficiency was added in the calculation. Two configurations were considered for the permeate side of the ethanol-selective membrane: a depressurized system (Figure 7a) and a sweep air system (Figure 7b). The operation temperature of both ethanol- and water-selective membranes was assumed to be 337 K. The stage cut-off, the fraction of feed permeating through a membrane, was assumed to be 90%. The retentate of feed was recycled to the fermenter, as shown in the figure. The ethanol recovery rate was calculated by considering the ethanol loss at the liquid/gas separator and through the water-selective membrane.

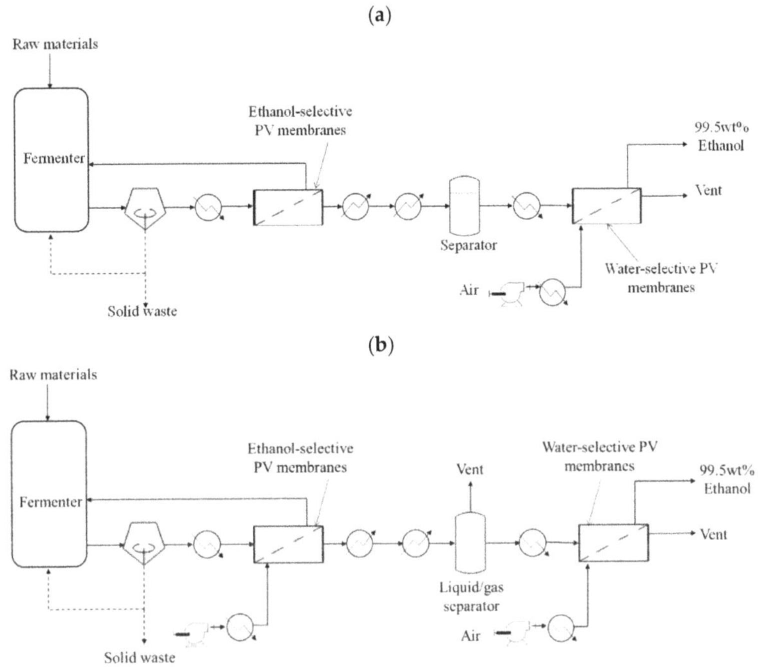

Figure 7. Downstream process consisting of two types of membranes: (**a**), a diagram of ethanol-selective membrane with vacuum followed by water-selective membrane with sweep; (**b**), diagram of ethanol- and water-selective membranes with sweep.

Table 3 shows the energy required to concentrate 10% ethanol to 80%, calculated with the process scheme in Figure 7. In the case of the depressurized permeate system, as shown in Figure 7a, the condensation of ethanol vapor was not possible with industrial water, which increased the duty on the chiller. As a result, the total energy requirement per kilogram of ethanol was almost equivalent to the distillation case. On the contrary, in the compressed air sweep system, as shown in Figure 7b, about 20 to 35% energy reduction was obtained from the distillation energy demand. Higher sweep gas flow rate requires more energy as it increases the compression energy, the heat required to bring the sweep gas to the membrane temperature, and the chiller energy to condense permeated vapor with the sweep gas. Too small sweep flow rate will limit the membrane flux as the driving force of the permeation gets smaller [50]. Accordingly, there is a range of optimum flow rate, which needs further investigation. In the case with a sweep gas, a liquid/gas separator is required after the chiller, as shown in Figure 7. The separator of liquid from gas reduced the ethanol recovery. About 0.5 to 1.3% of the ethanol permeated through the ethanol-selective membrane was lost at the separator, as some ethanol vapor was removed with the exhaust gas. The loss depends on the sweep gas flow rate; the higher the flow rate, the larger the loss. While an ethanol-selective membrane with a sweep flow system has the possibility to reduce energy consumption, the optimum process configurations need further study.

Table 3. Comparison of the energy required to concentrate 10 wt% ethanol to 80 wt%.

Process Configuration		Energy Demand (W/kg-EtOH)		
		Feed Treatment	Chiller	Total
Distillation		-	-	1803
Membrane separation	Vacuum at the permeate side	906	956	1862
	Air sweep at the permeate side (x 1.3 *)	930	270	1199
	Air sweep at the permeate side (x 2.2 *)	946	379	1325
	Air sweep at the permeate side (x 3.1 *)	962	488	1450

* Sweep flow rate compared to the flux through a membrane.

Various process configurations are possible to concentrate 10% ethanol solution to 99.5%. A few conditions are compared in Table 4. Replacing azeotropic distillation with membrane dehydration reduces the energy required for the separation to about half. Employing ethanol-selective membranes with compressed air as a sweep gas at the permeate side instead of distillation can further reduce the downstream energy. A membrane-based downstream process is an interesting choice, especially for small to medium scale applications, where the scale-merit of distillation is limited. The modular design of membrane units and the simple operation/maintenance are other advantages of the membrane process. In addition, membrane separation is environmentally friendly as no or very few additional chemicals are required [69].

Table 4. Comparison of the energy demand to concentrate ethanol from 10 to 99.5 wt%.

Process Configuration	Energy Demand (W/kg-EtOH)			Ethanol Recovery (%)
	10 wt% Ethanol to 80 wt%	80 wt% Ethanol to 99.5 wt%	Total	
Distillation +azeotropic distillation	1804	2339	4142	99.95
Distillation + water-selective membrane #	1804	287	2091	99.95
Ethanol-selective membrane with vacuum at the permeate side + water-selective membrane #	1862	287	2149	99.5
Ethanol-selective membrane with 1ir sweep at the permeate side (×1.3 #) + water-selective membrane *	1199	277	1480	99.4
Ethanol-selective membrane with 1ir sweep at the permeate side (×2.2 #) + water-selective membrane *	1325	277	1602	99.0
Ethanol-selective membrane with 1ir sweep at the permeate side (×3.1 #) + water-selective membrane *	1450	278	1728	98.7

* Sweep at the permeate side with flow rate 9.1 times higher than the membrane flux, # Sweep flow rate compared to the flux through a membrane.

One of the current challenges is the higher cost of inorganic membranes compared to conventional polymeric membranes. It is difficult to find the exact price of membranes as it depends on the production method, scale, and other factors. However, a few have reported that the inorganic membrane cost is about two orders higher than the polymeric membrane [70,71]. The robustness of the inorganic membranes may give the membranes competitiveness in long-term usage; however, stabilities of the membranes need to be evaluated under realistic conditions. The ceramic support cost occupies a major part of the fabrication cost of zeolite and zeolite-like micro-porous inorganic membranes [70]. Therefore, various ideas, using, for example, less expensive hollow fiber supports [71], re-usable stainless-steel supports [72], and other materials [73,74] are proposed. Improving the membrane flux is another approach because higher flux requires less membrane area and will reduce the capital cost of the membrane unit. The membrane synthesis conditions should be scalable in an economic and environmentally friendly way. The successful industrialization of dehydration zeolite membranes and modules [42,54] gives a guideline for the implementation of ethanol-selective membrane processes.

7. Ethanol-Selective Silicalite Membranes Applied to a Fermented Solution

Membrane properties are often evaluated with a single component permeation or a separation of binary synthetic mixtures. For example, the performances in Table 3 were measured with synthetic ethanol/water mixtures. On the contrary, bio-ethanol produced via fermentation contains various components. The influences of these third chemicals are not possible to predict at this moment. Nomura et al. studied the influence of yeast and salts in ethanol solutions. They reported that salts enhanced the ethanol selectivity due to the salt effect on the liquid-vapor equilibrium that enlarged the ethanol vapor pressure [58]. Other contaminants than salts, such as acids, can also influence the membrane properties. Offeman et al. used mixed matrix membranes consisting of hydrophobic zeolite particles and polydimethylsiloxane (PDMS) and applied them to separate fermentation broths [75]. They reported reductions of selectivity and flux in the broth due to a strong adsorption of, e.g., oleic acid to the zeolitic pores.

Figure 8 shows the flux and the concentrations of feed and permeate through a silicalite membrane. A distillate after a single distillation of fermented broth obtained by a HTF was used as feed. In this case, only volatile contaminants co-exist in the aqueous ethanol solution. The distillate had an ethanol concentration of 14 wt% and a pH of about 4. A synthetic mixture of 14 wt% ethanol solution was applied before and after the distillate test to check the change of membrane performance. All the tests were performed at 348 K. The selectivities of the silicalite membrane, measured with a synthetic mixture before and after the distillate separation, were the same, suggesting that the selective layer was stable. On the contrary, the flux became about half after the distillate separation. Both selectivity and flux decreased when a distillate was used as feed instead of a synthetic mixture. Acids or other volatile components in the fermented broth may be adsorbed into the membrane surface and block the zeolitic pores. The results suggested a contribution of reversible adsorption and irreversible changes. Pre-treating the fermented broth to reduce the concentration of contaminants, or modifying the membrane surface to prevent adsorption of contaminants, may reduce the reduction of the membrane property in a bio-ethanol separation. Another aspect of the ethanol-selective membrane process is the need to develop a cost-effective, large-scale membrane preparation method.

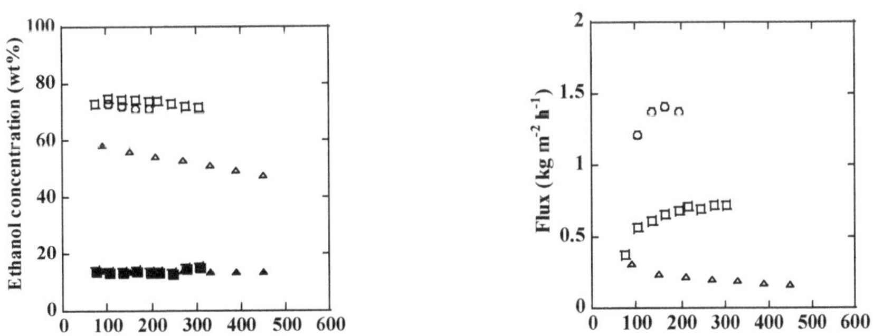

(a) (b)

Figure 8. Changes of silicalite membrane permselectivity with time in a synthetic mixture and in a distillate of fermented broth measured at 348 K: (**a**), ethanol concentration in the feed and in the permeate; (**b**), flux as a function of time; ○, a synthetic mixture of 14 wt% ethanol as feed measured before the distillate separation; △, distillate as feed; □, a synthetic mixture of 14 wt% ethanol as feed measured after the distillate separation; open keys, permeate concentration and flux; closed keys, feed concentration.

8. Conclusions

To meet the increasing demand and to improve the profit of using bio-ethanol, energy consumption at the conversion process should be limited. High-temperature fermentation (HTF) does not require temperature control. Cooling water is dispensable, making the fermenter configuration simpler and the energy demand smaller. This type of fermentation is particularly interesting in tropical countries.

Membranes can be integrated into the conversion process in different ways. Dehydration PV membranes are getting increasingly accepted as an energy-saving alternative to azeotropic distillation. To simplify the membrane separation process, replacing the vacuum line at the permeate side with a compressed air flow was considered. A preliminary simulation showed that the energy demand can also be reduced by applying water-selective VP membranes with sweep gas.

The preliminary simulation also showed that employing ethanol-selective membranes with sweep gas instead of distillation can reduce about 20–30% of the energy demand to concentrate 10 wt% ethanol to 80 wt%. Several research groups have reported that silicalite membranes, a type of nano-porous inorganic membrane, can concentrate 10 wt% ethanol to over 80 wt%. However, in this study, when a silicalite membrane was applied to a bio-ethanol solution produced by HTF followed by a single distillation, the flux through the membrane became about 10% of the flux obtained with an ethanol/water synthetic mixture. This result shows that it is necessary to pre-treat the bio-ethanol before applying a membrane or develop new types of membrane whose adsorption is influenced less by contaminants.

Combining HTF and membrane separation has the potential to realize a simple conversion process, which will facilitate on-site bio-ethanol production at farm areas. However, further developments in each of the technologies, investigation of a better configuration of the integrated process, and a scale-up study are required.

Author Contributions: Conceptualization, I.K.; methodology, M.Y. (Morihisa Yokota), R.T. and Y.S.; writing—original draft preparation, I.K., W.K., M.M. and M.Y. (Mamoru Yamada); writing—review and editing, I.K., W.K. and J.W.L.; funding acquisition, I.K., M.M. and M.Y. (Mamoru Yamada). All authors have read and agreed to the published version of the manuscript.

Funding: This research was funded by e-ASIA Joint Research Program, which was granted by Japan Science (JPMJSC16E5).

Institutional Review Board Statement: Not applicable.

Informed Consent Statement: Not applicable.

Data Availability Statement: Not applicable.

Conflicts of Interest: The authors declare no conflict of interest. The funders had no role in the design of the study; in the collection, analyses, or interpretation of data; in the writing of the manuscript, or in the decision to publish the results.

References

1. Portero Barahona, P.; Bastidas Mayorga, B.; Martín-Gil, J.; Martín-Ramos, P.; Carvajal Barriga, E.J. Cellulosic ethanol: Improving cost efficiency by coupling semi-continuous fermentation and simultaneous saccharification strategies. *Processes* **2020**, *8*, 1459. [CrossRef]
2. Park, J.-H.; Yoon, J.-J.; Park, H.-D.; Lim, D.J.; Kim, S.-H. Anaerobic digestibility of algal bioethanol residue. *Bioresour. Technol.* **2012**, *113*, 78–82. [CrossRef]
3. Pang, B.; Cao, X.-F.; Sun, S.-N.; Wang, X.-L.; Wen, J.-L.; Lam, S.S.; Yuan, T.-Q.; Sun, R.-C. The direct transformation of bioethanol fermentation residues for production of high-quality resins. *Green Chem.* **2020**, *22*, 439–447. [CrossRef]
4. Cesaro, A.; Belgiorno, V. Combined biogas and bioethanol production: Opportunities and challenges for industrial application. *Energies* **2015**, *8*, 8121–8144. [CrossRef]
5. García-Herreros, P.; Gómez, J.M.; Gil, I.D.; Rodríguez, G. Optimization of the design and operation of an extractive distillation system for the production of fuel grade ethanol using glycerol as entrainer. *Ind. Eng. Chem. Res.* **2011**, *50*, 3977–3985. [CrossRef]
6. Karimi, S.; Karri, R.R.; Yaraki, M.T.; Koduru, J.R. Processes and separation technologies for the production of fuel-grade bioethanol: A review. *Environ. Chem. Lett.* **2021**, 1–18. [CrossRef]

7. Solarte-Toro, J.C.; Romero-García, J.M.; Martínez-Patiño, J.C.; Ruiz-Ramos, E.; Castro-Galiano, E.; Cardona-Alzate, C.A. Acid pretreatment of lignocellulosic biomass for energy vectors production: A review focused on operational conditions and techno-economic assessment for bioethanol production. *Renew. Sustain. Energy Rev.* **2019**, *107*, 587–601. [CrossRef]
8. Alzate, C.C.; Toro, O.S. Energy consumption analysis of integrated flowsheets for production of fuel ethanol from lignocellulosic biomass. *Energy* **2006**, *31*, 2447–2459.
9. Google Scholar. Available online: https://scholar.google.com/ (accessed on 3 June 2021).
10. Barros, S. *Brazil Biofuels Annual*; United States Department of Agriculture (USDA): Washington, DC, USA, 2020.
11. Chase, M. *China Biofuels Annual*; United States Department of Agriculture (USDA): Washington, DC, USA, 2020.
12. Ankit, C. *India Biofuels Annual*; United States Department of Agriculture (USDA): Washington, DC, USA, 2020.
13. Prasertsri, P.; Chanikornpradit, M. *Thailand Biofuels Annual*; United States Department of Agriculture (USDA): Washington, DC, USA, 2020.
14. Ryan, B. *Philippines Biofuels Annual*; United States Department of Agriculture (USDA): Washington, DC, USA, 2020.
15. Arif, R. *Indonesia Biofuels Annual*; United States Department of Agriculture (USDA): Washington, DC, USA, 2020.
16. Kokkinos, N.C.; Nikolaou, N.; Psaroudakis, N.; Mertis, K.; Mitkidou, S.; Mitropoulos, A.C. Two-step conversion of LLCN olefins to strong anti-knocking alcohol mixtures catalysed by Rh, Ru/TPPTS complexes in aqueous media. *Catal. Today* **2015**, *247*, 132–138. [CrossRef]
17. Just-Auto.com. Thailand to Replace E10 "Gasohol" with E20 in 2020. Available online: https://www.just-auto.com/news/thailand-to-replace-e10-gasohol-with-e20-in-2020_id193299.aspx (accessed on 19 May 2021).
18. Tunapaiboon, N. Ethanol industry. In *Thailand Industry Outlook 2019-21*; Krungsri Research: Bangkok, Thailand, 2019.
19. DEDE. *Ethanol Data 2007–2017*; Department of Alternative Energy Development and Efficiency: Bangkok, Thailand, 2017.
20. AEDP 2018. Available online: https://www.dede.go.th/download/Plan_62/20201021_TIEB_AEDP2018.pdf (accessed on 3 June 2021).
21. Sakchai Preechajarn, P.P.; Maysa, C. Biofuels Annua. Available online: https://apps.fas.usda.gov/newgainapi/api/Report/DownloadReportByFileName?fileName=Biofuels%20Annual_Bangkok_Thailand_11-04-2019 (accessed on 2 June 2020).
22. DEDE. *Feasibility Study on Cellulosic Ethanol Production in Commercial Scale (In Thai)*; Ministry of Energy: Bangkok, Thailand, 2012.
23. Sakachai, P.; Ponnarong, P.; Maysa, C. Thailand Biofules Annual. 2019. Available online: https://apps.fas.usda.gov/newgainapi/api/Report/DownloadReportByFileName?fileName=Biofuels%20Annual_Bangkok_Thailand_11-04-2019 (accessed on 19 May 2021).
24. Matsushita, K.; Azuma, Y.; Kosaka, T.; Yakushi, T.; Hoshida, H.; Akada, R.; Yamada, M. Genomic analyses of thermotolerant microorganisms used for high-temperature fermentations. *Biosci. Biotechnol. Biochem.* **2016**, *80*, 655–668. [CrossRef] [PubMed]
25. Kosaka, T.; Lertwattanasakul, N.; Rodrussamee, N.; Nurcholis, M.; Dung, N.T.; Keo-Oudone, C.; Murata, M.; Götz, P.; Theodoropoulos, C.; Maligan, J.M.; et al. *Potential of thermotolerant ethanologenic yeasts isolated from ASEAN countries and their application in high-temperature fermentation. Fuel Ethanol Prod. Sugarcane*; BoD—Books on Demand: Norderstedt, Germany, 2019; pp. 121–154.
26. Rodrussamee, N.; Lertwattanasakul, N.; Hirata, K.; Limtong, S.; Kosaka, T.; Yamada, M. Growth and ethanol fermentation ability on hexose and pentose sugars and glucose effect under various conditions in thermotolerant yeast *Kluyveromyces marxianus*. *Appl. Microbiol. Biotechnol.* **2011**, *90*, 1573–1586. [CrossRef]
27. Charoensopharat, K.; Thanonkeo, P.; Thanonkeo, S.; Yamada, M. Ethanol production from Jerusalem artichoke tubers at high temperature by newly isolated thermotolerant inulin-utilizing yeast *Kluyveromyces marxianus* using consolidated bioprocessing. *Antonie Leeuwenhoek* **2015**, *108*, 173–190. [CrossRef]
28. Nitiyon, S.; Keo-Oudone, C.; Murata, M.; Lertwattanasakul, N.; Limtong, S.; Kosaka, T.; Yamada, M. Efficient conversion of xylose to ethanol by stress-tolerant *Kluyveromyces marxianus* BUNL-21. *Springerplus* **2016**, *5*, 1–12. [CrossRef]
29. Yuangsaard, N.; Yongmanitchai, W.; Yamada, M.; Limtong, S. Selection and characterization of a newly isolated thermotolerant *Pichia kudriavzevii* strain for ethanol production at high temperature from cassava starch hydrolysate. *Antonie Leeuwenhoek* **2013**, *103*, 577–588. [CrossRef] [PubMed]
30. Rodrussamee, N.; Sattayawat, P.; Yamada, M. Highly efficient conversion of xylose to ethanol without glucose repression by newly isolated thermotolerant *Spathaspora passalidarum* CMUWF1–2. *BMC Microbiol.* **2018**, *18*, 1–11. [CrossRef] [PubMed]
31. Arora, R.; Behera, S.; Sharma, N.K.; Kumar, S. A new search for thermotolerant yeasts, its characterization and optimization using response surface methodology for ethanol production. *Front. Microbiol.* **2015**, *6*. [CrossRef]
32. Talukder, A.A.; Easmin, F.; Mahmud, S.A.; Yamada, M. Thermotolerant yeasts capable of producing bioethanol: Isolation from natural fermented sources, identification and characterization. *Biotechnol. Biotechnol. Equip.* **2016**, *30*, 1106–1114. [CrossRef]
33. Techaparin, A.; Thanonkeo, P.; Klanrit, P. High-temperature ethanol production using thermotolerant yeast newly isolated from Greater Mekong Subregion. *Braz. J. Microbiol.* **2017**, *48*, 461–475. [CrossRef]
34. Prado, C.D.; Mandrujano, G.P.; Souza, J.P.; Sgobbi, F.B.; Novaes, H.R.; da Silva, J.P.; Alves, M.H.; Eliodório, K.P.; Cunha, G.C.; Giudici, R. Physiological characterization of a new thermotolerant yeast strain isolated during Brazilian ethanol production, and its application in high-temperature fermentation. *Biotechnol. Biofuels* **2020**, *13*, 1–15. [CrossRef] [PubMed]
35. Limtong, S.; Sringiew, C.; Yongmanitchai, W. Production of fuel ethanol at high temperature from sugar cane juice by a newly isolated *Kluyveromyces marxianus*. *Bioresour. Technol.* **2007**, *98*, 3367–3374. [CrossRef]
36. Lertwattanasakul, N.; Kosaka, T.; Hosoyama, A.; Suzuki, Y.; Rodrussamee, N.; Matsutani, M.; Murata, M.; Fujimoto, N.; Tsuchikane, K.; Limtong, S.; et al. Genetic basis of the highly efficient yeast *Kluyveromyces marxianus*: Complete genome sequence and transcriptome analyses. *Biotechnol. Biofuels* **2015**, *8*, 1–14. [CrossRef] [PubMed]

37. Murata, M.; Nitiyon, S.; Lertwattanasakul, N.; Sootsuwan, K.; Kosaka, T.; Thanonkeo, P.; Limtong, S.; Yamada, M. High-temperature fermentation technology for low-cost bioethanol. *J. Jpn. Inst. Energy* **2015**, *94*, 1154–1162. [CrossRef]
38. Kumar, S.; Singh, N.; Prasad, R. Anhydrous ethanol: A renewable source of energy. *Renew. Sustain. Energy Rev.* **2010**, *14*, 1830–1844. [CrossRef]
39. Khalid, A.; Aslam, M.; Qyyum, M.A.; Faisal, A.; Khan, A.L.; Ahmed, F.; Lee, M.; Kim, J.; Jang, N.; Chang, I.S. Membrane separation processes for dehydration of bioethanol from fermentation broths: Recent developments, challenges, and prospects. *Renew. Sustain. Energy Rev.* **2019**, *105*, 427–443. [CrossRef]
40. Sato, K.; Aoki, K.; Sugimoto, K.; Izumi, K.; Inoue, S.; Saito, J.; Ikeda, S.; Nakane, T. Dehydrating performance of commercial LTA zeolite membranes and application to fuel grade bio-ethanol production by hybrid distillation/vapor permeation process. *Microporous Mesoporous Mater.* **2008**, *115*, 184–188. [CrossRef]
41. Huang, Y.; Baker, R.W.; Vane, L.M. Low-energy distillation-membrane separation process. *Ind. Eng. Chem. Res.* **2010**, *49*, 3760–3768. [CrossRef]
42. Morigami, Y.; Kondo, M.; Abe, J.; Kita, H.; Okamoto, K. The first large-scale pervaporation plant using tubular-type module with zeolite NaA membrane. *Sep. Purif. Technol.* **2001**, *25*, 251–260. [CrossRef]
43. Khayet, M.; Matsuura, T. Pervaporation and vacuum membrane distillation processes: Modeling and experiments. *AIChE J.* **2004**, *50*, 1697–1712. [CrossRef]
44. Hubadillah, S.K.; Tai, Z.S.; Othman, M.H.D.; Harun, Z.; Jamalludin, M.R.; Rahman, M.A.; Jaafar, J.; Ismail, A.F. Hydrophobic ceramic membrane for membrane distillation: A mini review on preparation, characterization, and applications. *Sep. Purif. Technol.* **2019**, *217*, 71–84. [CrossRef]
45. Hietaharju, J.; Kangas, J.; Tanskanen, J. Analysis of the permeation behavior of ethanol/water mixtures through a polydimethylsiloxane (PDMS) membrane in pervaporation and vapor permeation conditions. *Sep. Purif. Technol.* **2019**, *227*, 115738. [CrossRef]
46. Nguyen, T.V.N.; Paugam, L.; Rabiller, P.; Rabiller-Baudry, M. Study of transfer of alcohol (methanol, ethanol, isopropanol) during nanofiltration in water/alcohol mixtures. *J. Membr. Sci.* **2020**, *601*, 117907. [CrossRef]
47. Kumakiri, I.; Yamaguchi, T.; Nakao, S.-i. Application of a zeolite A membrane to reverse osmosis process. *J. Chem. Eng. Jpn.* **2000**, *33*, 333–336. [CrossRef]
48. Holloway, R.W.; Childress, A.E.; Dennett, K.E.; Cath, T.Y. Forward osmosis for concentration of anaerobic digester centrate. *Water Res.* **2007**, *41*, 4005–4014. [CrossRef]
49. Shibuya, M.; Sasaki, K.; Tanaka, Y.; Yasukawa, M.; Takahashi, T.; Kondo, A.; Matsuyama, H. Development of combined nanofiltration and forward osmosis process for production of ethanol from pretreated rice straw. *Bioresour. Technol.* **2017**, *235*, 405–410. [CrossRef] [PubMed]
50. Kataoka, T.; Tsuru, T.; Nakao, S.-I.; Kimura, S. Permeation equations developed for prediction of membrane performance in pervaporation, vapor permeation and reverse osmosis based on the solution-diffusion model. *J. Chem. Eng. Jpn.* **1991**, *24*, 326–333. [CrossRef]
51. Nakao, S. Optimization of membrane process for concentrating alcohol solution. *Membrane* **1994**, *19*, 344. [CrossRef]
52. Li, L.; Dong, J.; Nenoff, T.M.; Lee, R. Desalination by reverse osmosis using MFI zeolite membranes. *J. Membr. Sci.* **2004**, *243*, 401–404. [CrossRef]
53. Lee, K.P.; Arnot, T.C.; Mattia, D. A review of reverse osmosis membrane materials for desalination—development to date and future potential. *J. Membr. Sci.* **2011**, *370*, 1–22. [CrossRef]
54. Kondo, M.; Komori, M.; Kita, H.; Okamoto, K.-I. Tubular-type pervaporation module with zeolite NaA membrane. *J. Membr. Sci.* **1997**, *133*, 133–141. [CrossRef]
55. Sano, T.; Yanagishita, H.; Kiyozumi, Y.; Mizukami, F.; Haraya, K. Separation of ethanol/water mixture by silicalite membrane on pervaporation. *J. Membr. Sci.* **1994**, *95*, 221–228. [CrossRef]
56. Ueno, K.; Negishi, H.; Okuno, T.; Saito, T.; Tawarayama, H.; Ishikawa, S.; Miyamoto, M.; Uemiya, S.; Sawada, Y.; Oumi, Y. High-performance silicalite-1 membranes on porous tubular silica supports for separation of ethanol/water mixtures. *Sep. Purif. Technol.* **2017**, *187*, 343–354. [CrossRef]
57. Ueno, K.; Yamada, S.; Watanabe, T.; Negishi, H.; Okuno, T.; Tawarayama, H.; Ishikawa, S.; Miyamoto, M.; Uemiya, S.; Oumi, Y. Hydrophobic* BEA-type zeolite membranes on tubular silica supports for alcohol/water separation by pervaporation. *Membranes* **2019**, *9*, 86. [CrossRef] [PubMed]
58. Nomura, M.; Bin, T.; Nakao, S.-I. Selective ethanol extraction from fermentation broth using a silicalite membrane. *Sep. Purif. Technol.* **2002**, *27*, 59–66. [CrossRef]
59. CAPE-OPEN to CAPE-OPEN (COCO). Available online: https://www.cocosimulator.org/ (accessed on 19 May 2021).
60. Van Hoof, V.; Van den Abeele, L.; Buekenhoudt, A.; Dotremont, C.; Leysen, R. Economic comparison between azeotropic distillation and different hybrid systems combining distillation with pervaporation for the dehydration of isopropanol. *Sep. Purif. Technol.* **2004**, *37*, 33–49. [CrossRef]
61. O'Brien, D.J.; Roth, L.H.; McAloon, A.J. Ethanol production by continuous fermentation–pervaporation: A preliminary economic analysis. *J. Membr. Sci.* **2000**, *166*, 105–111. [CrossRef]
62. Zhu, M.-H.; Kumakiri, I.; Tanaka, K.; Kita, H. Dehydration of acetic acid and esterification product by acid-stable ZSM-5 membrane. *Microporous Mesoporous Mater.* **2013**, *181*, 47–53. [CrossRef]

63. Zhu, M.-H.; Xia, S.-L.; Hua, X.-M.; Feng, Z.-J.; Hu, N.; Zhang, F.; Kumakiri, I.; Lu, Z.-H.; Chen, X.-S.; Kita, H. Rapid preparation of acid-stable and high dehydration performance mordenite membranes. *Ind. Eng. Chem. Res.* **2014**, *53*, 19168–19174. [CrossRef]
64. Zhang, Y.; Qiu, X.; Hong, Z.; Du, P.; Song, Q.; Gu, X. All-silica DD3R zeolite membrane with hydrophilic-functionalized surface for efficient and highly-stable pervaporation dehydration of acetic acid. *J. Membr. Sci.* **2019**, *581*, 236–242. [CrossRef]
65. Claes, S.; Vandezande, P.; Mullens, S.; Leysen, R.; De Sitter, K.; Andersson, A.; Maurer, F.; Van den Rul, H.; Peeters, R.; Van Bael, M. High flux composite PTMSP-silica nanohybrid membranes for the pervaporation of ethanol/water mixtures. *J. Membr. Sci.* **2010**, *351*, 160–167. [CrossRef]
66. Claes, S.; Vandezande, P.; Mullens, S.; De Sitter, K.; Peeters, R.; Van Bael, M.K. Preparation and benchmarking of thin film supported PTMSP-silica pervaporation membranes. *J. Membr. Sci.* **2012**, *389*, 265–271. [CrossRef]
67. Kujawska, A.; Knozowska, K.; Kujawa, J.; Li, G.; Kujawski, W. Fabrication of PDMS based membranes with improved separation efficiency in hydrophobic pervaporation. *Sep. Purif. Technol.* **2020**, *234*. [CrossRef]
68. Naik, P.V.; Kerkhofs, S.; Martens, J.A.; Vankelecom, I.F. PDMS mixed matrix membranes containing hollow silicalite sphere for ethanol/water separation by pervaporation. *J. Membr. Sci.* **2016**, *502*, 48–56. [CrossRef]
69. He, X.; Hägg, M.-B. Membranes for Environmentally Friendly Energy Processes. *Membranes* **2012**, *2*, 706–726. [CrossRef]
70. Caro, J.; Noack, M. Zeolite membranes–status and prospective. In *Advances in Nanoporous Materials*; Elsevier: New York, NY, USA, 2010; Volume 1, pp. 1–96.
71. Lin, Y.; Duke, M.C. Recent progress in polycrystalline zeolite membrane research. *Curr. Opin. Chem. Eng.* **2013**, *2*, 209–216. [CrossRef]
72. Zhang, F.; Zheng, Y.; Hu, L.; Hu, N.; Zhu, M.; Zhou, R.; Chen, X.; Kita, H. Preparation of high-flux zeolite T membranes using reusable macroporous stainless steel supports in fluoride media. *J. Membr. Sci.* **2014**, *456*, 107–116. [CrossRef]
73. Nguyen, N.M.; Le, Q.T.; Nguyen, D.P.-H.; Nguyen, T.N.; Le, T.T.; Pham, T.C.-T. Facile synthesis of seed crystals and gelless growth of pure silica DDR zeolite membrane on low cost silica support for high performance in CO_2 separation. *J. Membr. Sci.* **2021**, *624*, 119110. [CrossRef]
74. Achiou, B.; Beqqour, D.; Elomari, H.; Bouazizi, A.; Ouammou, M.; Bouhria, M.; Aaddane, A.; Khiat, K.; Younssi, S.A. Preparation of inexpensive NaA zeolite membrane on pozzolan support at low temperature for dehydration of alcohol solutions. *J. Environ. Chem. Eng.* **2018**, *6*, 4429–4437. [CrossRef]
75. Offeman, R.D.; Ludvik, C.N. Poisoning of mixed matrix membranes by fermentation components in pervaporation of ethanol. *J. Membr. Sci.* **2011**, *367*, 288–295. [CrossRef]

MDPI
St. Alban-Anlage 66
4052 Basel
Switzerland
www.mdpi.com

Processes Editorial Office
E-mail: processes@mdpi.com
www.mdpi.com/journal/processes

www.ingramcontent.com/pod-product-compliance
Lightning Source LLC
Chambersburg PA
CBHW061502210926
51850CB00008B/1792
9783725804191